TO BARBARY'S FAR SHORE

by
Michael J. Kozlowski

.. For Melissa, my Miracle and my Blessing

... For Mom, who taught all three of us a love of reading and learning

... and most of all for Sergeant Donald N. Kozlowski, United States Marine Corps

Semper Fidelis

The Author

Michael J. Kozlowski retired from the United States Air Force after a career that spanned three decades and saw the rebuilding of US forces after Vietnam, the fall of the Soviet Union, and the Allied triumph of Operation Desert Storm. He served as an aerospace munitions specialist with the 8th, 1st, and 20th Fighter Wings, as well as a tour with the 379th Bombardment Wing of the legendary Strategic Air Command. He was also selected on three occasions to serve as a flightline supervisor at the USAF's famed Red Flag exercises. He has been awarded the Senior Maintenance Badge, the Missileman's Badge, and the USAF Commendation Medal with two Oak Leaf Clusters. He has a Bachelor's Degree with Honors from the University of South Carolina and an Associate's Degree from the Community College of the Air Force. He resides in Columbia, SC, with his wife Melissa and a slightly odd cat named Eek!

Copyright Notice

Copyright © 2013 Lion by Lion Publishing ISBN 978-1-939335-31-9 No part of this book may be reproduced or transmitted in any form or by any means, electronic or mechanical including photocopying, recording or by any information and retrieval system without permission in writing from the author and publisher.

CONTENTS

Preface 1
One: Stroke Of State 3
Two: The Practices Of Presence 31
Three: The Philadelphia Problem 49
Four: Call Her *Intrepid* 69
Five: The Practices Of Politics 105
Six: Under Orders 131
Seven: Preparations 167
Eight: The Politics Of Presence 187
Nine: Good King Hamet And The House Of Rendezvous 207
Ten: The March 233
Eleven: A Wide Spot In The Road 253
Twelve: A Bit Of Colored Cloth 269
Thirteen: The Presence Of Politics 281
Fourteen: L'envoi 307
Epilogue 345
References And Biblography 385
Appendix 1 387

LION BY LION
PUBLISHING

Books by Michael J. Kozlowski

The Long Patrol: The CSS H.L. Hunley, Charleston, and the Civil War

To Barbary's Far Shore

To Barbary's Far Shore

PREFACE

The history of the US military, and the Marine Corps in particular, is highly documented, especially within the last one hundred and fifty years. Unfortunately, what came before is less well documented. In some cases, all we know are the bare-bones facts of the matter. The attack on Derna, forever immortalized in the words "To The Shores Of Tripoli," is one of those instances. We know more about the secret and ultimately fruitless diplomatic maneuvers in Washington and Tripoli than we do about the military attack that has since gone down in history. We know with certainty the names of only a handful of the four hundred-odd men who marched on the Dey of Tripoli's fortress; of the marines themselves, we can only be sure of three of them. Even the US Marine Corps Historical Center, a national treasure in every sense of the term, is not sure of all the details. A dozen learned authors and researchers have written as many superb works on the subject, and no two have the same facts.

And, in the end, that was what made writing this book such a challenge. The story of a handful of Americans, deep in hostile territory, pulling off an incredible victory against overwhelming odds, is one that has great relevance today, where the descendants of those Americans are doing the same thing all over again. But this time the enemy is not merely capturing ships on the high seas and holding their crews for eventual ransom. This time, they murder thousands of innocent Americans. This time, they seek ways to kill millions at one blow; their ultimate goal, world domination. This is, quite simply, a story that needs to be retold; to remind us that what we have done before, what we had to do before, we can do again.

The trouble is, how does one tell a story where one isn't sure of exactly what happened? After considerable thought, I decided to try and tell the story the way it would have come down through the voices of the men who were there. As nearly as can be determined, the historical facts, that is to say, times, dates, and places, are accurate and cross-referenced. The conversations and dialogues, however, are fictional, but conform to events as we know them. I have done my level best to have the characters speak and act as their historical prototypes did, and in this I believe I've been successful.

What I do wish to stress is that there will be differences between this account and others. Two hundred and eight years downrange, no two accounts will agree. What I wanted to portray is the way the story would have been told

Preface

down through the years, from one marine to another at the campfire or the tavern. There has been artistic license taken with some aspects of the story, hopefully to the benefit of the reader's entertainment. I rest confident that the reader will be able to determine the difference between license and historical fact.

The sharp-eyed reader will note that the prefix 'USS,' or United States Ship, does not appear in this work when referring to specific vessels. The term was not officially brought into use until President Theodore Roosevelt so directed it in Executive Order 549 in January of 1907. Prior to that, vessels were referred to as 'US (ship type),'' for example, 'US schooner *Enterprise,*' or 'US Brig *Siren.*" As the terminology was therefore not in use, it is accordingly set aside.

I am, of course, indebted to everyone who helped in the writing of this book, and whose work provided the basis for it. The staff at the USMC Historical Center offered priceless assistance and information. Three books provided the foundation for telling the story: Professor William Fowler's Jack Tars and Commodores, Richard Zacks' The Pirate Coast, and a remarkable little book called Our Navy And The Barbary Corsairs by Gardner Allen, written in 1905. The staffs and crews of the US frigate *Constellation* and USS *Constitution* were unfailingly kind and generous in answering my questions.

My wife Melissa was a light and a joy throughout. She occasionally does not understand my love of and fascination with history, but she has always understood how much it means to me. As always, Bob Dedmon was a steadfast support during the seeming eternity it took to get this one done, offering advice, assistance, and the occasional threat if I didn't get the doggone thing finished. Debbie Ciannella kept a determined devotion to the project, for which I will be eternally and affectionately grateful. My friends at the Board were a constant source of advice and information as well as encouragement at one point long ago when personal dilemmas threatened to overwhelm the entire project. They are as much family as friends, and they are cherished accordingly. And I am especially indebted to Stuart Slade for giving this book and my previous work, The Long Patrol, a chance. I have considered him a friend and teacher for many years and look forward to the future.

As always, any errors are mine and mine alone.

M.J.K.

ONE: STROKE OF STATE

Americans tend to think of Washington, D.C., the way ancient Romans used to think of Rome itself; all marble and nobility and grandeur, with everything just far enough out of normal scale to overwhelm. Of course, in the 21st century, we've added some uniquely American touches. There are purposeful convoys of black limousines carrying Important People, all escorted by grim phalanxes of motorcycle cops. Flocks of protesters of one stripe or another spin about; some carrying handmade signs, while others totter about on stilts and manipulate grotesque puppets, seemingly unable to tell the difference between 'protest' and 'surreal theatre.' Politicians rush to and fro with endless speeches, sound bites and photo ops, sometimes against things and sometimes for them, but more often than not desperately trying to avoid being pegged one way or the other. And flying above it all, haloed in a halogen glow, is the Media; made up and blow dried within an inch of their lives, while telling us in ponderous terms What It All Means. Welcome to Washington, D.C., at the beginning of the millennium; an (arguably) more dignified Las Vegas, where a President, a Supreme Court, and five hundred and thirty five professional politicians gamble with our money. We think, we believe, we know with all our hearts that it has always appeared thusly. And we'd be wrong.

It is only within the last seventy-five years that Washington has come to resemble an Imperial 24/7 circus with marble tents. For instance, many of the monuments and buildings that we have come to regard as eternal as the stars themselves simply didn't exist until the 1930s. They were make-work projects during the Great Depression. In fact, within living memory, Washington was considered something of a hardship post for politician, diplomat, warrior and reporter alike. Until the late 1940s, it was a hot, humid and segregated backwater considered by many as thoroughly unfit for the grand purpose it was intended to fill.

Stroke of State

Go back a hundred and eleven years and Washington becomes far less grand than we know it today. There are working farms within sight of the Capitol, with their attendant sights, sounds, and smells. The Unknown Soldier has yet to take up his majestic and lonely post. Arlington Cemetery itself cradles the sleeping fallen of the Civil War, with only a few newer headstones from the Spanish-American War. Pennsylvania Avenue is not far past being a rutted, semi-solid river of dirt, mud, and manure. It was finally paved in 1872 with wooden blocks, which were cheaper than bricks or macadam, and the local political bosses could pocket the savings. They should have saved themselves the larceny; the blocks were rapidly turned into flinders by horses' hooves and then pounded back into the eternal mud. Presumably, if one digs deep enough, one will find a few splinters still there awaiting future archaeologists. The Jefferson and Lincoln memorials are nowhere to be seen; indeed, they will not appear for almost another three decades. The Mall is heavily forested, with gardens running through it.

Back another fifty years. The Washington Monument abruptly loses half of its height, unfinished due to funding problems and civil war; from 1858 to 1876 it will stand like a whitewashed chimney. To this day, the dividing line between the original construction and the later one is still clearly visible. The White House will mourn three presidents and hear distant artillery fire as Confederate forces flash up and down the Potomac Valley. The precise rows of white stones at Arlington will gradually recede, until Robert E. Lee's home once more stands alone atop a hill with a clear view of the leafy shores of the Potomac. No offramps for the Pentagon, no Metro trains rattling past, no jetliners shrieking into the air from Reagan National Airport, no mobile missile launchers near the Washington Monument. The Mall is very nearly overgrown wilderness, with a malodorous ditch, the Tiber Creek Canal, running along its northern edge, what is now Constitution Avenue. The canal has become an open sewer, and it will trickle sluggishly from the base of Capitol Hill to the Potomac, almost to the base of the Washington Monument, until 1874.

Another few decades and we are almost to our destination. The Mall is forest, and the Tiber Creek rolls pleasantly along its north side, until it floods. Then the Mall and present day Constitution Avenue become pestilential swamps that breed malaria, yellow fever, and God alone knows what else. The Capitol's soaring dome, with Freedom herself atop it, is gone. In its place is a low, modest hemisphere. The White House and the Capitol show black scorches from each window and door, then suddenly blossom fire; the calling card of the British Army in 1814. The President's House is no longer white, but now a warm brown-gold sandstone. It sits with quiet nobility beside the narrow dirt track that has just been christened Pennsylvania Avenue. The view eastward to the Capitol is almost unobstructed, except for a few trees and fewer buildings, but this is not the Capitol you and I recognize.

To Barbary's Far Shore

Only the North Wing stands erect on Capitol Hill now, its marble glowing in the Virginia sunlight. Just eleven years ago its cornerstone was laid by President Washington himself, may God bless his soul. The Senate meets in the solid, castle-like structure, but the House meets one hundred and fifty feet away in a temporary round brick building known only half jokingly as 'the Oven.'.Legislators traveling to and from the North Wing must walk through a rickety wooden passageway to go about their business. The buildings and masts of the Washington Navy Yard in Anacostia are just visible from Capitol Hill, as are the tents of the temporary headquarters of the US Marines at what will become the intersection of Eighth and I Streets. The glass and chrome towers of Crystal City, the soulless designs of modern government buildings, the grandeur of the Smithsonian and the Supreme Court, are all gone. In their places are scattered handfuls of farmhouses, homes, taverns, lots of taverns, boarding houses, merchants and markets of various stripes and reputation, and the occasional brothel; all tied together with a spider's web of dirt roads and trails.

Welcome to Washington, District of Columbia, in the Year of Our Lord One Thousand, Eight Hundred and Four.

You can see Washington quite some distance away today, whether or not you're on I-95, the nightmarish tangle of traffic and madness known as the Beltway, or the mighty Potomac River itself. In 1804, one had to get a lot closer to see it. For the most part, as you came upriver from Chesapeake Bay, you didn't really see anything until you were coming around that last bend in the river south of town. Given how densely populated the region is now, the southern end of a megalopolis that stretches from Washington to Boston, it is difficult, if not impossible, to imagine the area being almost uninhabited for huge stretches at a time. And the fastest way there wasn't by land, either. Only two or three dirt roads led into the District, and they were a horror to travel over. It took days to get from the mouth of Chesapeake Bay to Washington by land, a trip one can make now in about four hours. It still took a couple of days to sail the same distance as well, but it was a much more safe and comfortable trip.

William Eaton would have stood in the bows of the frigate *Constellation* as she crawled slowly up the Potomac on the fifteenth of March, 1804; watching the trees begin to blossom and turn deep, rich shades of green, far more lushly than they did in his native Connecticut. To see a large warship gliding up the river today would be a sight to back up traffic for miles along the river's shores; but in 1804 it would be common, and would remain that way until the bigger and heavier steel warships became too heavy and unwieldy to routinely traverse the Potomac. *Constellation* was headed to the Washington Navy Yard for refit and crew recruiting, and Eaton was basically hitching a ride home. He would have enjoyed the smells of the blooming green Potomac shores after spending four years in dry and often malodorous North Africa, and he would have stood at attention as *Constellation* passed President Washington's home at Mount

Stroke of State

Vernon earlier that day and rendered honors to our first President, gone only these five years now.

Eaton was thirty-nine that cool March morning on the Potomac, and he'd already led a life that could be described as fairly eventful. His father was a schoolmaster and farmer in the little hamlet of Woodstock, CT when William was born. From the beginning, he didn't seem to be very interested in a quiet, sedate way of life. At the age of sixteen, when our children are concerned about grades, status, clothes and dating (not necessarily in that order), he enlisted in the Connecticut Militia. He served in the closing years of the American Revolution, rising to the rank of sergeant in just two years, a fairly notable accomplishment in those days, all the more so because of his age. Eaton seems to have had a genuine affinity for military life and service. Today, it's difficult to get young adults in their teens and twenties to even consider joining the military in peacetime, but that was a different world. Records of his Revolutionary service are sketchy, but it looks as if he acquitted himself well. When the Revolution ended, he worked as a teacher, and then went to Dartmouth College. There being no GI Bill then, one can assume that he paid his own way, with assistance from his family. He did well at Dartmouth, and graduated in 1790. He briefly dabbled in politics before deciding he wasn't happy with that either. Missing the excitement and camaraderie of the service, he contacted some old friends from his militia days and got himself a commission in the tiny Regular Army as a captain.

One might think that his service as an enlisted man would have given him a solid foundation for his service as an officer. One would be wrong. Captain Eaton didn't do anywhere near as well as Sergeant Eaton did, for a variety of reasons. First, Will Eaton, for all his capabilities and intelligence, had a personality that unfortunately would have tried the patience of a saint. He was at turns arrogant, bombastic, egotistical, sarcastic and high-strung to the point of mania, in addition to the fact that he apparently never entertained the possibility that he could be in error about anything.

Ever.

Mind you, in and of themselves, these are not necessarily career-killing qualities. Any number of fresh new shavetails from the various academies share them, until the forces of reality, and their NCOs, set them straight. If one didn't know George Custer's stellar record before that particularly bad afternoon above the Little Big Horn, one could read his some of his writings (he wrote several books about his military experience as did his wife, to whom we owe, for better or for worse, a great deal of the Custer legend) and wonder if they were reading the memoirs of a particularly prissy schoolmaster/naturalist/warrior god who was eternally saddled with a regiment of dolts far beneath his intelligence, leadership, and social status. Douglas MacArthur made it to five stars and immortal legend with an ego the size of Mount Rushmore and a self-image that seems to have been inspired by Shakespearean hams. But even Custer and MacArthur knew when to shut up and follow orders in their early careers. Will Eaton didn't. In an era when histrionics, flourishes and boundless self-confidence were de rigueur for military officers, Eaton apparently stood out from the crowd.

This might not have been so bad. As I mentioned, in those days that sort of thing was the rule, and had Eaton been a bit more careful, he would simply have been regarded as an unusually good example of the genre. However, he had another problem on top of that; the tough veteran NCO of the Revolution was, frankly, a lousy officer. He questioned just about every order, regulation, suggestion, and other directive that ever came down the pike at him; most unusual, when one remembers his earlier service. He was apparently also rather indecisive about things and usually had to be badgered by higher authority into making a decision. This could at least be tolerated in some East Coast garrison, where the worst risk one faced was getting back past the sentries if one stayed out past curfew. But Eaton was on the old Northwest Frontier, where Indian warfare would continue in one form or another for nearly another fifty years.

Today's Ohio, Michigan, Indiana, and Kentucky were vicious, bloody killing grounds where both sides fought without quarter or mercy; one to protect fellow citizens from natives they saw as red barbarian hordes, the other to save the only homes and only way of life they had known for millennia. Foolish and hesitant conduct by the paleface here could, and often did, get your scalp handed to you. Assuming, of course, that they left you any hands, as the native tribes there were rather keen on mutilation of defeated enemies. Eaton may, stress the 'may,' have also had a drinking problem. This wouldn't have been at all uncommon among officers and NCOs in the wilderness in those days. When the locals weren't trying to separate you from your scalp, life was deadly dull and drinking was often the only recreation anyone had.

Alcohol abuse in the US Army was at levels we would consider shocking from the Revolution through the Civil War, and on the frontiers it was epidemic. Evan O'Connell, in his masterpiece <u>Son Of The Morning Star</u>, presents convincing evidence that the majority of officers west of the Mississippi during the 1870s and 1880s most likely fit the clinical definition of an alcoholic. We know that Eaton did develop a drinking problem after 1805; but to what extent it was evident before, we don't know for sure. Let us leave it at this. Even if he was a moderate to heavy drinker, which would have put him in the same category as most US Army officers of the day, it wouldn't have done him any good and could only have made things worse.

But, worst of all for Will Eaton, his senior commanding officer was General Anthony Wayne. Yes, that one; 'Mad' Anthony Wayne, and they meant

Stroke of State

it. Wayne was a battle-scarred, fire-breathing legend whose Revolutionary record was carved out in the Northwest against the British and the natives. His nickname actually came from one battle in which he had his troops climb a cliff behind British forces in order to attack them. His Majesty's forces had assumed that they were invulnerable from that direction, and lived just long enough afterwards to regret that assumption. Ascending the cliffs was considered insane, or 'mad;' henceforth the sobriquet. And, in keeping with such an image, Wayne had a volcanic temper, was prone to physical violence upon his subordinates during his rages, and possessed a zeal for combat that today would be, at best, faintly disturbing. Obviously standards were different then, but Wayne was enough of a standout that the 'Mad' could easily have covered all of those too.

It could probably go without saying that someone like Anthony Wayne was less than forgiving with officers who didn't measure up, and it could not have been long then before Captain Eaton's failings were brought to the desk of the Commanding General. Remember too, in those days, there were a lot fewer officers between even a lowly captain and a commanding general. Even if his superiors didn't bring the subject up the chain of command themselves, old Mad Anthony probably had a chance to witness Eaton in full flower at least once or twice. We don't know how many, if any, chances Captain Eaton was given to straighten out. Based on my own military experience, I would guess that Wayne probably gave Eaton even less slack than other officers because of his prior service as an NCO. Put simply, Will Eaton should have known better. But the Old Army was a forgiving one, and it took a great deal of effort to get oneself in a position where one was standing before grim justice in the form of a court-martial or summary dismissal.

Eaton, apparently, exerted himself to that extent.

In 1795, Eaton was formally charged under the Articles of War for a 'misunderstanding' between himself and a superior officer. The nature of the charges is at best unclear, as record-keeping on the frontier was, to put it gently, erratic. It almost certainly involved a physical confrontation over certain actions taken by the captain, which appear to have included profiteering and allowing a prisoner unauthorized liberty. Eaton was found guilty; not a terrible thing, as a court-martial conviction was not unusual. The punishment of 60 days confinement wasn't that bad either, except for the mind-numbing boredom that would have accompanied it. Eaton remained on the frontier for two more years and then decided that his military career had come to a conclusion. He may well have wanted to stay longer, but it's likely that, if he did, he may have very well been taken aside and had it strongly suggested to him that he needed to look at a career change; and, if he wouldn't, Mad Anthony would do it for him.

Not really knowing what else to do, he went home. In those days, that in and of itself would have been an adventure, traveling alone across nearly half a continent in a time when a good portion of that distance would have been contested by the US Army and the natives. Arriving back east, he was 'at liberty' for a short time, until he ran into a friend of long acquaintance who was willing to help him out. The friend was Timothy Pickering, Secretary of State under Presidents Washington and Adams. Pickering knew Eaton as a friend as well as an educated, sophisticated gentleman, and decided that he would be perfect for an opening in the State Department's Corps de Diplomatique. And as if the Gods had not found enough pleasure in this, Secretary Pickering sent him to the Barbary Coast to deal with the assorted deys, beys, bashaws, and other satraps who ran the place.

Given the propensity these days for Congress to investigate government appointees within an inch of their lives, it seems safe to say that Eaton wouldn't have made it today. Once his record in the Northwest Territory came out, that would have pretty much been the end of it. It also seems that then, and most definitely afterwards, Eaton had a tendency to inflate his military rank, knowledge, and experience to people who didn't know better. Today, when senior officers have been publicly pilloried for wearing the wrong ribbon, Eaton would have been run up a flagpole. However, things were done differently then. Pickering knew and vouched for Eaton and, in those days, that was enough. Accordingly, Eaton was named Consul to Tunis in 1797, but, for reasons unclear, did not report to his consulate until March of 1799.

The official record says that Eaton was there from 1799 to 1803. His official mission was to keep the Bey of Tunis happy and off American shipping. How he managed to last forty-eight months is a mystery to this writer, because, if there was ever a man more constitutionally and temperamentally unsuited to be a diplomat than Will Eaton, he has yet to be found. Eaton paid a courtesy call to the Dey of Algiers, the always charming and affable Bobba Mustafa, on the way there. Eaton was utterly shocked at what he saw as American groveling to the various rulers of the North African coast, and, being who he was, had little inclination to keep his opinions to himself. The nature of diplomacy is to express the attitudes and needs of one's nation, no matter how outrageous, in polite, formal tones, listen politely while the other side does the same thing and then repeat the dance until both sides have come to some kind of compromise they can both call victory. That was most definitely not Will Eaton's style. He made his feelings known most clearly to his superiors, subordinates and diplomatic counterparts alike, and there was little diplomatic about those feelings. Eaton had nothing but disgust and repulsion for Bobba Mustafa, and his attitudes about the other royalty in the area were little better and often worse. He regularly sent ponderous missives back to the State Department demanding that military action of a massive and overwhelming nature be unleashed upon the Barbary States, and PDQ at that.

Stroke of State

These dispatches were just as regularly ignored. At the time, the US Government had no desire or, for that matter, no capability to start a war with anyone, least of all the Barbary Pirates. As far as Congress was concerned, it was still cheaper to pay them tribute than to build an army and navy; one frigate alone cost more than $100,000 just to build. Manning, maintaining, and actually deploying it tripled or quadrupled that, and you needed a force of at least six to eight frigates, plus all their associated support ships, crews, and other impedimenta. And a large, standing army, capable of deploying overseas in sufficient numbers to effect a victory? Absolutely out of the question. That was one of the imperial horrors that lurked in the minds of the Founding Fathers, and no one was about to try and change that decision any time soon. On the other hand, for just $275,000 a year, the Dey of Tripoli would be quite happy, and probably wouldn't ask for any more until next year.

No one in Washington liked the situation; who possibly could? After all, Americans were being captured or killed, held hostage, their property stolen, their lives disrupted and bankrupted and their commerce halted while the Deys, Bashaws and Co. were making a profit on the whole thing. But . . . was it worth a war? Was it worth the cost? Surely, nothing could be worth the bloodshed and the pain, surely it would be far better to just go along, wouldn't it?

We still hear these questions today about other rulers in that part of the world and their actions. Is it worth it? Should we anger others just to prove a point? Is anything worth a war? William Eaton thought so, and, give him his due, he never wavered from that position for the rest of his life. He believed that the only language the deys would understand was force; to make the game far more painful than it could ever be worth. Other American diplomats would face the same choice in other times and other places, and sadly, far too many would believe that it was better not to stand up and draw the line, but far better instead to smile politely and pay the money . . . or worse. What they always seem to forget is that, as Rudyard Kipling would point out eighty years later,

> It is always a temptation to a rich and lazy nation,
>
> To puff and look important and to say --
>
> "Though we know we should defeat you
>
> We have not the time to meet you.
>
> We will therefore pay you cash to go away."
>
> And that is called paying the Dane-geld;
>
> But we've proved it again and again,
>
> That if once you have paid him the Dane-geld
>
> You never get rid of the Dane.

To Barbary's Far Shore

The deys probably didn't know or care who the Danes were, but they would have understood the concept quite nicely; they certainly applied it well enough. Every year, the Dane-geld went up, and the pirates would comply for a little while. Then the attacks would start rising again, and the complaints would be made, and the deys would offer to halt the attacks completely this time, for just a little more money, and the gavotte would begin again, ad nauseum. As long as no one was shooting at anyone else on an organized basis, Congress and the Administrations were perfectly willing to let this state of affairs go on indefinitely. Needless to say, this infuriated Eaton to the point of distraction. Not only did the letters keep flowing from that side of the world, but it was becoming painfully clear that Eaton was not at all interested in keeping his contempt of the Barbary rulers hidden. He was not alone in his feelings, mind you. His fellow consuls, Richard O'Brien at Algeria and James Cathcart at Tripoli, shared his attitudes, but did a much better job of keeping them under wraps. He also had serious problems getting along with the men who were sent out to try and provide some muscle for our policy there, the Commodores of the Mediterranean Squadron.

First was Commodore Thomas Truxtun, who had made a name for himself commanding *Constellation* during the war with France. Truxtun was the second most senior officer in the USN at this point, behind only Richard Dale, of whom we will hear again momentarily. He brought over a four-ship squadron in 1801 that was nowhere near sufficient to deal with the Barbary threat. When they arrived that spring, Eaton was horrified to discover that Truxtun was more interested in personal glory than in backing up diplomacy. And, since he had neither the ships or the men to achieve such glory, Truxtun spent his time showing the flag and trying to get back home to command one of the big 74-gun ships-of-the-line that were being planned just then. Eaton fairly exploded in correspondence; while the Secretary of the Navy realized that Truxtun wasn't doing anyone any good and recalled him upon the arrival of Richard Dale.

In the meantime, Eaton, Cathcart, and O'Brien had their hands full. Relations were deteriorating badly with Tripoli, as the Dey seemed to have no real interest in getting his lads under control. Cathcart, probably the most capable diplomat of the three, did his best to hold things together until Dale arrived, but it was no use. On May 14th, 1801, a curious and terrifying ritual took place at the American Consulate in Tripoli. A group of the Dey's senior ministers and advisors, followed by a huge crowd shrieking dire imprecations, showed up at the consulate. The Dey's men then marched into the courtyard and, before the horrified eyes of Cathcart and his staff, cut the flagpole and its banner down. The crowd went berserk and dragged the flag through the streets of Tripoli, making sure to run it through whatever piles of garbage and camel dung they could find. Bringing it back to the consulate, they then burned it, rising to transports of ecstasy. One of the Dey's men kicked on the door of the

Stroke of State

consulate until the mortified, and furious, Cathcart answered, then told him that this is how we declare war in Tripoli. Laughing, he strode away, leaving Cathcart and his men surrounded by the mob for another few hours until they tired of their sport and went home. Cathcart was no fool, and he evacuated the Consulate as soon as it was humanly possible to do so.

Dale knew none of this as he sailed from the States on June 2nd, still with only four ships, the frigates *President*, *Philadelphia*, *Essex*, and the schooner *Enterprise*. On the face of it, he was an excellent choice to take things over. He had plenty of combat experience against the best in the world, the Royal Navy. Dale had shipped as John Paul Jones' first lieutenant aboard *Bonhomme Richard*, as well as commanding a privateer. The first inkling Dale had that anything was wrong was when he stopped at the Royal Navy's base at Gibraltar, finding two Tripolitian frigates there, commanded by Murad Reis himself. The British were, as always, helpful and willing to do just about anything to assist, but here their hands were tied. They were neutral in this fight, and Reis and his task force were there quite legally under international law.

There were a few spectacular fights, most notable of which was when Lieutenant William Sterrett, commanding *Enterprise*, captured the Dey's flagship *Tripoli*, following a magnificent example of the 'false flag' trick that would capture *Mastico*. Sterrett ran up the Red Ensign, and the *Tripoli's* captain, talkative as well as incautious, proceeded to hail the 'Englishman' and tell him that they were looking for Americans to capture, and did he know where they might find any? Sterrett, who once ran one of his own gunners through when he thought the man was about to desert his post during a battle, must have smiled like the Devil Himself when he answered. At almost the very same moment, *Enterprise* let loose a broadside at almost point-blank range, and the fight was on.

The pirates followed their doctrine to the letter. They closed and tried to board, which led to some anxious moments aboard *Enterprise*. Sterrett's men were able to keep the grappling hooks off, and *Enterprise* backed away to pound them into scrap wood. The pirates then used a stratagem that has, sadly, made reappearance in that part of the world, a false surrender. Twice within ninety minutes, the pirates struck their colors and, as Sterrett closed to board, ran them back up again and opened fire. It was only on the third 'surrender' that it took. Sterrett wasn't talking any chances; he kept firing right up to the moment he sent the boarding crews over. Needless to say, that brought the casualties aboard *Tripoli* to an unusually high level; fifty dead or wounded out of a crew of eighty. *Enterprise* hadn't lost a single man.

However, instead of sending *Tripoli* to the bottom, or, better yet, hauling her into a port with a prize crew, just to prove a point, Sterrett's orders from Dale required him to dismast her, send her cannon over the side . . . and let her go. In fairness, *Enterprise* would have been hard pressed to provide a prize crew, but why Dale gave orders like that is a bit tough to figure out. It seems that he may have interpreted his orders to indicate that they were not to actually sink the Tripolitans, but rather just teach them a lesson: "This is what we can do to you, now leave us alone."

Unfortunately, this idea didn't work any better with the Tripolitans in 1801 than it did with the North Vietnamese in 1965. They didn't see it as a sign of American restraint, but rather as a sign of American weakness. Mind you, *Tripoli*'s captain probably didn't see it that way. When he got his crippled ship back to Tripoli, the enraged Dey gave him five hundred with the bastinado, then paraded him through the streets of Tripoli seated backwards on an ass, with the entrails of a sheep draped around his neck. If nothing else, that would be tend to be conducive to far better command performance in the future.

But, in any event, Eaton was distinctly unhappy that the *Tripoli* was allowed to get home, no matter what her condition. His demeanor only got worse when a series of problems began to dog the squadron. First, *Enterprise* came up short on supplies after taking care of Tripoli, so Dale had to detach *President* to replenish her at sea. Then, there was a run of bad or spoiled supplies in the squadron as a whole, plus a sick list that was disturbingly large, and getting larger. The US Navy had fairly enlightened policies for that era in trying to keep its crews healthy enough to fight, but for decades to come yet there would still be only so much they could do, and it looks as if the surgeons on each of the ships had their hands full.

Dale had been trying to maintain something resembling a blockade, but with the handicaps he was now laboring under, there was simply no way he could keep it up. So, on September 3rd, without consulting any of the US diplomats in the Mediterranean at the time, Dale had *Philadelphia* stay on station off of Tripoli, apparently to just remind the Dey we cared, ordered *Essex* to stay near the Rock, and then started out for home himself in *President*. Or at least tried to. A local pilot ran *President* aground while trying to get her underway, and damaged her keel so badly that any chance of a passage across the North Atlantic was finished without spending some time in a proper dockyard. The nearest one was the big French base at Toulon, so Dale gingerly guided her into drydock there, and there she stayed until February 1802. The frigate *Boston* came out to replace her, but, for all practical purposes, that left only *Philadelphia* and *Boston* on station in the Med.

Dale's early successes did have the effect of sending the Dey's fleet into harbor for a few months; just long enough, as it turned out, for people to get the impression that the danger had passed. Two centuries later, we would throw the occasional cruise missile or airstrike at other malefactors and assume that, because they were briefly silent, then the crisis was past. We would live long

Stroke of State

enough to regret that assumption, and so would the Jefferson Administration. To be fair, Jefferson seems to have understood that the situation was worse than it appeared. In his first State Of the Union message, he made it clear that he felt more military force might be needed. However, Jefferson didn't push the issue and neither did Congress. In the meantime, William Eaton's feelings on the matter may be imagined. More letters, written in high dudgeon, crossed the stormy winter Atlantic in the fastest ships Consul Eaton could find.

The Navy had taken the opportunity of the winter stand-down to prepare a new squadron to go over in the spring of 1802. Built on the frigates *Chesapeake*, *Adams*, and *Constellation*, with *Enterprise* in trail, it was to have been commanded by Tom Truxtun. Truxtun, however, was already of a very doubtful mind on this particular deployment. *Chesapeake* had been tied up at Norfolk since early 1801 with only a single caretaker officer aboard, and it showed. She was in miserable condition, but Truxtun rolled up his sleeves and turned to. He had just started to make some headway when the Navy Department threw another spanner into the works.

Truxtun had agreed to command the squadron on the condition that he would have a flag captain aboard *Chesapeake* to relieve him of the day-to-day drudgery of running that vessel, so he could devote his full attention to the squadron. The Navy Department agreed that this was a most sensible idea, and assigned Captain Hugh Campbell to the job. Campbell, a long time Navy veteran, was an exceptional commander and excellent administrator. In addition, he got along well with the notoriously prickly Truxtun, so everything seemed to be looking up for a late April deployment. Unfortunately, as time went by, the Navy Department kept putting Truxtun off about getting a full crew aboard *Chesapeake*, and there was no sign of Captain Campbell. Truxtun was getting progressively more suspicious of the department, and it showed in his letters.

Finally on February 23rd, the boom, so to speak, was lowered. The Secretary of the Navy let Truxtun know that Campbell would not be reporting to *Chesapeake*, and that no one was going to be assigned to take his place either. That did it for Truxtun. He wrote an extremely angry and intemperate letter to the Secretary of the Navy, telling him that either he got a flag captain for the *Chesapeake*, or he was resigning as Commodore. The Secretary, Robert Smith, was cut from the same cloth as Will Eaton, so the final collision must have been as impressive as it was terminal. The Secretary accepted Truxtun's resignation with 'painful regret.' One doubts the truth of that, but never mind. There is the strong possibility that Truxtun was bluffing about the resignation; but, if so, then it raises the interesting question of why he never tried to make amends and get his position back. But never mind. The Secretary then appointed a new commodore to take the squadron out, and Tom Truxtun would never command so much as a rowboat in the USN ever again.

To Barbary's Far Shore

Commodore Richard Valentine Morris, United States Navy, was . . . well . . . possibly not the best choice for the job. He was given a much clearer set of orders than Dale had been, and there were high hopes for his tour there. The gods, however, apparently having begun to tire of their sport with Eaton as a diplomat, decided to amuse themselves with Commodore Morris.

Morris ended up arriving in Gibraltar on May 25th, making good time from Norfolk. *Constellation* and *Enterprise* were already there, and the *John Adams* finally showed up on July 21st, commanded by Hugh Campbell. Campbell dropped anchor and reported to Morris to find that apparently, preparing for combat had a low priority in his squadron. Morris had brought with him, on a combat deployment, his wife and son, plus their maid, Sal. Morris had already formed a tight little clique of officers who seemed to feel that their mere presence in the Med was sufficient to deter the Dey from further misbehavior. Actual patrols, on the other hand, were an unpleasantness to be avoided at all costs.

Constellation, under Alexander Murray, actually ended up going in alone within a few days of Morris' arrival. Murray got to the Tripoli coast in good time, to discover, of all things, a Swedish squadron on blockade duty there; some of their merchants having been annoyed by the Dey. Murray was glad to have the help, as *Constellation* drew far too much water to go in close, and Murray, though a good man, was a less than aggressive commander. He then proceeded to assure every American merchant skipper he met that there was no cause for alarm, and they could go happily about their business. Some, to their eternal regret, believed him and were most discomfited a few days later when the Dey's men went a-raiding again. One merchantman, the *Franklin,* was captured by the pirates and then actually towed past a stunned Captain Murray, who did nothing because he feared taking *Constellation* in close enough to recapture her, lest he run aground.

Commodore Morris was angry about Murray's performance, but apparently not angry enough to actually leave Gibraltar until mid-August, and then only to visit Malta to socialize with the Royal Navy leadership there. The strains of such an active schedule apparently caused him to lose track of his ships, because without his knowledge Murray took *Constellation* to Malta to pay his respects to the Commodore. All this time, *Adams* and *Enterprise* were still anchored at the Rock, waiting for orders. So there were now four warships in the Med, not a single one of which was covering their enemy's only base.

Cathcart and O'Brien were apoplectic, but Eaton was apocalyptic, and, for once, with good reason. At Gibraltar, the Navy's routine seemed to be nothing but parties and receptions, followed by hangovers and recoveries. On the rare occasions when they left the safe shelter of the Rock, they simply headed for the nearest friendly port and resumed their entertainment schedule again. Cathcart caught up with Morris in Livorno, Italy, to try and talk some sense into him, as well as catch a hop to Algiers to see if the Dey there could be convinced to use his good offices in America's favor. At the end of January 1803, Morris set sail for Algiers with *Chesapeake, John Adams,* commanded by Lieutenant John Rodgers, and *New York,* Captain James Barron, commanding. Morris was expecting a quiet crossing. He didn't get one.

The weather had turned remarkably ugly; combined with an unexpected supply shortage Morris, decided to put in at Tunis. Unknown to Morris, who really was not minding his shop, *Enterprise* had captured a small Tunisian ship named the *Paulina*. The Dey of Tunisia, only marginally friendlier than the Dey of Tripoli, was not at all happy, and made his feelings clear to Commodore Morris. Will Eaton, already on duty in Tunis and being something of an expert in dealing with infuriated Berbers, counseled restraint and good feeling, and to avoid going ashore. Morris, however, decided he knew better and went ashore anyways, taking Cathcart and Lieutenant Rodgers with him. The real problem here was that Morris and Eaton cordially loathed one another. In doing so, each most cheerfully ignored advice that might have made a difference. So it was with the utmost in self-confidence that Morris showed up at the Dey's palace and demanded, not asked, not respectfully requested, but demanded, an audience with the Dey.

And did he ever get one.

Morris broke just about every one of the myriad and complex rules of etiquette in the Muslim world during his audience with the Dey. The climax came as the meeting ended. Instead of waiting for the Dey to dismiss him, Morris simply rose and turned his back on the Dey as he left. This was literally the straw that broke the camel's back. The Dey was almost beside himself with fury. Commodore Morris, on the other hand, seems to have been almost blissfully unaware that he did anything wrong. In a culture where merely showing the heel of one's shoe to someone is considered near to a mortal insult, remember the video of gleeful Baghdadis slapping the fallen statues of Saddam Hussein with their sandals, Morris had pretty much gone completely beyond the pale.

Worse was to come. As Morris, Cathcart and Rodgers headed back to the waterfront, they were headed off at dockside by approximately three hundred enraged, and heavily armed, Tunisians, and placed under arrest for the crime of insulting the Dey and, by extension, Tunisia. The Americans were confined for about a week or so, while Will Eaton did his level best to get them out. Morris, serene in his righteousness, refused to even discuss compromise until Eaton and Cathcart pointed out that since the United States Navy, represented by the person of Richard Valentine Morris, seemed to be so loath to even try and protect American merchant ships worth millions of dollars, why did the

To Barbary's Far Shore

Commodore think that the rest of the Navy would bestir itself to rescue two officers and a junior diplomat? One can believe that the two Consuls made their point, and Morris finally decided that discretion was the better part of valor. Sending a suitably contrite note to the Dey, Morris admitted that he may indeed have spoken rashly and impolitely, and perhaps the good ship *Paulina* was better off with her original owners. The Dey, ever magnimanious in triumph, agreed and decided that twenty-three thousand dollars would be a suitable palliative for his distress.

What happened next isn't exactly clear. It appears that Eaton, who had been exceptionally well behaved during his efforts to get the three Americans released, was in the process of delivering the ransom, excuse me, settlement, to the Dey when his self restraint finally cracked in a most spectacular fashion. Eaton proceeded to deliver a blistering, acid-tongued speech to the Dey regarding the whole sorry affair that had to have been impressive even by Eaton's own high standards. When he was finished, there were four men headed out of Tunisia: a much-chastised Richard Morris, a relieved James Cathcart and John Rodgers, and a thoroughly unrepentant Will Eaton.

Eaton had been, in modern diplomatic terms, 'PNG'd,' or declared *persona non grata* and sent home. However, even in disgrace his amour propre was unruffled. He refused to sail with Morris on the comparatively luxurious and roomy *Chesapeake*, instead asking Rodgers if he might have a small corner aboard *Adams*. Rodgers, whose opinion of Morris was very nearly as high as that of Cathcart and Eaton, happily agreed and the flotilla left Tunisia behind on the southern horizon. They finally arrived in Algiers on March 19th, where the Dey of Algiers refused to even talk to poor Cathcart. Richard O'Brien, the US Consul in Algiers, was now the only accredited diplomat on the entire Barbary Coast. The very next day, Morris raised anchor and headed for the safety and order of Gibraltar.

The Commodore seems to have dimly realized by this point that he was in deep trouble. Washington was also realizing it too, due to the fact that Morris had not sent a single report back to the Secretary of the Navy for more than four months. That spring, Morris tried to get his act together once more for an attack on Tripoli. It almost misfired when *New York* suffered a near-catastrophic powder explosion that took her out of the line and killed fourteen of her crew. Pressing on despite recurring maintenance and supply problems, Morris actually accomplished quite a bit, capturing or destroying several Tripolitan ships and finishing with an amphibious raid west of Tripoli on June 1st. There was some talk that he might have changed his ways, until he suddenly pulled himself and the fleet away from blockading Tripoli to return to Valetta, where his wife was about to give birth to their second son.

Stroke of State

Morris always denied that was the reason, but it mattered not. Navy Secretary Smith had been getting mounting reports through other channels about the confusion and incompetence of the Mediterranean Squadron, and he was not likely to let it go on forever. Therefore, exit Commodore and Mrs. Morris, their children, and the beloved Sal, stage left. John Rodgers, who was routinely called many things by his men, but never unaggressive, was placed in charge until Edward Preble could get there. After arriving in Gibraltar after the unpleasantness in Tunisia, Eaton left almost immediately aboard *Constellation*, which was headed home for a badly needed refit. On the way home, he spent most of his time writing his report to the Secretary of State.

It was not a rosy one. Eaton reviewed everything that had happened since his arrival four years before. With the exception of a few standout victories, it had pretty much been one long litany of disaster, mistake, and confusion, leavened with just a touch of utter incompetence. Eaton stated for the record that he felt that whatever small good the United States had accomplished there had been completely undone by the repeated military and diplomatic failures. He also felt that unless something was done, and soon, the US might be forced to evacuate the Med altogether.

Will Eaton was by nature an optimist, but it had been quite a while since he'd been optimistic about anything. For all his faults, he took his job and his responsibility very, very seriously. Even with his unquenchable faith in himself and his abilities, he was doubtful about what might come next. A review of the facts on the ground, as a later generation would say, would hardly give any cause for hope.

There was almost no American diplomatic presence on the Barbary Coast, save poor, lonely Richard O'Brien at Algiers, trying to juggle a half dozen recalcitrant deys, beys, and etc. The Danish consul at Tripoli, Nicholas Nissen, did the best he could to assist, often placing himself at immense risk to do so. The military effort had sputtered from marginal to hideously ineffective under three commanders whose abilities were hindered by personal quirks that varied from quests for glory and reluctance to push one's orders to advanced military and diplomatic socializing.

Whatever military and political gains had been made, and there had been damned few, hadn't impressed the Dey of Tripoli at all, other than to give his negotiating skills a workout. And worst of all, the signals from Washington were mixed, weak, or nonexistent. All Congress had advised anyone on an official basis was that they 'recognized' a state of war between the US and Tripoli, but did not declare one in return. Given the confusion in the area at that point, it's hard to see how that could have possibly helped matters. In fairness, a great deal of that could be laid to the problems inherent in a minimum eight to twelve week time for a message to get there and come back. But other nations, the

To Barbary's Far Shore

British most notably, had the same problem and worse, yet seemed to be able to do things much more smoothly. It might have just been more experience on the British part. On the other hand, the United States had some genuinely impressive talent working in the infant State Department in the form of Thomas Jefferson, James Madison, the present Secretary of State, and Tim Pickering, the Secretary who had appointed Eaton.

However, things had not been smooth at State for some time. Jefferson, as the first SecState, had set a standard that has been dauntingly tough to meet ever since. But after that, President John Adams, who rightfully suspected that Pickering was serving two masters, had fired Pickering. The second master in this case had been Alexander Hamilton, who was Adams' mortal political enemy. In those days, when political parties were more numerous, early Presidents had to occasionally work in almost a coalition fashion in order to secure legislative support, so they ended up appointing Secretaries from other parties or who supported other politicians. This is what happened with Pickering, who was apparently doing less than his best to support Adams' policies and rather more than he should have been to support those of Hamilton.

Under Jefferson, James Madison got the job and proceeded to put his own unique stamp on things. Madison had been one of that pantheon of immortals we know today as the Founding Fathers, the men who had laid down the bedrock that our nation has been built on. Madison was a physically small man, but made up for it with a ferocious aptitude for combat both political and military. He had served with distinction alongside Washington in the Revolution, and in the vicious political knife fights that had surrounded the writing of the Constitution he had held his own and more. The Bill of Rights in particular was Madison's handiwork, and, for that alone, we should be eternally grateful to him. In the meantime, he was throwing himself into his duties at State with a gusto that has been rarely seen since. Madison was what we would call today a hawk; willing to wield American power to achieve American ends, and not ashamed to admit it. The trouble with this was that, unlike later periods in American history, the United States essentially had no real military power that could deploy outside of the United States and its coastal waters. A quick look at what the US actually had to work with is probably in order here.

There really was no army worthy of the name. The official beginning of the US Army was on June 14th, 1775. It consisted of just ten rifle companies, or about one thousand men. Everything else, everything, came from the state militias. These were basically every adult male from age eighteen up, and they came together as needed for very limited periods of time, usually around a year or so. Needless to say, the retention problems made planning anything resembling a coherent campaign absolute hell for General Washington, but, as history shows, he did a magnificent job.

Stroke of State

The old Continental Army eventually rose to about twenty thousand men; all but a handful of whom simply went home after the signing of the Treaty of Paris. A very few state militia units were kept on duty at certain military strong points, such as West Point or Fort Pitt in Pennsylvania; at Federal expense, of course. In fact, Congress had actually ordered the Continental Army to disband in 1783 when the Revolution ended, but quite a few of the remaining troops . . . well, remained. Congress, always mindful of the Founding Fathers' wishes regarding a standing army, not to mention the fact that the recalcitrant troops were costing money that they were averse to spending, gave the disband order again on June 2, 1784; apparently just noticing that they still had an army. Congress stated in the order that "standing armies in time of peace are inconsistent with the principles of republican government." Noble sentiment indeed, and truly in keeping with the attitude of the Fathers. Said sentiment lasted just forty-eight hours. For, on June 4th, Congress performed an about face breathtaking in its speed and scope.

In that short span of time, someone seems to have convinced Congress that the best thing to do was make a virtue out of necessity. They proceeded to order the formal creation of a single regiment, to consist of ten companies: eight infantry and two artillery. Designated the First American Regiment, they served with distinction up until the point where they went into combat against the Indian tribes in the old Northwest Territory. The First American's commander, General Josiah Harmar, wasn't the best Indian fighter in the world and the First American had its head handed to it on several occasions. The worst of these was a three-day stretch in October of 1790, when Harmar managed to lose, through poor planning, coordination and possible drunkenness on his part, two hundred and twenty-three men out of five hundred and forty; forty-one percent of his force. The First American would take a further beating and finally withdraw to rebuild and refit, while General Harmar would be officially exonerated of any blame. On the other hand, it had to have been a grim victory on Harmar's part; for although the court-martial cleared him George Washington did not, firing him from his post as the Senior Officer in the United States Army.

The First American would later, barely, survive near annihilation a year later at the hands of the Miami chief Little Turtle, in what is still the worst defeat the US Army would ever take at the hands of Native Americans. Eight hundred and thirty two US soldiers and camp followers were left to return to the dust near present-day Fort Recovery, Ohio on November 4th, 1791. Put into perspective, George Custer and just two hundred and sixty seven others would lie on the dusty bluffs above the Little Bighorn eighty-five years later.

As with any major disaster suffered by any branch of the service, the first thing the Government decided to do was reorganize things. General Harmar, somehow managing to survive all of this, retired in 1792 and went home to Pennsylvania, where he lived in relative comfort. His replacement, and the man who led the First American into the deadly embrace of Little Turtle, was General Arthur St. Clair. He too survived his encounter with the Miamis, and although he was officially exonerated, it seems reasonable to believe that whatever his faults as a commander, he knew which way the wind was blowing. Taking retirement not long after Harmar, he served as Governor of the Northwest Territory before it became the state of Ohio in 1803. St. Clair wanted very much to see the Northwest Territory become two states and worked hard to stop a single-state solution, something that apparently did not endear him to the leadership in Washington. With that, President Jefferson relieved him of his duties and he retired to Pennsylvania, dying there in 1818. The men Harmar and St Clair left behind were reformed into a new organization, the Legion of the United States, a combined-arms (in this case, infantry, artillery, and cavalry) unit of approximately brigade size. And, in the first truly intelligent decision anyone had made up to this point, the Government put Major General Anthony Wayne is charge, which is how William Eaton met him in 1792.

With Wayne in charge, things changed for the better, and right quickly. Wayne improved training, laid out an intelligent supply system based on a series of forts, and then went about securing all that Harmar and St. Clair had lost. It was a brutal, vicious campaign with no quarter asked or given. It finally ended on August 20th, 1794, at an anonymous little place in present-day Maumee, Ohio, known then as Fallen Timbers. Wayne not only lived up to his nickname, he improved on it, launching a bayonet charge against the Miamis. The Miami war chief, Blue Jacket, tried to escape with his braves into a nearby British fort, but the commander there decided that this was not a good day to involve His Majesty's Government in a war with the United States, and barred the doors. The Miamis scattered, and the Northwest Territory, though still a grimly dangerous place for some years to come, was finally under the control, however tenuous, of the United States of America.

However, at about the same time, the Whiskey Rebellions, where local refusal to pay excise taxes on home brew rapidly turned into a widespread revolt against Federal authority, scared the living daylights out of the Federal Government, which had to borrow state Militia troops to put it down. The rebellions had been simmering for nearly three years by that point, and when they finally boiled over, President Washington himself had taken command, the only time a sitting President has led troops in the field, and then turned it over to General Henry 'Lighthorse Harry' Lee III (father of Robert E. and the sitting Governor of Virginia) to execute the actual campaign. The militia turned in a reasonably decent performance; with Washington and Lee at its head, it would have been difficult to do otherwise.

The problem, as it turned out, had been in getting the militia into battle in the first place. First, the Federal government wanted to call up approximately thirteen thousand men; a number that exceeded what Washington usually had in the field during the Revolution, and a number that Congress had been notoriously bad about actually paying for when the nation's independence was at stake. But, when the call-ups went out, very few men actually stepped forward. The Federal Government, in its infinite wisdom, then decided that a draft, America's first, was in order. The result was a disturbingly high number of men who evaded the draft or worse, led riots against it. The new army's first task, therefore, was to put down a rebellion caused by its own establishment. Finally, the miserable performance of state militias in fighting still-belligerent Indians in what would be Ohio, Michigan and Indiana drove the message home that there was going to have to be a regular military force under Federal control.

Congress, always willing to spend less when more was needed, authorized a single infantry regiment of seven hundred men. Not much, in present terms, but the concept of a standing army was anathema to the Founding Fathers, and anything larger was considered a threat to the very existence of the Republic. That single regiment, the sapling from which would emerge the most powerful and capable army in history, would be it for quite some time to come. The idea of deploying it outside the United States was quite unthinkable. For the time being, it would have no other purpose but to hold the line in on the frontier against the Indians.

The Navy wasn't really in much better shape. Initially organized on October 13th, 1775, the Revolutionary Navy had little success on the high seas. A Congress reluctant to establish and fund a proper army wasn't willing to do any more for a Navy that had cost far more than it seemed to have been worth. By 1784 the USN consisted of exactly two near-derelict ships, the *General Washington* and the *Alliance*, both of which were actually working as chartered freighters at the time to help pay their way. As the Barbary pirates started their depredations at about that time, Congress repeatedly refused to fund a navy capable of stopping them, and finally decided to sell both the *Washington* and the *Alliance* in order to save money.

When the new Constitution was debated in 1787, the idea of a Navy was a noisily argued one. More than a few delegates felt that a navy was an invitation to conflict with other nations. Many of the rest felt that America needed to look inward, to the seemingly endless space that unrolled west of the Alleghenies, instead of to the Old World. More practically, there were serious regional differences that were blocking a navy; most notably the fact that the North, with most of the wood and sailors and almost all of the good harbors, would profit the most.

It was James Madison, though, who made some of the most effective arguments in favor of a navy. He pointed out that, without a strong navy, we were simply inviting trouble in the future. It worked, but again not by much. It would not be until 1794 that any serious efforts were made at actually

To Barbary's Far Shore

assembling a navy. A report done that year called for four 44-gun frigates and two smaller 24-gun ships, all intended to deal with the Barbary threat. Seemingly reasonable enough, Madison, now a congressman, took a look at the figures, which called for building those six ships, manning them, and maintaining them for one year at $850,000. Madison realized those figures were, to use a more modern term, lowballed, and said so.

The result was to end up going to Joshua Humphreys, the Philadelphia naval architect and shipbuilder who had come up with the concept of a 'superfrigate,' capable of taking on and defeating any other frigate in the world. The result was the *Constitution* and her sisters, possibly the finest sail warships ever built. He would build them for a much more realistic $100,000 each. But, once Congress got loose in the process, things spiraled out of control. Their construction was spread out between Boston, New York City, Philadelphia, Norfolk, Portsmouth, N.H., and Baltimore. Each ship had a different officer overseeing its construction. For instance, Tom Truxtun had *Constellation* at Baltimore and Richard Dale had *Chesapeake* down at Norfolk. That meant that although the ships were built to the same general design, each one was fitted out according to that officer's wishes. The rough equivalent today would be telling the prospective commander of a missile cruiser that as long as it was a certain length and displacement, he could put whatever he wanted on it.

In addition, the amount of 'live oak,' that indispensable shipbuilding material of the age of sail, that would be needed for construction was badly underestimated; though this seems like it would be more the error of a War Department bean-counter than the experienced Humphreys. As the frigates slowly and painfully took shape, their expenses quite understandably began to spiral into that rarefied atmosphere we call today the cost overrun, which led to eager efforts on the part of some congressmen to scuttle the entire project. And as if that wasn't bad enough, an even more dangerous threat appeared.

There had been a clause in the initial funding authorizations, Section Nine, that said in the event that peace was achieved between the United States and the pirates, any and all naval building and funding would be shut off. And sure enough, in late 1795, with none of the frigates even close to completion, a treaty was achieved with the Algerians. For a one time payment of one million dollars, the Algerians would happily leave US shipping alone forever and ever, amen; while we would save even more by scrapping those damned frigates which we didn't need anyways. The fate of the US Navy hung in the balance for the better part of 1796, but in the end furious negotiations resulted in funding to complete the first three ships, while the other three, *President*, *Congress*, and *Chesapeake*, would be suspended to save money.

The first one, USS *United States*, wouldn't even be launched until May 10th, 1797, at Philadelphia. *Constellation* was launched at Baltimore in early

Stroke of State

September, and *Constitution* herself wouldn't get into the water until September 20th at Boston. Well, actually, she tried to get into the water on September 20th; she made it halfway down the ways towards the Charles River, and then stopped cold. It took thirty-one more days and two more tries to get her in, but she finally slipped into the muddy Charles on October 31st, and her commander, Captain James Sever, toasted her with the best Madeira he could find. But once *Constitution* overcame her reluctance to get wet, all that did was put three unfinished hulls in the water; hulls that would take further months to fit out and get ready for sea. In the meantime, the USN was relying on converted merchant hulls to hold things together, and it wasn't working that well. The pirates kept capturing US ships and holding their crews for ransom, and the USN seemed to be powerless to do much about it.

Fortunately, the American gift for finding the right man at the right time made a fortuitous appearance when President Adams named Benjamin Stoddert as Secretary of the Navy. His predecessor, James McHenry, had been a well-intentioned but utterly incompetent political appointee, one who was further crippled by his connections with Timothy Pickering and Alexander Hamilton. Stoddert, a wealthy Georgetown merchant who had also held a number of high posts under the old Continental Congress, was also an exceptional administrator who understood matters of shipping and had some solid business and political connections. He had been a little reluctant to take the job; in his late forties, he was actually looking at quiet, well-funded retirement. Fortunately, he decided to take the job, and just in time. On the very same day Stoddert accepted the post, Congress authorized hostilities with France in what has been called the 'Quasi War'.

France was still in upheaval after the nightmare of the Revolution, and their behavior was as aggressive now as it had been helpful in the closing days of the War for Independence. Now at war with the British and Spanish, the French interpreted the old Revolutionary War treaties with us as entitling them to buy and outfit privateers in American ports, as well as organize and launch assaults against Spanish territory. Needless to say, the British and Spanish were less than pleased, and American shipping, doing business with both sides, was caught in the middle. The French interpreted a new US treaty with the British as a direct move against them, and proceeded to start stopping, and frequently taking, US shipping.

Now, as we have learned to our repeated surprise on several occasions, the French are nothing if not practical. Their envoy to the US, Edmund Genet, seems to have suggested as much on several occasions, but, being the simplismé types that we are, it didn't really sink in until early 1798. A group of French 'agents,' who have come down to us only as Messrs. X, Y, and Z, came to the US and said, very discreetly, but quite directly, that all would be well, if only President Adams would deliver a groveling apology and a suitable bribe, excuse me, 'compensational payment,' was made to the French. It is at least possible that the messieurs were acting on their own behalf and not that of the French government, that sort of thing was not unknown, but no conclusive evidence of that has ever come to light.

That did it for Congress, once they saw the reports of the 'XYZ Affair' in April of 1798. The muted fury of American reaction to French diplomacy in 2003 was nothing compared to the thunderstorm that boiled out of Congress and public opinion. Suddenly, the idea of a strong navy didn't look so bad or so expensive, as it was the only effective way we could hit back at the French. Accordingly, Congress appropriated not quite one million dollars, probably the largest single military appropriation in US history to that date, to get the Navy up to speed. Three days after that, on April 20th, 1798, the Department of The Navy, with a Secretary of Cabinet rank, was established.

On the whole, the US Navy's war against the French went pretty well. The first ship to sea was *Ganges,* a converted merchantman. None of the Humphreys frigates was ready yet, though Stoddert was bearing down hard on their captains. Tom Truxtun was having fairly serious disciplinary problems with the crew he had put together for *Constellation,* which he cured in his own inimitable fashion: he mustered the crew on the main deck, read them the Articles of War, then flogged a marine private for the crime of 'insolence,' a very wide-ranging and flexible offense in those days. The object lesson seems to have done the trick, and Truxtun's crew settled down and got back to work.

There were a few solid victories. Stephen Decatur Sr. aboard *Delaware* nailed the French privateer *Croyable*; an odd victory in that the *Croyable's* captain wasn't aware that the two nations were at war yet. Brought into Philadelphia, the first prize of the French War was recommissioned into US service as *Retaliation* and given to the young Will Bainbridge. However, a pair of ugly losses followed this early victory.

First, Bainbridge lost the *Retaliation* back to the French on November 20th; though in doing so he enabled two other ships, *Norfolk* and *Montezuma,* to escape. On the other hand, as was previously mentioned, Bainbridge's conduct was such that even the French had to admit he had acted in a competent and correct manner, so no blame was attached there. Would that the same could have been said for the sad case of *Baltimore,* Captain Isaac Phillips commanding. Phillips came up on the short end of a confrontation with a Royal Navy commodore bent on impressing some extra crewmen.

The RN, for the most part, kept its press raids confined to merchant ships, but there were a few RN commanders who felt that anything afloat was fair game. Commodore James Loring, RN, was one of them. And the game was truly afoot that day. Loring originally took fifty-five of *Baltimore*'s crew off the ship, but in the end only kept five. Phillips made the unfortunate mistake of giving in,

Stroke of State

without any effort to resist. In fairness, Phillips' orders may not have been entirely clear on what to do if hit by a press gang; after all, we weren't at war with the British, so Phillips had no reason to be expecting trouble. It must be remembered though that honor was all two hundred years ago. Even if Phillips was outnumbered, outgunned, and outmaneuvered, he still had a responsibility to make some effort at resistance. Secretary Stoddert moved fast and hard to make sure this sort of thing was nipped in the bud. Phillips was cashiered from the Navy, and Stoddert laid down an order that as long as they were able to fight, no captain was to ever allow a single man to be taken from their decks ever again.

The Humphreys frigates finally started entering service, and gave very good accounts of themselves. *United States*, *Constellation*, and *Constitution* all scored impressive victories over French warships, and although *Congress*, *Chesapeake* and *President* were finally completed, they didn't get a great deal of action in. The war with the French proceeded in fits and starts until July of 1800, when the French diplomatic and political positions started to change fairly suddenly in America's favor. It wasn't solely because of the string of essentially unbroken American victories at sea. The French were also dealing with the Royal Navy as well, and in those days, the RN was the eight hundred pound gorilla of naval combat. French morale was also headed downwards as well, as much as from leftover problems from the Revolution as much as anything else, and that was an absolutely unacceptable situation in a time when a ship was the sum of its men more than its technology.

Unfortunately, an old American tradition reared its lovely head at this point, disarming after a victorious war. When Thomas Jefferson became President in March of 1801, his Treasury Secretary, Pennsylvania's Albert Gallatin, took a meat cleaver to the Federal budget, which in those long gone innocent days could be reckoned in sums of less than twelve figures. The budget he submitted to Congress that summer for fiscal 1802 was less than half of the last Adams budget. Imagine the screams of outrage if next year's Federal budget was around 45% of this year's. As usual, everything else was more important than the military, which we wouldn't really need again since we'd won the war, hadn't we? If it hadn't been for the last minute political shenanigans of the Adams administration, which 'reduced' naval strength by permitting disposal of all the converted merchantmen that made up the bulk of the fleet, Gallatin probably would have disposed of the entire fleet, and he would have probably gotten away with it. Jefferson was nowhere near as anti-navy as he has often been made out to be, but he did feel that a large navy would encourage expansionist and interventionist sentiment; that, Jefferson felt, would have made us just like the British. Seeing as how the Revolution had been fought only thirty years previously, no one wanted to go imperial just yet. That would come in just about a century, but never mind. On the other hand, Jefferson did believe that we needed a fleet of some kind, if only to protect shipping. The question in his mind, and one that he seemed to be willing to discuss without end, was what kind of fleet should be assembled to do that job.

The sleight-of-hand pulled off by Adams saved the USN, but just barely. By the time Gallatin was done, the USN had an effective strength of thirteen frigates. Seven of those would be placed into mothballs, and the other six would remain on active duty, but only at about seventy percent manning. The Navy's officer corps was cut to nine captains, thirty-six lieutenants and one hundred and fifty midshipmen. Today, thirty-six lieutenants would barely meet the manning requirements of a single aircraft carrier and one could probably find nine captains in the Pentagon cafeterias at lunch hour; but, by late 1801, that's all there was.

What saved the USN from dwindling away to nothing was Will Bainbridge's ordeal in command of *George Washington*, when he was detailed by Bobba Mustafa to take his diplomats and menagerie to Istanbul. This was considered a flat-out insult to the United States. Even the normally calm and composed James Cathcart wrote a message to Secretary Madison that "whatever answer he gave to the Dey should be accompanied by two frigates."

The other arm of the US military in 1803 was that small band of brothers who would carve themselves an eternal place in history, and, in doing so, lay the foundations for a legend: the United States Marines. Congress says they created the Marines on July 11th, 1798, authorizing a strength of a single major, four captains, twenty-eight lieutenants, forty-eight sergeants, and seven hundred and sixty-eight lower ranks. The Marines, as might be expected, have a different take on the matter. They observe their birthday on November 10th, for that is the day in 1775 that the doors to the Tun Tavern on King Street in Philadelphia were opened to recruit a small group of men to serve as marines on American shipping. This also enables them to modestly point out that they are actually nine months older than the nation they have served so well, but never mind. The first few good men numbered two hundred and sixty eight, led by Major Samuel Nicholas and Captain Robert Mullan, who, by fortuitous happenstance, was also the tavern's owner.

The Revolutionary Marines served well, given the limited capability of the old Continental Navy. Growing to an eventual strength of 3,124, they executed a number of small amphibious raids, harbingers of what would come in the future; but, for the most part, they provided security for ships and their officers, as well as other, primarily ceremonial, duties. After the Revolution, as the Navy shrank, so did its need for marines, and it is unclear if there were any formal marine detachments at all on the handful of ships that unevenly bridged the years between Yorktown and the Quasi War with the French. The looming conflict with the French not only enlarged the Navy, but also made the formation of a

Stroke of State

proper marine unit on every warship a necessity. Marines were usually about a tenth of the crew of any given ship, but this could vary in either direction. Detachments were commanded by an officer, a captain or lieutenant, along with a handful of NCOs.

The long-running rivalry between marines and the sailors they serve alongside started here. The marines had very specific and limited duties, regardless of the type or size of ship they were assigned to. They served as ship's security in the event of an attempt at mutiny, protecting the captain and the command staff. They guarded the ship's magazine, the area where powder, matches, and other explosive materials were maintained; the one place where a single individual could blow even the most massive man-o-war into eternity. They provided men for the boarding parties; teams of men armed with everything they could carry whose job was to jump from their ship to another, to take by brute force what had already been pounded into insensibility by the gunners. They served as a ship's honor guard; often a crucial, if ceremonial, duty in the days when ship's captains were also de facto ambassadors. They were lethal sharpshooters in battle whose job was to knock down enemy officers and gunners in combat, and do it through smoke, flame, steam, and in darkness or light. In so doing, they created the foundation for one of the present day USMC's most cherished values: solid, well trained marksmanship. They stood, armed and ready, when the old rum ration was passed out every day; though there seems to be no indication of who stood guard over them when it was their turn.

They were not, repeat, not, permitted to share in the hundreds of mundane details that made up the care and feeding of a sail warship, and this seems to be the cause of the difficulties with ships' crews since. They did not scrub decks, except in their own areas; they did not paint, they did not climb the rigging to handle sail. Ships' officers tended to look down socially on marine officers, but in the close confines of shipboard life, relations there were always at least correct, if not cordial. On the other hand, most sailors of that day regarded marines as good-for-nothing and overpaid passengers, and most marines regarded sailors as a drunken rabble who were unfit to be entrusted with the responsibility that the marines were given.

For some reason, this led to the occasional friendly dispute.

The one thing that everyone did agree on regarding marines was their priceless utility as seagoing and amphibious raiders. Marines trained in standard infantry tactics, but they also trained in going ashore either from dockside or by small boat 'over the beach.' The result was the ability of a ship's captain or commodore to send ashore a highly disciplined and well-trained trained elite in any size from a squad to a company (depending on the size of a given ship), and rely on them to be able to handle whatever they ran across without intervention from the quarterdeck. This ability was put to use infrequently, as it took men away from their shipboard duties, but knowing it was there gave officers and diplomats one more option.

President Thomas Jefferson

It was, however, another facet of this ability that needs to be examined. The marines were able, on short notice and with minimal equipment, to get on a ship and sail to wherever they were told, and then go ashore to do their fighting if it became necessary to put muskets and boots on the ground. The United States

Stroke of State

Army of that time was utterly unequipped and unprepared to do so; but let this author make clear that it was through no fault of the men who served and those who led. Political beliefs and financial constraints made it impossible for them to do anything else. The Navy had ships that could sail anywhere on the planet and deliver a highly respectable punch, but they could not always get in close enough to do the kind of damage that convinces an enemy that discretion is the better part of valor. In the end, only the marines had that ability and training in 1803.

It's not always clear whether or not President Jefferson's reputation for brilliance encompassed an ability to understand and use the military options he had. He was a brilliant political thinker, but men of that stripe do not always bridge the gap between what is diplomatically possible and what must be done by force. He would have been quite happy to have left the rest of the world alone and turned westward, so that America could start its long, majestic march to history.

All Tom Jefferson wanted was to get along, to learn and exchange ideas, and in time, bring liberty to the world. In that, a bit of Tom Jefferson lives in most Americans. We really don't ask much more than that, when one looks at history. But as other presidents since have learned, be it through the smoke of a sinking fleet or that of a shattered skyscraper, our adversaries have tended over and over again to mistake a desire for peace as submission; a dislike of war as cowardice. And as other, more recent enemies have learned, after you hit us enough.

We hit back.

As if to reinforce that point, a few months later the good people of Tripoli paid that charming social call to the American Consulate, and the war was on.

TWO: THE PRACTICES OF PRESENCE

Aboard US Schooner Enterprise, *eight miles off the shore of Tripoli, December Twenty-Third, Eighteen Hundred and Three*

Enterprise capturing Barbary Corsair. Source: U.S. Navy History and Heritage Command

Most people tend to think of the Mediterranean Sea as a lovely spot that is always warm and friendly year round. And, since most people have never been to the Mediterranean, most people would be wrong – in the winter, it's cold in that part of the world, and the men who manned their action stations aboard the sloop of war *Enterprise* knew it. The northerly winds that time of year send a chill through even the most hardened soul; knowing that one could face one's maker that day makes it even colder.

Combat at sea two centuries ago was a very personal thing; not the lightning-quick duel of rocket propelled homicidal robots, lethally invisible submarines, omnipotent electronics and sensors, and terrifyingly fast aircraft that it has become today. You were often, far too often, close enough to see the gunners who were trying to kill you. Sometimes you were able to even see the brutal cannonballs that could turn you from a living, breathing human being in God's own image into a side of poorly butchered beef in a mere heartbeat. And, if by some miracle you survived the hurricane of cannonballs and knife-like wooden splinters they trailed behind them, then you had to face the boarding

Practices of Presence

parties. Often men just as terrified of combat as you were, but determined to make sure your ship was the one that paid the ultimate price, and prepared to hack and shoot their way through every last man Jack of you to do it.

Now those dangers were bad enough. But, human nature being what it is, you could always convince yourself that you had a chance against them. The cannonball might take the lad beside you, and splinters might shred another into bloody ribbons, but you would be able to avoid them. Someone else might not be able to fight off a boarding party, but you were pretty damned good with a blade, if you said so yourself, and you would be able to hack your way out of a fight.

But the marines . . . The marines were another matter entirely. Their job was to lurk with their muskets in the fighting tops like vultures, except that was probably an insult to vultures, who at least had the decency to wait for something else to kill you before they did their work. The marines would perch up there . . . hunting you. You couldn't see them for the smoke from the guns and the web of rigging and the forest of masts, spars and yardarms that surrounded them. You could be looking right in the direction of the one who'd get you and, in the god-awful meleé of combat, you might see the flash and fan of smoke, but you'd never hear it. Not even the most eagle-eyed tar could see the ball that would come at you faster than the speed of sound, a concept that wouldn't even be understood by science for another hundred years.

That was probably just as well, because a musket ball was truly hated by its targets for the way it killed you. A cannonball or splinter might turn you into something that not even your sainted mother could recognize, or leave you as something that couldn't even be buried in consecrated ground. A blade might sever an artery, but it was generally accepted that it would at least be fast. But a musket ball took its time, like it was some mindless monster with an urge to make you suffer and in no particular hurry to do it. They were frequently irregular lead spheres that averaged a half inch or more across; slowly tearing and rending their way through your body, leaving a gory tunnel from where it entered to wherever you were unlucky enough for it to stop. If there was anything good to be said about it at all, it was that the marines were not prejudiced at all about who they targeted; everyone within sight was a potential corpse, though they did show a definite preference for officers. A French marine would demonstrate that two years later at Trafalgar, when he put one through Horatio, Lord Nelson, severing his spine and condemning him to a slow, miserably painful death.

Enterprise's marines were kneeling on the deck beside her masts, ready to get to their positions in the fighting tops when they were given the word. Low to the deck like this, they're invisible to anyone on another vessel; that's exactly the way Master Commandant Stephen Decatur, *Enterprise's* captain, wants it.

Sergeant Solomon Wren, the senior enlisted marine aboard, is stalking the deck like a caged tiger, snarling at the marines and ready to take out any of his own men faster than the enemy would if they even looked as if they were going to shirk their duties. That was usually not anything they had to worry about, as the marines were generally far more afraid of their own officers and sergeants than they were of any foe. As long as Wren is out there, Decatur knows he has nothing to worry about.

Stephen Decatur lives for combat, one of those men of that era who sought adventure and challenge at every turn of his life. He came by it honestly enough; his father, Stephen Senior, had been one of the first captains of the US Navy when it was reconstituted in 1794. His sons, Stephen Jr. and James, were tough, rambunctious boys who thrived on the glory and thrills of life at sea. They followed in his footsteps almost as soon as they could set foot on a deck, and they both went from midshipmen to lieutenants in just a few years each; a quick amount of time in a day when it could take a decade or more to become an officer. Stephen went first, quickly building a reputation for himself as smart, aggressive, perhaps a little too much so, but in the early 19th century that is a very subjective thing, and extremely capable. It should be pointed out, however, that one writer suggests Decatur may have gone to sea to avoid problems after being acquitted of the murder of a woman of 'doubtful integrity.' There does not seem to be much, if anything, to back this up, and nothing in Decatur's early life or the way he was lionized after the War of 1812 seems to indicate that there is anything this potentially unseemly in his past.

Master and Commander Stephen Decatur. Source: U.S. Navy History and Heritage Command

At the age of twenty-six, he is already a Lieutenant Commandant in command of his own ship, with a glorious career ahead of him. When his brother James became a middie, he too came up just as fast and just as brilliantly; command of *Gunboat No. 2* was given to him just a few weeks before. And, in seven months, James would be dead, almost within sight of his brother; shot through the head by a Tripolitanian captain who was feigning surrender. His brother will lead an epic assault on the captain and ship that killed him; one ending in a man-to-man swordfight that will end with the ship, and the pirate skipper's head, in the hands of Stephen Decatur. That, though, was in the future; a future that would hold glory, triumph, and tragedy unimaginable as yet.

Practices of Presence

Right now it was a cold morning off North Africa, and they had work to do.

A few hundred yards away from *Enterprise*, the frigate *Constitution*, under the command of Commodore Edward Preble, is making for a set of sails that had been spotted a few minutes before, low on the horizon and making way towards the Tripoli shore. *Constitution* is probably the most advanced and capable ship of her type in the world; designed by a genius named Joshua Humphreys, who wanted to produce ships that could not only outmaneuver and outrun any other frigate in the world, but outgun it as well. Humphreys succeeded magnificently with *Constitution* and her sisters *President, United States, Chesapeake, Congress* and *Constellation*, the first capital ships of the new United States Navy.

Edward Preble is a young, forty-two, tough, no-nonsense commander, aggressive and demanding. He requires absolute perfection in his men; firmly believing that the more one sweats in training, the less one bleeds in combat. Preble's record goes back to the Revolutionary War, where he served as a midshipman aboard the Massachusetts State Navy vessel *Protector*. The *Protector* does quite well, capturing several British vessels, but eventually Preble pays a high price for his bravery and patriotism. When *Protector* is captured in May of 1781, Preble is sent to a British prison hulk in New York Harbor, the *Jersey*.

His Majesty's Ship *Jersey* had been a sixty-gun ship-of-the-line when she was built in 1736; but, by the time Edward Preble is put aboard her at bayonet point she is no longer a sleek fighting machine. She has instead become a floating graveyard, known to her inmates simply as 'Hell.' Built to carry a crew of approximately five hundred and thirty men and boys in conditions that could be called cramped, there were now more than a thousand American prisoners crammed shoulder to shoulder beneath her rotting, pestilential decks. Eight to ten Rebels a day were carried dead to her forecastle, then lowered over the side to be buried in shallow graves on a nearby riverbank. Those who escaped Hell feet first joined almost eight thousand others from other hulks on that riverbank. In time, their remains shall be placed in a vault in 1808, near what would become the Brooklyn Navy Yard. *Jersey* herself was abandoned where she sat and quietly decayed into history.

In any event, Edward Preble was one of the lucky ones; he contracted, and survived, typhoid fever. Typhoid in those days was a dreaded killer, and some kindhearted British officer became convinced that Preble was at death's door and allowed him to be released, to spend his last days with his family. Preble was made of tough Maine stock and made it home to recover. After a long recuperation, Preble promptly went back to sea aboard the sloop *Winthrop* and became a terror to the British coastal trade.

To Barbary's Far Shore

A spell as a merchant captain followed independence. When the Quasi-War with the French began, Preble went back to the newly reestablished United States Navy and was given command of the frigate *Essex*. In the course of an epic voyage, Preble took *Essex* into both the Pacific and Indian Oceans, a first for any US warship. While fighting Biblical storms, an occasionally mutinous crew, and, every now and again, the French, Preble brought his ship and eleven merchant ships back home intact. There was, however, yet another toll demanded; Preble contracted malaria, from which he will suffer for the rest of his life.

In 1802 he was given command of the frigate *Adams* as she is fitting out in New York, but his health was so bad that he was unable to take her out to join the Mediterranean Squadron. Preble had up to this point done a pretty good job of keeping up with his duties; but this time, it is so bad that it cannot be hidden or overlooked and he submits his resignation. Fortunately for the future, the Secretary of the Navy simply sends him home to get his strength back. In early 1803, he is sent out to take over the Mediterranean Squadron as Commodore aboard the frigate *Constitution*. He was clearly expecting the orders and eager to get to sea. Whipple pointed out that Preble reported to *Constitution* within twenty-four hours of receiving his orders and got the frigate refitted and ready for sea in record time.

Preble is an extraordinarily capable administrator and planner, who will not launch an operation unless he has maximized every advantage he can in conjunction with maximum firepower. He will rehearse an operation until every officer and sailor knows their part in their sleep. He demands constant readiness among his crews. He is not often a cheerful man; indeed, he is known for being something of a martinet, but he takes immense pleasure in bringing up promising midshipmen and young officers, giving them their heads and training them how to be skilled, highly capable commanders. He also knows now that he is dying, very slowly and very painfully, of tuberculosis; but, as if defying God, he refuses to even acknowledge it. It is starting to show now; long coughing fits and spitting blood when Preble thinks no one can see it. They do see it, though; and it is tearing the hearts out of the men who proudly call themselves 'Preble's Boys.'

Whatever Preble and Decatur are tracking is fairly small and moving fast; that almost certainly spells a blockade-runner. Blockade duty, such as that the United States Mediterranean Squadron is conducting off Tripoli, can be mind-numbingly dull, but one good catch is all it takes to make the whole thing worth it. Two centuries ago, when a ship is captured, each member of the crew gets a pay bonus, plus a share of whatever cargo the ship is carrying. Now mind you, the method used for deciding who got how much is, to be polite, murky; more fitting for determining the Federal budget than fairly dividing rightfully earned prizes. Officers of course get far more than the ordinary seaman, and frequently

Practices of Presence

the marines don't get any at all. But even at that rate, one or two good captures could equal two or three years pay in an era when an ordinary seaman only made $108.00 a year, so it was a prize worth waiting for.

Constitution and *Enterprise* are a few hundred yards apart, the frigate to the south, the sloop to the north and slowly opening the distance, knifing through the waves at about nine and a half miles an hour. Slow today, yes, but respectable in 1803 and far from their highest possible speed under full sail. The idea is that, since their target doesn't yet appear to have seen them, they will bracket the stranger, cutting off any chance it will have of escaping. None of the three ships are flying any colors yet; not unusual in those days, ships of all kinds and nations had very similar appearances at a distance, and the longer you could keep the other captain guessing, or ignorant, of your identity, the better. Warship skippers in particular were remarkably inventive with opportunities like this, as will be seen.

Preble is on *Constitution's* quarterdeck, breathing hard through his damaged lungs as he tries to get a good look at the ship bucketing in from the north. He has no qualms at all about Decatur and *Enterprise* getting out of hailing range; the young Lieutenant can be something of a loose cannon, but he is always an effective one. He'll know what to do when the time gets there. The sea is choppy; not overly so, but with a target as small as the oncoming ship, it's enough to hide it effectively from an observer on the ships' decks. Preble is relying on his lookouts aloft to get him the earliest possible warning. It comes just a few minutes later.

"She's a ketch, sir! Low in th' water and movin' fast, no colors!"

Preble had to think fast now, but that was what he was paid to do. No colors, small, fast and apparently heavily laden. Without question in his mind, she was now a runner, but how best to catch her . . . Preble coughed once, then smiled.

"First Officer!"

"Sir?"

"Time we let our friend there know who we are."

Constitution's First Officer looked at Preble, ready to do whatever he was ordered but not knowing exactly what the Commodore had in mind. Preble just smiled grimly and said, "Run up the Red Ensign, lad."

Now it was the First Officer's turn to smile. The Red Ensign, one of three flags flown by Britain's Royal Navy, would convince the runner of their noble intentions, or at least allay his fears a bit longer. Today flying another nation's flag would be patently illegal, although that never seems to stop nations avoiding arms and oil embargoes; but never mind. In 1803, it was considered more-or-less fair play. When actually challenged by hail, a ship generally had to identify herself honestly, but, until that moment, anything went.

On *Enterprise*, Decatur's First Officer tapped him on the shoulder and pointed towards *Constitution*. A huge red banner, with the Union Jack in the upper right corner, was suddenly whipping backwards from the frigate's mainmast, snapping like a cat o'nine tails.

"I'll be damned, gentlemen," Decatur laughed heartily. "We're back under the flag of His Majesty King George the Third! Well then, let's do our best for King and Country!" Waving his hat over his head, Decatur led his crew in a mock cheer.

A gunner called, "Here's to hopin' that His Majesty pays better than His Excellency the President!" Everyone laughed at that, but everyone also knew now that there was going to be shooting that day; but not knowing whether it would be mere target practice or a brutal, vicious fight that would leave dozens dead and maimed on both sides.

"Captain," Decatur's First Officer shouted over the wind, "Shall I run it up as well?"

Decatur thought for a moment, and then shook his head. "I don't think so, Lieutenant. After all, we're one happy Royal Navy here, and I have no problems letting Commodore Preble speak for us. Maintain our course and keep up our speed."

"Aye aye, sir!"

Five nautical miles away, the ketch *Mastico*, inbound for Tripoli with a hold full of silks, spices, liquors and other luxury goods for the Dey of Tripoli and his favorites, had been keeping an eye on the two ships that had suddenly appeared to the south, one to either side of *Mastico's* course. Because of the prevailing winds, *Mastico's* skipper couldn't veer north away from the oncoming ships, and the unknown to starboard certainly seemed to be big enough and fast enough to cut them off if they tried to go south and hug the shore the rest of the way in to Tripoli Harbor. *Mastico's* skipper, a veteran of the on-and-off seaborne sniping that made up the Barbary Wars, had been through this before; so far, he saw no reason to panic but certainly plenty of reason to be prudent. Having his crew lay on all the sail they could, *Mastico* dug her heels deeper into the rough seas, and kept straight ahead, as inoffensively as possible.

"How do they sail?" asked *Mastico's* first mate.

"Parallel to us; one on either side and opening the distance," answered the captain, folding his spyglass. "What do you think?"

Practices of Presence

The first mate peered through his glass for a moment, swiveling from side to side before answering his captain. "They don't seem to be moving to cut us off, Effendi, though they certainly must be able to see us. British, or perhaps French?"

The captain stroked his chin thoughtfully. "Perhaps Melikan, American."

The first mate gave a derisive snort of laughter. "After what we did to their *Philadelphia*? I think not, Effendi. They will be most circumspect with the Dey's ships for some time to come."

Scanning the horizon again, the captain said, "They did it to themselves, my friend. That can make one humble before God and man, or very, very angry indeed . . . But they do seem to be taking their time; wait . . . "

The captain looked hard at the ship to starboard, not sure if he'd seen a flash of color or not at the ship's mast as it dipped its bows into the rolling Mediterranean once again. Then he saw it again, crimson and a sliver of blue against the gray African skies. *Allah be praised*, the captain thought to himself, though careful not to let his men see his relief, *an Englishman*. The captain nodded to the first mate and motioned toward the frigate. The first mate looked through his spyglass for a moment, then nodded and called back to a crewman, who raced up a mast with a colorful cloth wrapped around him.

"Got him!" Decatur said triumphantly, as a square yellow-and-red banner unfurled from the runner's mast, with a matching pennant over it.

"From Tripoli, by God!" Preble called. "Helm, bring us to pass along his starboard, but do it slowly, slowly!" *Constitution* began to slowly pivot to her starboard, lining up so that they would pass close enough to hail, but not so close that one would fear a broadside; at least, not right away. In the meantime, Decatur maintained his course, trying to keep himself aimed directly at the runner's bow.

Mastico's captain motioned for his helmsman to maintain his course. At this rate, the smaller ship would pass across their bow, and the larger, looking to be a frigate, would pass comfortably to their starboard. "Have the gun crews stand down," the captain said. "Get them below and start getting things ready for port." The captain started to walk away, but noticed the first mate looking with concern at the oncoming ships.

"What's wrong?"

The first mate continued looking at the oncoming frigate. Through his spyglass he could see the Red Ensign all right, and an officer waving his hat in a most friendly manner. But the frigate was changing course; slowly to be sure, but it was changing course . . . and the smaller one, a sloop from the looks of it, wasn't.

"Effendi, " the mate said gently, "I do not wish to question you, but –"

"Then why do you?" The mate bit his tongue at that. Captains were absolute rulers on their ships the world over, but on a Barbary ship, they were even more absolute than most. A misconstrued look could have you beaten or flogged; a truly significant mistake could mean becoming acquainted with the bastinado, a thin, hard stick usually made of ebony or some other dense wood. You would be held down and beaten on the soles of the feet until you saw the error of your ways . . . Or, more properly, until the captain felt you had. The penalty for outright defiance was entirely too unpleasant to contemplate. The mate knew better than to push the matter, and bowed. "A thousand pardons, Effendi. I meant no disrespect."

The captain looked at him with that sleepy expression that meant he was tired of a discussion and wanted it ended. "Hul kalast, Are you finished?"

"Yes, Effendi."

"Good. Get to work." The mate bowed as he stepped nimbly backwards for a few steps before turning to herd the handful of gun crews back below.

Constitution continued to bear down on *Mastico*, and Preble was waving towards the oncoming runner, with an unctuous grin on his face he didn't mean.

"First officer?"

"Sir?"

"As soon as they're close enough, hail 'em. The damned marines aren't in the tops, are they? I'd hate to go through all this nonsense to have some trigger happy Leatherneck spoil the party."

Captain Daniel Carmick, commander of *Constitution*'s marines, heard Preble and called back over his shoulder, "Fear not, Commodore. The boys are all below until you give the word."

"Good. Keep 'em there."

There is a time when one must confront an unpleasant reality, and for *Mastico's* first mate, that time was rapidly approaching. The frigate was still closing the range, but everything looked absolutely normal on deck; gunports shuttered, crews going about their work, no sharpshooters in the masts, officers waving to them as if they were their oldest and dearest friends . . . But it still didn't feel right. And that sloop that was accompanying it . . . it was still bearing down on them, almost as if on an intercept course, but seeing it bow on made it impossible to tell what they were up to.

"You still worry." The captain had stepped quietly beside him as he watched the ships approach.

Practices of Presence

Well, thought the first mate, *as well for a lion as for a sheep* . . . "Yes, Effendi, I do. For ships that are not interested in us, they pass close. Too close."

"Even the British have trouble handling their ships sometimes. And with the winds behind us, it is not easy to head away from the coast."

"I have seen them before, Effendi. They can tack into the wind, away from the wind, or wherever they wish to. Something does not feel right, not at all." The frigate was close now; close enough to see men's faces. The sloop would be that way in a moment, still without colors.

The captain considered this for a moment, and a dawning expression of doubt began to creep over him. Picking up a speaking trumpet, the captain called in heavily accented English.

"Ahoy the British!"

"Commodore, he's hailing."

A cough. "Let him wait."

"Captain Decatur, he's hailing *Constitution*."

"Keep an eye on him, gentlemen. Here's where it gets interesting."

There was no answer from the Englishman. That finally got through to *Mastico*'s captain. *Stay calm*, he thought. *You have the wind and you have the ability to get away from him. Stay calm.* He picked up the trumpet once again and called.

"You tell me what ship, damn all!"

Edward Preble grinned wickedly. "Gentlemen, answer him."

It took only a second for a seaman to haul down the Red Ensign and send a red, white, and blue banner skyward, unfolding in the wind like an eagle spreading its wings to strike. At almost the same moment, the shutters over *Constitution*'s cannon pivoted open with hollow wooden thumps, letting forty-four cruel black snouts run out through the hull of the ship. There was the crash of boots as Carmack's marines crashed across the deck and up into their perches in the rigging. Preble grabbed his trumpet to answer *Mastico*'s hail.

Mastico's captain and first mate stood in stunned shock for just a heartbeat before the first mate screamed, "Tawarek, tawarek, emergency!!" and ran for the alarm bell.

"This is the United States Frigate *Constitution*, and you will stand by to be boarded, you damned heathen pirates!" The roar took a lot out of Preble, and he had to brace himself as he started hacking . . . but it was worth it.

To Barbary's Far Shore

Mastico's captain was no fool. Once he'd recovered from his initial shock he actually made the best decision he could have under the circumstances; he gave the helm hard a starboard, turning towards *Constitution*. At first glance an almost suicidal gesture, it actually cut down on the number of guns *Constitution* could bring to bear. There was certainly no way the frigate could turn tightly enough quickly enough to get back into its broadside position. At the same time, if Preble decided to try for a snap broadside as he turned, he stood a very good chance of hitting *Enterprise* with anything that missed *Mastico*. As long as the wind held, and there seemed to be every possibility of that, *Mastico* could run like mad for the shallows where only *Enterprise* could follow, and that only in a stern chase, inshallah.

Had it been someone with less experience than Edward Preble, a ship other than *Constitution*, and someone less aggressive than Stephen Decatur, he just might have gotten away with it.

Decatur snarled, "Helm, hard a port!" *Enterprise* suddenly heeled hard to the left, her rail hissing along the waves as she started to come broadside to *Mastico*. Holding on for dear life, Decatur's first officer called, "Sir, they're going for their guns!"

Decatur shook his head in cruel mock sorrow. "Sergeant Wren," he shouted. "Let them know I disapprove!"

Wren nodded, then looked up and gave a sharp whistle. It was answered by the sound of two musket volleys from the fighting tops.

Mastico's captain saw the flashes from the fighting tops on the sloop. He started to shout a warning when he saw the gun crews on his two bow chasers suddenly leap backwards from their posts. Red spray hung in the air for the flicker of an eyelid, then settled gently on the deck as they lay there, terrifyingly motionless. He turned to the first mate to tell him to get more men on the bow chasers. The first mate was simply standing there, looking down in confusion at the spreading crimson stain on his shirt. He slowly, almost reverently, sank to his knees and was in Paradise before his head was pillowed on *Mastico*'s rough-hewn deck. Turning to try and get under some kind of shelter, the captain realized too late that the sloop was far more heavily armed than he ever could have imagined. Her guns were just a few yards away, aimed at him.

"You may fire when ready, Master Gunner!" Decatur called.

The Master Gunner hesitated not. *Enterprise's* broadside cut loose with a thunderous roar of sparks, smoke, and flame. Knowing they'd be far closer than any prudent skipper would go before they'd open fire, Decatur had the starboard side cannon loaded with canister. An awful reaper's scythe of iron balls packed inside paper containers that disintegrated on firing. it let the balls fly in a deadly cloud that wouldn't sink a ship, but would turn every man in their way into bloody scarecrows.

Mastico jumped as she broached a wave. That saved the captain's life; he stumbled and fell face first onto the deck with a whump that knocked the wind out of him. He actually heard the deranged whirr of the canister shot going over his head, cutting down the crews from the main guns as they tried to get away to the hatches. The helmsman staggered and slipped, but held on grimly to the polished spokes of the wheel. The captain felt the bodies falling onto the deck, saw the looks of shocked terror on their faces and the thin trickles of blood that started to creep slowly out from under their bodies.

Mastico's crew was, to put it gently, rattled. The captain was pinned to the deck by the fire from the American ship. The lead gun crews were dead or dying, and the rest of the deck crew was frantically trying to find some sort of cover on the small, open deck. That was compounded by the fact that *Mastico* was, not to put too fine a point on it, a bully of the sea. Exceptionally good at stopping and overrunning unarmed merchant ships and then terrorizing the crews, but when they were up against well-trained men and purpose-built warships it was a different story. Other bandits have learned this lesson over and over again in recent history and don't seem to remember it, but never mind.

Enterprise was now sailing almost parallel to *Mastico*; perhaps just a touch slower, as she came out of the turn almost close enough to throw cannonballs at the runner, much less shoot them. Solomon Wren stood with his musket on one hip, watching for someone to be foolish enough to move in a threatening manner towards him or his ship. His marines had reloaded in respectable time, though never fast enough to suit him; now they were covering *Mastico*'s deck to make sure no one even thought about trying to get to their stations. *Constitution* was now turning behind the runner, far more tightly than *Mastico*'s captain would have liked to believe possible. When the turn was finished, Preble would be along *Mastico*'s starboard, with a broadside that would be quite capable of pulverizing her. Decatur would still be alongside her, to port. There would be no way that the pirate could possibly outrun or outmaneuver the Americans without being swatted out of the water, as surely as if she had never existed.

Checkmate.

Mastico's captain realized he had to do something, quickly, if he wanted to see the sun go down that day, but he wasn't at all sure what. He had crewmembers lurking in quivering fear just below the hatch coamings. They were outnumbered and outgunned, and any effort to fight back would probably result in glorious, if unpleasant, martyrdom.

"Ahoy the ship!" Decatur's voice pierced the captain's thoughts. "You have to the count of three to strike your colors and heave to! One . . . "

Some of the runner's crew, brave, courageous, and utterly foolhardy, made a run for a small cannon facing *Enterprise*.

"LA!!" The captain screamed. "NO!" He tried to get to his feet to stop them.

With reflexes born of years of bitter combat experience, Solomon Wren spun his musket down, lined up his shot and fired; in more of a reflex than a planned action. The ball flew approximately true; we say 'approximately' because, though Wren had planned to put it squarely through the captain's trunk, even the best of shots in those days had to submit to the vagaries of powder, ball, and breeze. It slapped *Mastico*'s captain in the right thigh. Ripping through it with blunt force that tore and rent rather than cleaved, it shattered the captain's right thighbone in the process of passing completely through and then burying itself in the deck. The captain spun down to the deck again. This time, he stayed there screaming in agony. His men stopped dead in their tracks and then whirled back for the safety of the hatch, screaming in horror and wishing that they could be anywhere else besides that bloody patch of teak, almost in sight of home.

Decatur frowned at Wren, but not in anger. He'd prefer to take everyone alive, and a ship's captain most of all. But Wren was only doing what he'd been trained to do, and if he thought there was any threat to his ship the sergeant would swim over himself and push the ball in with his bare hands if he had to. Turning to his first officer, Decatur asked, "Where was I?"

"'Two,' sir!"

"Thank you, First." Raising the trumpet again, Decatur called, "Two!"

At that, hands appeared over *Mastico's* rail, waving in a universally understood gesture. One crewman, trying to bob-and-weave, bow in respect, and wave frantically at the same time, sidestepped to the mainmast and the flag halyards. He cut them with a terrified smile. The red and yellow flag, topped by its slender pennant, shot down the severed halyards, as if it was trying to seek shelter as well. With *Enterprise*'s marines, and now those of *Constitution* as well, covering every movement, *Mastico*'s crew slowly moved to their posts, studiously avoiding anything resembling an aggressive action while they reefed the sails and brought the careening ship to a halt. Cheers rose skyward from the two American ships; as much in gratitude that there would be no vicious gunfight and boarding action as the fact that the runner was surrendering.

It took a few minutes to get all three ships stopped. When they did, every gun that could be brought to bear covered the *Mastico,* as the boarding parties went over the side and prepared to take their prize. The marines went up the rope ladders onto *Mastico* first; tall and intimidating in their blue-and-white uniforms, with towering shakos, bayonets flashing on the muzzles of their

Practices of Presence

muskets, and the high cowhide collars intended to protect against a saber cut but more remembered for the nickname it gave them: leathernecks.

This was the most dangerous part of the entire operation. If a crew suddenly decided they didn't want to stay surrendered, you would probably lose most of the boarding party. Even the most ferocious commander would think twice before firing at point blank range into a ship that now carried an appreciable fraction of your crew as hostages. As a rule, though, by the time a ship was boarded, even the most veteran crew was near or in a state of stunned terror from the sustained roar and damage of broadside fire at close range, the shredding effect of splinters and shrapnel, and the unseen fingers of death the marines would send outward without warning from their aeries. All together, those things would be called 'shock and awe' by a later generation, and it would have a very real effect in those days.

The marines took their positions, the quarterdeck, the bow, along the rails and the main battery, with cool precision, led by Sergeant Wren. Others would manhandle the crew into an appropriately submissive position until the ship could be secured. It was not at all uncommon for them to try and render first aid to particularly hurt crewmembers. A team would head below and into any visible holds and cabins, making sure there were no diehards lurking about who wanted to become dead heroes. That was a less dangerous task than one would think. Today, no boarding party wants to go aboard a ship without having memorized its blueprints; but, in a time when ships tended to look alike, they tended to be designed alike and a sloop or frigate from one nation was laid out very much like a sloop or frigate from any other. The belowdecks team was still wary though, weapons at the ready.

Mastico's captain was rolling about in the deck in pain; an ugly purple/red splotch on either side of his right thigh and what looked to be a broken end of his thighbone sticking out. His crew had gathered quite obediently, kneeling in a few ragged rows that ran fore and aft; the usual motley collection of Tripolitanians, Egyptians, Tunisians, Moroccans, Algerians and assorted thieves, cutthroats and other nefarious individuals that comprised a typical Barbary crew. Wren and his men did their best to look as menacing, unpleasant and, in general, as potentially psychotic as possible for the first few minutes of the boarding. It always worked. He reviewed the prisoners prior to signaling Captain Decatur that it was clear to come aboard. Something, though, struck Wren as a bit odd as two of the below-decks team came stomping back up the ladder and reported that the ship, was indeed, secure.

Wren looked curiously at a little knot of men who were trying to stay inconspicuous towards the rear of the formation, then motioned one of his corporals over.

"Corporal?" Wren murmured.

"Sir?"

"Without making it too damned obvious, take a look at those poor sods along the port rail and tell me what you think."

The corporal nodded as if listening intently, and looked at them as best he could out of the corner of his eye. There were five men there; all on their knees, with their hands above their heads. Each one was tanned, with luxuriant moustaches and nicely kept clothing; not at all dirty or worn, like every other man kneeling on that deck. He continued to nod as if listening to Wren, then looked directly back at the sergeant.

"Awfully well done up for Berber crew, Sergeant. A bit too much so. Officers?"

"Very good, lad; you're beginning to convince me you might be trainable. And if there are that many officers aboard, there's likely to be a great deal of interesting things aboard that they don't want to let go of. Take a couple of the lads and discreetly, *discreetly*, y'hear, make sure you've a few muskets on 'em."

"Aye aye, Sergeant Wren."

Sure to his word, the corporal moved slowly over the deck, not making any great show of anything. He quietly nodded to four marines, tilting his head towards the port rail. Each one in turn nodded, understanding exactly what was needed. They lowered their muskets just enough to insure that bayonets jutted out towards the five. It wasn't terribly obvious, because none of the pirates could see all the marines, but the combined effect was to strongly discourage any thoughts of resistance or escape. At that, Wren waved his shako in a circle over his head three times, the signal to *Enterprise* that all was well.

That set off another round of cheers aboard the sloop, with much handshaking and backslapping before Decatur could get into one of the boats and set off for the *Mastico*. From *Constitution,* Commodore Preble was on his way, followed by a prize crew; the frigate having a much larger crew, it was considerably easier for them to spare the extra bodies than it would have been for *Enterprise.*

As Wren and his marines watched the parade from the other ships wend its way to them, the below-decks team was doing a more thorough search of the ship. Marines are a naturally curious lot, and mere locked holds and closed doors were but minor impediments. One private used his musket's metal plated butt to knock a lock and hasp off a barred gate that led into the main hold. Sergeant Wren would have had words, and, quite probably, more than just words, with the private had he seen the man abusing his weapon like that, but the private worked on the assumption that what the Sergeant didn't know about wouldn't annoy him.

Practices of Presence

The gate swung open and admitted the private and another marine into the hold, where the treasures of Croesus awaited them. Rich silks from all over the Middle East and the Orient, leathers and beautifully carved woods, and small wooden kegs with wax seals, something that every marine on duty off the Barbary Coast soon learned meant gold and silver coin. That meant prize money enough to keep them all drunk and in pleasant female company until the end of their enlistments and beyond. The private who'd forced his way in leaned his musket against the bulkhead and began to rummage through the piles of goods as his partner came in behind him.

"'Ere now, Wren'll be lookin' for ye."

"Sergeant Wren told me to do a thorough search down here, and 'tis exactly what I'm doin'. Don't want anyone hidin' in this lot, do ye?"

The private lifted bolts of magnificently colored silk, feather light and smoother than anything he'd ever wore. Holding up one bolt of glorious crimson, he asked his partner, "Did ye ever see the likes o' this stuff? Old Hannah in Philadelphia would feed me for a year for just a few yards . . . D'ye think I could cut a bit off with the blade and wrap it under the jacket then? . . . What d'ye have there, lad?" The private had turned to show his partner the silk, only to see him rummaging around two battered wooden boxes that had been covered with a filthy, pestilential blanket.

The private walked over to his partner, just as he held up what was unmistakably a Holy Bible, with a sailing ship engraved on the tooled leather binding. Remaining inside the boxes, the private could see other books, as well as clothing, a sextant, swords, and pistols.

"Looks like these boys had been busy, busy indeed. Wonder whose they were?"

The private saw one book inside that was smaller than the rest, with a bright red cloth binding, and the word Diary embossed on it in gold leaf. "Well," said the private, "one way to find out." He opened it, flipped a page, and then froze, his expression slowly hardening.

"What's wrong, then?"

The private lowered the book, his eyes fixed on nothing, and then he looked up and motioned for his partner to hand him the Bible. He obediently gave it over, and the private opened it to the frontispiece, read the inscription, and then slowly closed the cover.

"Lad, let's get topside and see Wren. He'll want a look at this."

Decatur and Preble were aboard now, surveying their prize. *Mastico*'s captain was propped against a deck hatch coaming as *Constitution's* surgeon tended to his mangled thigh. There wasn't much he could do for the crewmen who'd taken the musketry and canister but cover them up decently until the chaplain could get here and say a few words over them, heathens though they were. Decatur fell in on Preble's left as they walked over to the captain and looked down at the writhing, weeping man.

"Surgeon," Preble said, "I see you are doing your usual excellent work. Will he live?"

Constitution's surgeon, a bookish doctor from Maine, didn't look up from his efforts to acknowledge Preble; something that would have gotten a kick in the pants or worse were it anyone else, but healers in the military have traditionally had a certain leeway that others do not. After all, one never knows when one will need a doctor, and having him angry with you is not the wisest of things. "Well," the surgeon said affably, "it doesn't appear any of the major vessels were severed, though I don't believe that thighbone will ever be right again. I swear, with the kind of damage these do, you'd be better off saving the expense of a musket and going back to Indian war hammers and tomahawks."

Decatur laughed harshly and replied, "I shudder to think what marines would do with tomahawks, Surgeon. Besides, it would take into the next millennium to train them." Squatting down, he leaned over the captain and said, "I am Leftenant Stephen Decatur, United States Navy. You are my prisoner, sir, and your ship is –"

"You lie me, you lie!!" The captain howled tearfully. "You not British!"

Decatur raised one eyebrow. "Of course I'm not British, you ass. I'm American. Am, err, ick, can, understand, savvy?"

"You fly red flag, red!" The captain was able to thrust an accusing finger at Preble, waving it like a sword.

Decatur looked up at Preble with an expression of exaggerated shock and horror. "Commodore, did you fly His Majesty's Red Ensign?"

Preble looked thoughtful for a moment. "You know, Decatur, I believe I might just have, at that!"

"Sir, I'm shocked, shocked indeed, that you could do such an awful thing! How could that have happened?"

Preble coughed once and then shrugged. "Chalk it up to old age, Decatur. My memory just isn't what it used to be. I suppose there's really only one way to make it up to the poor man."

"And that would be?"

Practices of Presence

Preble smiled slowly, like a wolf that just spotted lunch. "We'll just have to take him back to Sicily with us as our guest. I promise you, Mister Decatur, I shall write a most sincere letter of regret to the Dey as soon as possible!"

The surgeon stood, wiping his hands. "If you two are finished with your music hall play," he said with a look of faint disdain; no more than faint, because with Edward Preble, even a doctor could push too far. "I'd like to get this man back to the *Constitution* and into sick bay. He should be off that leg for a while and under medical supervision. Trust me Commodore, he won't be trying to swim back to Tripoli."

Preble nodded. "Make it so, then. Decatur, let's see what else your men have found . . . " The sight of Solomon Wren striding towards him, his face as dark as a storm cloud and two marines behind him, each bearing a wooden box, told him that they may have found something more than they'd bargained for.

It took several hours to get *Mastico* fully secured and underway again, with her crew now under heavy guard belowdecks. They'd have protection from the elements, plus food and water, but not much else for the long sail against the wind back to Sicily, where the Mediterranean Squadron normally operated. The dead had been given a decent burial at sea, closed up in weighted linen shrouds by *Constitution*'s sail maker, remembering as he did so the words of the Apostle reminding them that they would all someday undergo a sea change.

As the sun set on the western horizon, the American crews could take a great deal of pride and pleasure in the day's work. They'd nailed an honest-to-God blockade runner with a hold full of riches for the Dey, and they'd done it without so much as a scratch between the two ships. Even better, Mister Decatur, as the skipper responsible for the capture, made a magnimanious little speech before he left *Mastico,* telling everyone how proud he was of them, and that all of them, even the marines, would get a share. The marines were, in the main, quite happy with that. After all, the pay for a private was only nine dollars per month; not per hour or per day, but per month, which averaged out to something like thirty cents a day. The sailors had something to say about it, as they felt they had done all the real work and all bringing the marines in did was lower everyone else's share. Needless to say, there would be some very spirited discussions aboard *Enterprise* regarding this matter, but that could wait for tomorrow.

The evening routine had started, and the ships were slowly winding down for the night. They never completely rested, but rather stood down into a less demanding schedule.

THREE: THE PHILADELPHIA PROBLEM

Sadly, for Edward Preble, duty never wound down to a less demanding level. His dinner sat in front of him; boiled beef and potatoes getting progressively colder, a tall goblet of Madeira almost untouched beside it.

As the Commodore of the Mediterranean Squadron, he was both the administrative and tactical commander of the unit; not to mention, by default, the senior American commander in this part of the world. Now, as many a soldier, sailor, airman, or marine has discovered since, gallivanting about the globe in defense of Liberty can be a great deal of fun. It's the paperwork that's annoying, and Preble was well and mightily annoyed that evening. The tidal wave of requisitions, orders, discharges, letters, enquiries, and assorted nonsense is the same for the commander of today's Sixth Fleet aboard a nuclear carrier; the only difference is the sheer magnitude.

The one consolation was that, unlike his four hundred-odd shipmates aboard *Constitution*, he was able to do it in reasonable quiet and isolation. The Captain's cabin aboard *Constitution* is remarkably small to those of us used to the open horizons of life ashore; not much bigger than the living room of a typical suburban house, with the only real view astern. There would have been a little bit of furniture; nothing terribly well carved or assembled, all of it severely functional. In fact, it's quite likely that the largest single object there would have been the massive oak cabinet that covered the top of the rudderpost, which protruded about three feet straight up from the deck.

Preble was scratching away with a quill pen, trying to figure out how his ships had gone through several metric tons of suet, for the love of God, when there was a knock at the cabin door.

Cough. "Come."

Captain Daniel Carmick, United States Marines, poked his head in. "G'd evening, Commodore; you wanted to see me?"

The *Philadelphia* Problem

Preble never looked up from his writing, and instead motioned Carmick to a chair beside his desk. Carmick, a big, red-haired Bostonian, nodded a thank-you and sat down.

Preble inclined his head towards the carafe of wine on his desk. "Help yourself, if you'd like."

Carmick smiled graciously. "Thank you, sir; was feeling a bit parched."

Carmick reached over, taking care to stay out of Preble's way. Pouring a goblet, he sat back down and took a sip. Preble put down his pen and pushed his spectacles up on to his forehead, rubbing the bridge of his nose. Leaning back slightly in his chair, he picked up his wine and toasted Carmick.

"Confusion to the damned Berbers," Preble said, tapping his goblet against Carmick's.

"Confusion."

They both took deep swallows, and Preble grimaced as he put his down. "Dear God," he said with disgust. "I don't know what we're paying for that swill, but it's entirely too much. I take it you know about the treasure trove we recovered from that runner this morning?"

Carmick sipped his wine. "Aye, sir. A great many personal effects of one sort or t'other, I'm told."

"Among other things. The gear's over there." Preble motioned to two wooden boxes beside the rudderpost cover. Carmick stepped over and dropped to one knee to get a better look at the contents of the boxes. The books and other items that Sergeant Wren had spotted were there, along with a massive triangular leather case. Carmick opened it, knowing what he would find just from the weight of the case.

Inside was a beautiful brass sextant, that indispensable navigator's tool in the days before inertial guidance systems and Global Positioning Satellites. It was in absolutely beautiful condition; brass gleaming, perfectly cut and ground glass optics, and a lovingly engraved plate on the inside of the cover:

FROM THE PEOPLE OF THE CITY OF PHILA., PA

TO THE CREW OF THE UNITED STATES FRIGATE *PHILADELPHIA*

"A STAR TO SAIL HER BY"

Carmick shook his head distastefully, as if unsettled simply by handling it. He continued to look through the boxes; swords, clothing, a Bible presented to the ship's chaplain. Picking up the diary, Carmick leafed through it quickly, then looked at the name written in careful, spidery script on the frontspiece: William

To Barbary's Far Shore

Ray, Private, United States Marines. Willie Ray, short, belligerent, but smart as a whip; one of the better marines Dan Carmick had ever worked with. Tossing the diary back into the box, Carmick stood up, looking down at the piles in sadness.

Turning back to Preble, he asked quietly, "Divvying up the spoils, then?"

Preble sipped his wine. "It looks that way. We still don't seem to have heard anything from the Dey, so we aren't sure if Will Bainbridge and his men are dead or hostages."

Carmick shook his head with a gentle but bitter laugh. "The Dey is a very practical man, Commodore. A corpse is worthless, but a live hostage is money in the bank."

"And he's got three hundred and eight of them."

The three hundred and eight in question were the officers and crew of the frigate *Philadelphia,* captured and in the hands of the Dey of Tripoli since October 31st, 1803. Her skipper, William Bainbridge, had been guarding the approaches to Tripoli harbor along with the sloop *Vixen* when he got word of two corsairs, the long, lethal pirate ships that made up the bulk of the Dey's fleet, operating one hundred and fifty miles to the north off Lampedusa. Exactly what Bainbridge was thinking has never been all that clear when he sent the shallow draft *Vixen* off into deep water and kept the deep-drafted *Philadelphia* in notoriously shallow waters, but it apparently seemed to be a good idea to him at the time.

As events would prove, it wasn't.

Will Bainbridge was a good man and a highly skilled captain, spoken of highly by all who served with him. Loyal, and aggressive, to a fault, he unfortunately had a reputation as something of a black sheep among the USN's commanders. He had been in command of a small US vessel, the *Retaliation,* during the naval war with France a few years earlier, and had been caught one morning, outnumbered and outgunned, in a position that really allowed no other option than surrender.

As difficult as it may be to believe, at that time surrendering one of your Navy's finest warships was not the career-ending move that it would be today. In a world of chivalry and honor among warriors, it was understood that occasionally the Gods of War would not be smiling upon you for whatever reason, and you simply might have no other choice, as long as you had made a genuine and honest effort to defend yourself and your nation's honor. There would be, of course, a Board of Inquiry upon your return. But, as long as everyone swore to your valor and courage, all would be forgiven, once, and only once. After that, it got much harder. Now in Bainbridge's case the French

The *Philadelphia* Problem

themselves even testified to his skill, daring, and bravery, so the Board easily found him blameless, and returned him to his duties.

A few years later, Bainbridge had command of the sloop *George Washington* on a trip to Algiers to deliver our nation's blackmail money, a thousand pardons, tribute money, to the Dey of Algiers. In those days, it was routine to pay off the various Deys, Bashaws, Emirs, Viziers, and other assorted titled criminals who ruled their little city-states along the North African coast. From Congress' point-of-view, it was cheaper to make the payoffs than organize, build, and operate a competent, standing Navy. The *Washington* was an excellent example of that particular line of reasoning. She was a converted merchantman with a reasonable punch of twenty-four guns, but certainly not a true frigate, and light years away from the capabilities of the six masterpieces that were getting ready to join the Navy back in the States.

In any event, Bainbridge did this distasteful duty with the bearing and skill expected of him, right up to the point where he anchored *Washington* in Algiers harbor. For reasons never fully clear, Bainbridge allowed his ship to anchor almost directly under the guns of the largest fortress on the harbor, the personal keep of the Dey of Algiers himself, a huge and utterly unpleasant character named Bobba Mustafa. Compounding the error, no one else apparently realized it either as Bainbridge went ashore to give the Dane his dane-geld.

Actually, that's not true. Bobba Mustafa seems to have realized it rather quickly, and when Bainbridge presented the booty and tried to leave, Mustafa none too gently explained it to Bainbridge. Mustafa put forward his belief (a previously unknown one) that, since Bainbridge was paying him tribute, he and all his crew were now Mustafa's slaves; therefore Mustafa was going to put him to work. He had a shipload of cargo that needed to be sent to Constantinople, and Bainbridge would take it there. Otherwise, as Mustafa's translator pointed out with undisguised glee, the guns of the castle would open fire and turn the United States Sloop *George Washington* into . . . oh, what was the word in your language . . . kindling?

Will Bainbridge protested to the best of his ability, which was formidable, but there really wasn't much he could do. Before he could even tell his men what was going on, the heavy batteries of the fortress would destroy the *Washington*, and the crew would be prisoners or dead. With a heavy heart, Bainbridge marched back to the *Washington* and told his crew what they would have to do. Without question, Bainbridge was mortified beyond belief at this point, and understandably so. What came next would probably have killed any other man. Bobba had *George Washington* brought to dockside and loaded the most bizarre cargo ever carried by a warship of the US Navy.

Under Bainbridge's amazed eyes, the Dey's functionaries marched aboard the following: one hundred and fifty sheep, one hundred Negro slaves, twenty-

To Barbary's Far Shore

five bulls, twenty 'gentlemen,' sixty women, four lions, four antelopes, four tigers, the new Algerian ambassador to Constantinople, and an unknown number of ostriches. Poor Will Bainbridge must have thought he had been put in charge of a new Noah's Ark, by way of the Marx Brothers. Somehow, he gathered his dignity and made it out of port, proceeding at his best possible speed to Constantinople. On arrival there, the Turks treated him with the utmost decency and politeness; all the more pleasant of them because they had apparently never heard of the United States of America and were quite mystified at the concept. On the other hand, they had no preconceived notion of things and Bainbridge took the opportunity to negotiate a treaty of friendship between the US and the Turks, a relationship that still exists. Bainbridge then headed straight home for Philadelphia.

Needless to say, none of this, with the exception of the Turkish treaty, went over well at home. William Eaton laid it out most bluntly:

"Genius of my country! How thou art prostrate! Hast thou not yet one son whose soul revolts, whose nerves convulse, blood vessels burst, and heart indignant swells at the mere thoughts of such debasement? Shade of Washington – behold thy orphan'd sword hang on a slave – a voluntary slave, and serve a pirate!"

Mister Eaton was obviously less than pleased about the situation.

The attitude of the US Navy's leadership was even less understanding. Although technically Bainbridge hadn't lost his ship, it had been an embarrassment of the first order, so it was once again to the Board of Inquiry. In Bainbridge's defense, the US Consul to Algiers, Richard O'Brien, by the book, his superior on the scene, had counseled him to go along with it. There was also the very practical point that, had Bainbridge sailed out of sight of Algiers and Bobba Mustafa then kicked the entire menagerie over the side, the Algerians would have turned their corsairs loose on American shipping in the Mediterranean. After overwhelming the tiny USN presence there at the time, they would have wiped out everything he could find. The Board was furious at the way things had turned out, but Bainbridge really had been in a no-win situation, and he was cleared of responsibility. But they were going to keep an eye on him.

Two years later, Will Bainbridge had command of the frigate *Philadelphia*, a beautifully designed and built vessel presented to the US Navy as a gift from the people of the City of Brotherly Love. Exactly how a warship would represent the Quaker ideals that the city was founded on escapes us today, but that was a different world. It had been a good cruise so far. *Philadelphia* had not only recaptured a US merchantman that had been taken by a Moroccan ship, but captured the Moroccan as well. On October 22nd, 1803, Bainbridge got that

The *Philadelphia* Problem

report of corsairs operating off Lampedusa, and sent *Vixen* north to see what they could do while he stayed off Tripoli with Philadelphia.

Nine days later, on Halloween, it was only fitting that all Hell broke loose.

At around nine o'clock that morning, Bainbridge's lookouts saw a sail moving fast and westbound, close to the shoreline. That meant a runner, and the chase was on. The runner, sure enough, ran up the yellow-and-red of Tripoli and made a run for it.

At eleven, with a strong wind behind her, *Philadelphia* was within firing range, and within sight of the fortifications at Tripoli Harbor. That meant that the waters were getting progressively more shallow and dangerous. Bainbridge knew that much, and had three sailors on the leads forward, calling out the depth every few seconds, so he knew he would have time to do something if the bottom suddenly started rising. But thirty minutes later, still behind the runner and getting closer to the formidable harbor defenses, even Bainbridge had to face the fact that the runner was going to get away. It had been a good chase and certainly livened up the day, but it was time to get back to work.

Bainbridge ordered *Philadelphia* brought about to the north and headed back to open water. Seconds later, *Philadelphia,* moving at not quite ten miles an hour, ran hard aground on Kaliusa Reef, a gravel bank known only to local pilots. Bainbridge had to have been stunned, but he kept calm. There were no other enemy vessels in the vicinity, but that wouldn't last long. Setting a crew into a boat, they quickly determined that there was still sufficient water under *Philadelphia*'s stern to get her floated again if they worked fast. To their credit, they did their best. The crew rigged the sails 'aback,' that is, backwards, threw the anchors overboard, started throwing their fresh water over as well, cut the foremast down and tossed it overboard, and then, as a last resort, sent most of the main battery over as well.

Nothing. Not so much as a quiver astern. The Tripolitanians were, to put it mildly, in transports of ecstasy once they realized what had happened, jeering and gesturing from the battlements of Tripoli Harbor. It seems to have taken them a bit of time to realize that *Philadelphia* wasn't going anywhere. At that point, some brave skippers hustled their crews aboard and got underway, wanting to take advantage of the incredible opportunity that Allah had dropped in their laps. The gunboats that headed out after the stricken frigate had every possible advantage, from the wind to sheer numbers. They had to stand out a bit farther away from *Philadelphia* than some of them would have liked lest they run aground themselves. But, by four in the afternoon, the end was in sight, and it would not be a pleasant one for Will Bainbridge and his men.

Calling his staff together, they reviewed the situation, and everyone there had to face facts. *Philadelphia* was not getting off Kaliusa Reef, and if the

To Barbary's Far Shore

pirates got aboard after a fight . . . well, that wasn't something anyone wanted to think about. Almost certainly, the marine detachment under Lieutenant William Osbourne would have put up a fight, but even he saw that it would have been glorious . . . and final. Accordingly, a little after four o'clock, Bainbridge ordered her colors struck and surrendered *Philadelphia* to the Tripolitanians. Some effort was made to scuttle her, but it seems that they waited a just a bit too long to do so. A small hole was punched in her thick oaken hide, but the only effect this had was to give her a distinct but nonfatal list to port.

The effect of losing *Philadelphia* was roughly equivalent to losing a guided missile cruiser today. People today ask, as they did then, why Bainbridge didn't blow his ship and incidentally, himself, into eternity and take a goodly percentage of the Tripoli pirate force with him. Bainbridge's answer was straightforward: " . . . I thought such conduct would not stand acquitted before God or Man, and I never presumed to think I had the Liberty of putting to death the lives of 306 Souls simply because they were placed under my command . . "

It is easy, sometimes, for those who have never had to face a grim, implacable enemy, to suggest that one should take one's own life and as many of theirs as possible rather than surrender. More than a few people, many of who should have known better, said the same thing when the Red Chinese captured a US Navy surveillance plane in 2001. The difficulty is that when you are the man, or increasingly these days, the woman, who has to look people in the eye and then give the order to push the little red button that you never talk about, or dive your aircraft into the water, or light that fuse and simply sit there and watch it burn down, taking the last seconds of your life with it, the entire subject becomes more than an abstract matter of duty and authority. You are responsible, and you will be the one who has to take that responsibility to eternity. William Bainbridge decided he and his men had a fighting chance to survive this, no matter what happened, and that was the path he followed.

The pirates weren't at all sure what to do first; ransack the ship or capture/beat their prisoners. In the event, they did both, some of the booty finding its way into those two wooden boxes on the *Mastico*. And as soon as they had the crew clear, the Dey's naval engineers, no mean talents, set about getting *Philadelphia* patched up enough to refloat her. Almost immediately, the Dey sent out his ransom demands. He was a reasonable enough man, and felt that a peace treaty, and three million dollars, would be sufficient to ease his mind.

As Bainbridge was alone when captured, it's not entirely clear how Commodore Preble got the word. It's possible that the first word may have come from the Danish Consul to Tripoli, Nicholas Nissen. Consul Nissen, may his name be forever honored, made it his personal struggle to see to the safety and care of Bainbridge and his men. He frequently stood up to the Dey about their

The *Philadelphia* Problem

treatment, paid for food and clothing for them, and on at least a couple occasions passed letters to and from Bainbridge, at considerable personal risk; at the very least, had the Dey found out, Nissen would have joined the *Philadelphia*'s crew.

In any event, Preble decided to head for Malta, where the British were traditionally pleasant, gracious hosts and quite probably would have more current information through their own diplomatic network. Sure enough, when *Constitution* arrived there on November 27th, he sent Lieutenant John Dent ashore to pay his respects and see what information they might have. Dent arrived back aboard just before dinner, and after that, it was likely Preble lost his appetite for the rest of the day. Dent had been given a couple of eyewitness reports as to *Philadelphia's* capture, and, the icing on the cake, a letter from Bainbridge himself, most likely smuggled out by Consul Nissen, explaining what had happened. In an era when senior naval commanders were notoriously bad about handling bad news, Preble's response must have been volcanic, but the details have not come down to us. But once he had settled down, his next decision was a clear one; have the squadron form up at Siracusa, on Sicily's southeast shore, and then try to figure out just what in God's name to do next.

One could only imagine what Preble was thinking as they raced north that cold November night. He knew what kind of a ship *Philadelphia* was, he knew what she was capable of under a good, solid crew, and worst of all, he knew that the Tripolitanians were more than capable of getting her back in the water. If she was refitted with even a relatively light battery, she would still be capable of savaging any merchantman, and it would take at least a frigate, or two, to bring her to bay, if they could find and catch her. And to make that thought even more pleasant, Preble knew what kind of punishment *Philadelphia* could dish out if she was cornered. That was a fight that simply did not bear thinking about.

When they got to Siracusa, the Mediterranean Squadron at least had the pleasure of a kind, affectionate welcome from the governor, a Sicilian nobleman named Marcello de Gregorio. Nothing was too good for his American friends, nothing at all, and whatever they asked for would be provided; at quite reasonable prices, of course. In truth, altruism wasn't His Excellency's entire motivation. The corsairs had a tendency to play rough with his shipping as well, and the navy of the Kingdom of Naples, which then controlled Sicily, was less than ineffectual in stopping them.

Preble stayed there for a bit more than two weeks, taking advantage of the time to repair, refit, and replenish, especially *Constitution*. The repairs and refits were easy enough. The US Navy has always shown a positive genius for being able to keep itself at sea and mission capable, no matter how tight the money or resources. What was a problem were provisions. Some supply ships had been sent out under escort from home and were now called for, only to find that a disturbingly high percentage of what they carried was rotten beyond salvage.

To Barbary's Far Shore

His Excellency the Governor was able to provide some items, but once the good merchants of Siracusa became aware of the Squadron's plight, prices suddenly rose dramatically. Preble could only do so much with that, so he reluctantly sent word back to the United States that he needed supplies and ships; to be precise, at least three frigates. He wanted two to replace *Philadelphia*, and the third to be stationed off Gibraltar to keep an eye on any pirates who might try breaking out into the open ocean. They had been known to on occasion, there are confirmed accounts of raids by the Barbary pirates as far north as Ireland, but with *Philadelphia*, the task would be a great deal easier.

By December 17th, Preble had gotten things together well enough to hoist sail in *Constitution* and, in company with Stephen Decatur in *Enterprise*, set sail for Tripoli Station badly outnumbered and outgunned. Preble knew all too well that if somehow they lost *Constitution*, either to damage that would knock her out of action or an outright loss, the remaining lighter warships of the Squadron would be utterly unable to keep the Tripolitans from breaking out and overrunning anything flying Old Glory. Six days later, Preble's lookouts spotted *Mastico* trying to slip past them, and suddenly the game had a new look to it. Lieutenant Dent was detailed to take the prize crew and bring *Mastico* up to Siracusa, where her former crew could be suitably lodged, while Preble and Decatur would go back to Tripoli and see what other mischief they could find.

It would be a short trip; the weather changed drastically, more fitting for New England than North Africa. Edward Preble and Stephen Decatur were aggressive skippers, but they weren't crazy ones. That kind of weather could kill ships on its own without having to worry about the pirates, not to mention the unnerving possibility of one, or both, of them being run aground. Christmas 1803 was spent in transit across the Mediterranean, and by New Year's Eve they were back in Siracusa.

Two centuries ago, ships simply could not fight year round. The weather was usually enough to keep even the pirates in, and as weather got worse, so did the general health of the crew. It was with that in mind that Preble decided to leave Lieutenant Dent in command of *Constitution*, and sail himself to Malta. For a man with Preble's illnesses, the climate was a great deal more congenial, and the Royal Navy commanders stationed there would certainly be the kind of company that would keep him cheerful and happy until he could take the Squadron out again in the spring. For their part, the Royal Navy could not have been kinder or more pleasant hosts, welcoming Preble to a round of dinners, parties, and assorted boys' nights out to trade stories, toasts, and opinions. It was at one of those that Preble was told that as far as His Majesty's Navy was concerned, *Constitution* was a welcome visitor anytime, and that Maltese sailors and pilots would always be permitted to sign on.

The *Philadelphia* Problem

Preble had a small office/billet set up in a building overlooking the famed Grand Harbor, where he could see any dispatch boats coming in from the squadron or from home, that place weeks away across the water that sometimes seemed more fondly remembered legend than truth. It was someplace quiet where he could work, think, and try to figure out what to do next. He had every intention of going on the offensive when the weather cleared, but he had to face the likelihood that he would have to do it without any help from home, or help delayed at best. The trick, he knew, would be to utilize every possible advantage he had; and, if he didn't have one, make one.

It was with that thought in his mind that there was a knock at the door of his room.

"Come," Preble wheezed, fighting back a coughing fit.

The door opened, and a young RN lieutenant who had been assigned as Preble's secretary, aide-de-camp, and general assistant, poked his head in. "Morning, sir. A moment of your time?"

Preble nodded, downing a glass of water to still his cough. "What is it, Leftenant?"

The lieutenant looked thoughtful for a moment, then spoke. "In truth, Commodore, I truly don't know . . . Lord knows I've never seen a circus like this before. We have some gentlemen here from the Dey of Tripoli to see you."

Preble raised one eyebrow. "Leftenant, you are joking."

The Lieutenant shook his head in true sincerity. "No, sir. The senior of them states his name is Mohamet Kemal. He speaks a fair approximation of the King's English, and he says he represents His Most Gracious and Serene Majesty Hamet Karamanli, the rightful Dey of the Tripolitian States."

Preble simply blinked. "My apologies, Leftenant. Indeed, you aren't joking. I've met the honorable Mister Kemal before. He is indeed quite a piece of work; a cross between a diplomat, a priest, and an opera singer, as well as an extraordinarily bad actor. What's he doing here? His Most Serene et cetera, et cetera was in Egypt last I heard, hiding from his dear brother."

"Some things one can't make up, sir. Shall I send them in?"

"What exactly do they want?"

"Mister Kemal says it is a matter of the utmost importance that may impact, and I quote him directly, Commodore, 'the future of the world and the fate of thousands.'"

Preble frowned. "He needs a loan of some kind then?"

"When they talk like that, sir, they usually do. However, if I may, he does seem to have an unusual sincerity to him."

"That, Leftenant, would be a miracle of the first water. Send him in and let's see what he wants."

"Aye aye, sir." The lieutenant closed the door as Preble straightened his jacket, flicked off the dust, and prepared to meet his visitor. In a moment, the door opened.

The lieutenant snapped to attention, announcing, "His Excellency Mahomet Kemal, Chancellor and Aide-de-Camp to His Most Gracious and Serene Majesty Hamet Karamanli, By The Grace Of Allah and the Prophet, Peace be unto Him, The Rightful Dey and Bashaw Of The Associated Tripolitian States."

The lieutenant wouldn't have been human had he not rolled his eyes as a slim, swarthy, bearded man, wrapped in a spectacular rainbow of silks and topped with a shimmering white turban sailed in, arms wide, a smile of unctuous insincerity and goodwill spreading across his bird-like features.

"Ah, Commodore," the visitor piped, "what a wonderful sight you are being to this poor servant of Allah! I am hoping that indeed you are well!" Throwing his arms around the reluctant Preble, Kemal planted a full kiss on each of the Commodore's cheeks.

Preble smiled as best he could, and replied, "Good day to you, sir; it is, as always, a pleasure to see you."

"But better to see the my dear friend the Commodore, whose name is known throughout the world for his bravery, honor, and kindness."

Preble bowed slightly in acknowledgement. "My reputation is far overrated sir, but I thank you for your kind words. Please . . ." He motioned to a seat. "May I offer you something? Some coffee or tea, perhaps?"

"Your hospitality is magnificent, my Commodore! Truly, I do not deserve this kindness! A little tea, perhaps." Preble reached over to a small hand bell on his desk and gave it a quick shake.

The lieutenant poked his head in. "Sir?" The look on his face indicated that he wasn't sure if he was being called in to serve or protect.

"Tea for our guest."

The lieutenant looked at Kemal skeptically, but nodded. "Very good, sir."

As he ducked back out, Preble turned to Kemal with as polite a smile as he could muster. "So, Your Excellency, you are well?"

The *Philadelphia* Problem

Preble was, to a certain extent, used to this sort of thing. He was the senior US military commander in the Mediterranean region, and, although there were US diplomats there as well, they usually didn't have the kind of information or mobility that Preble and his men did. Accordingly, like Will Bainbridge negotiating that treaty with the Turks, Preble and his captains often served as de facto diplomats with full powers of negotiation. On the other hand, they were expected to think like diplomats, rather than military commanders, which didn't always work out. For every successful treaty like the one Bainbridge negotiated, there were some spectacular failures, like the captain who claimed the Falkland Islands for the US. In the meantime, however, since much of the rest of the world considered their diplomats and warriors interchangeable, they expected the United States to work the same way. We know that Preble hated it, and he was far from alone, but it was something that had to be dealt with. If he wasn't a warrior, he was expected to be an ambassador, and that was that.

Kemal smiled broadly, and launched into a long monologue that touched not only on his health, but also on that of his family, his distant relatives, several old friends, and a long dead uncle. Preble nodded politely, doing his best to interrupt but failing miserably, until the lieutenant knocked once and brought in the tea. Kemal stopped cold, less to take a breath than surprised by the lieutenant's entrance. Preble took his opportunity.

"I am indeed grateful to hear that all is well, Mister Kemal," Preble quickly interjected.

Preble understood that in Arab cultures, one was required to spend hours on the various social niceties, even if one only needed to take care of a few minutes worth of business. On the other hand, Preble wasn't an Arab and his demeanor could be most politely described as brisk; a combination that usually required a bit of work on the Commodore's part. "I assume, though, that your visit extends beyond a very welcome social visit. How is your sovereign?"

Kemal raised his hands skyward in tribute to his God. "Praise Allah, he is safe and well, despite the efforts of his cruel and fiendish enemies, may their bellies be roasted! He continues his efforts to depose the bitter knaves who have stolen his family and his birthright."

Preble held up a hand. "His family?"

Kemal nodded vigorously. "His poor family; his wives and his children. The pretender, let his name be thrice cursed . . .," Kemal spat onto the red tile floor, " . . . has taken His Majesty's family hostage!" Kemal's indignation was insufficient to keep him from swallowing the silver-jacketed glass of tea in one gulp. "Even after His Majesty kindly offered to renounce his throne and any claim to it, the fiend whose name I shall not speak sent his minions to take His Majesty's wives and children into vile durance."

To Barbary's Far Shore

Making a mental note to find out exactly what a durance was to be considered vile, Preble took a sip of his tea. "I am truly hurt to hear of His Majesty's distress." What he had to say next did truly hurt, but it was required by diplomacy, no matter how much Preble wanted to avoid it. Swallowing hard, Preble gave his best smile and asked, "How can I be of assistance?"

Kemal looked at Preble in adoration. "Truly, my Commodore, you have been sent by Allah to assist His Majesty in his quest to regain his throne! That, in fact, is the very reason I have been sent here: to ask your assistance in returning the rightful heir to his rightful place! His Majesty has been understanding, but no more! And now, with the capture of your glorious ship, the . . . ah . . . a thousand pardons, Commodore, what is the lovely name of your ship once more?"

"*Philadelphia*."

"Of course, *Philadelphia*, a name that literally sings! With your magnificent ships . . ."

Preble smiled and held up a hand to try and interrupt. "Your Excellency . ."

"Your thousands of brave men and your marines, whose ferocity and courage are legendary . . . "

"Your Excellency, I am flattered beyond words that . . . "

"His Majesty has directed me to assist you in . . . "

"Mister Kemal . . . "

"Yes, my Commodore?"

Preble held a finger to his lips and said "Shh." Before Kemal could get started again, Preble spoke. "Your Excellency. I am very aware of His Majesty's wishes to regain his rightful throne. Although my country is not a monarchy, we could never support an illegal action on anyone's part to usurp power that is not theirs . . . "

Kemal leaped from his chair to throw his arms around Preble once more. "I knew you would understand, my Commodore! His Majesty will be so very pleased . . . "

"His Majesty will be nothing of the kind!" Preble's voice was almost sharp, and he had to remember that for a few minutes, he was an Ambassador, no matter how much he hated it. "If you are requesting that my country aid and assist His Majesty in regaining his throne, that is one thing. I am very much in sympathy with his position, and I suspect that many others would be as well. But I am not in a position to commit my nation and its treasure to such an effort, you must understand that."

The *Philadelphia* Problem

Kemal looked as if he had been slapped in the face. "But surely, my Commodore, you are able to grant us some help? All we need are a few hundred brave warriors . . . "

Preble tried to look sympathetic. "Your Excellency, if I had that many warriors, I would do it on my own. But I am short of men, ships, supplies, and money myself, not to mention trying to take care of my men who are held in Tripoli now . . . "

"My Commodore, if you could help His Majesty, your men would be free in a few weeks, a few days . . . "

Preble leaned forward, gently laying a hand on Kemal's shoulder. "Your Excellency . . . Tell me, how many men can His Majesty provide, even if we could help?"

At that, Kemal hesitated for just a moment, but no more than that. "Ah, my Commodore, thousands of brave warriors would flock to your side . . . "

Preble raised one eyebrow, like a schoolmaster gently rebuking a student. "'Thousands would,' Your Excellency? Does that mean that there are none now?"

Kemal was about to launch into another statement averring how many brave soldiers of the Rightful Dey would happily stand by him, but Preble's gently skeptical look stopped him. There was a pause, then Kemal cast his eyes to the ground. In soft, quiet tones, he said, "My Commodore, there are a few of us who would follow you . . . would follow you anywhere, anytime, to take back what is ours. What is rightfully ours. But . . . only a few. We have nothing to give those who could help us."

Preble nodded understandingly. "I do wish we could help, truly. His Majesty would be an honored and valuable ally. No, far more than an ally; an honored and valuable friend. But, Your Excellency, I can do nothing on my own. I have great leeway to do many things on my own account, but this is simply not one of them. I mean no disrespect, I mean no insult . . . but this is something I cannot do."

Kemal asked quietly, "Can you not ask your president? Can you not tell him of the help we could give you?"

Preble nodded. "I cannot ask him myself; our system simply does not work like that. I can pass this on to our senior consul here, Mister Cathcart, and he can pass it on to Washington. I shall most assuredly urge him to do so. But, Your Excellency . . . my hands are tied."

To Barbary's Far Shore

Kemal nodded slowly. "I see, Commodore. I do indeed appreciate your honesty. I shall tell His Majesty of your sincerity." Kemal bowed slightly, and stood, mustering what dignity he could.

Preble found it difficult to terminate the conversation, but not through any desire to continue. Rather, it was out of pity and sympathy for the man, knowing that he would have to go home empty-handed and ashamed. Preble could face defeat after honorable combat; that was something you had to always take into account, because there could come a day when the Gods Of War would frown on you, Will Bainbridge could attest to that. But to have to go begging for help, and then be told to go away, no matter how sympathetically, that would have sent a shiver down the spine of a man like Edward Preble.

Preble did the only thing he really could do. Coming to attention, he saluted Kemal, then offered his hand. Kemal's expression was crestfallen, but he shook Preble's hand, and slowly glided out of the room, looking far smaller than he did when he entered.

As Kemal left, the lieutenant entered the room. "Not good, sir?"

Preble looked out the window, watching Kemal and his companion fade into the crowds that always surrounded the Grand Harbor. "Not at all, Leftenant. They want us to try and help Hamet get his old job back."

The Lieutenant considered this for a moment. "In and of itself, not such a bad thing, sir."

Preble coughed, then gave an ironic smile. "Very true, Leftenant. But even if I had the resources to do it with, I don't have the authority to commit my country to war, unlike Lord Nelson. It's moments like this that make me envy a monarchy."

The Leftenant smiled. "Never too late, sir. Lord Nelson could always call some friends of his in Whitehall."

Preble looked at the lieutenant in mild confusion. "About helping Hamet?"

"Heavens no, sir. About bringing the United States back under the Crown."

Given the grin on the Lieutenant's face, Preble wasn't sure whether or not to throw something at him, and instead went back to his desk and pulled out a piece of stationery. As he wrote, Preble told the lieutenant, "Please send my compliments, and this, to Lord Nelson, and ask if I may see him at his earliest convenience."

The lieutenant almost turned white. "Sir, I was simply jesting about returning to the Crown."

The *Philadelphia* Problem

Preble smiled wickedly. "No offense taken, Leftenant. Your mention of His Lordship gave me a thought, that's all. Off with you, now." Folding the stationery in three, Preble handed it to the lieutenant, who clicked his heels and trotted out. Preble coughed once, watching the lieutenant move out into the bustling street and head for the Governor's Castle.

Horatio, Lord Nelson, always had time for another of his Band of Brothers, regardless of where he was from, and within a few hours he was shaking hands with Preble at the Governor's Castle. Nelson was not a big man to begin with, and the end results of his injuries, a blinded eye, a lost arm, showed hard on a man in those days before proper medical care and therapy. But no injuries could hide Nelson's personality, a bright, cheerful one that would contribute to his victories.

"Commodore Preble," Nelson smiled, stepping forward with his hand extended. "Always a pleasure to see you, sir."

Preble doffed his hat and shook Nelson's hand. "As always with you too, your Lordship. Thank you for taking time out of your hectic day to speak with me."

"Not a sacrifice at all, Commodore. Was simply doing paperwork, paperwork, and more paperwork. After a few weeks ashore of that," Nelson smiled, "one looks forward to getting out to sea and into a good fight . . . which unfortunately always leads to more paperwork. May I offer you something?" Nelson motioned for a Maltese servant to come closer.

"Thank you, Your Lordship. A glass of Madeira, if I may."

"As good as done." Nelson held up two fingers and the Maltese glided off into the Castle. "Now, sir, to what do I owe the honor? All is well on the harbor?"

"Couldn't ask for anything more pleasant or helpful. What I did wish to talk to you about was a visitor I had earlier today."

The Maltese had reappeared in complete silence, holding two chilled glasses of Madeira on a silver tray. Nelson and Preble each took one and clinked them together. Nelson smiled like a cat that had just spotted the day's canary. "You mean Mohamet Kemal?"

Preble was visibly taken aback for a moment before Nelson shook his head. "No, the leftenant didn't say anything. We do not spy on our friends, Commodore, please believe that. On the other hand, we do keep very close track of those whose loyalty and friendship are, shall we say, less than certain. His Excellency actually arrived here last night on a Turk, and certain . . . well, 'gentlemen' in His Majesty's employ have been keeping track of him since."

"I'm wondering if you haven't already read my mind," Preble said with an uncertain smile as he sipped his Madeira, making a mental note to find out where Nelson got his.

"I'm quite sure I haven't," Nelson replied. "If I had that ability, I'd know where Bonaparte would strike next. In any event, please, go on about His Excellency Mister Kemal."

Preble collected his thoughts for a moment. "Mister Kemal wishes us, the United States, to assist Hamet Karamanli return to the throne of Tripoli."

Nelson nodded. "In and of itself, not a bad thing. Hamet is, shall we say, far less aggressive than his dear brother Yusef, and much more likely to be helpful and cooperative."

"Indeed. And much less likely to be making a living through piracy, though that is a far from sure conclusion."

"Agreed, Commodore. I'm not sure anyone could get the pirates completely under control at this point without being more than willing to hang more than a few of them from a convenient yardarm."

"The problem, however, is this . . . " Preble paused for a moment to try and frame his words properly. "First, all we have is Mister Kemal's word that Hamet wishes to return to the throne. With that in mind, may I ask, if possible, what those 'gentlemen' in His Majesty's employ know about that?"

Nelson smiled with another sip of wine. "It is my understanding that Hamet Karamanli, even after the arrest of his family, really has no great interest in returning to Tripoli. Now, his ministers, advisors, and other assorted hangers-on are another matter. They seem to be most eager to return to their old positions."

"With all their old power and privileges . . . "

"Exactly. For his part, Hamet tends to vacillate between the occasional urge to reclaim what is, in all honesty, rightly his, and a much stronger urge to self-preservation. I am aware of at least one attempt on his life already since his exile to Egypt, which followed an announcement by Mister Kemal that Hamet was preparing his march back to Tripoli."

Preble nodded. "A corpse or two can have most persuasive abilities."

"Quite. My guess is that Mister Kemal and friends managed to browbeat him into considering another run at Yusef, but that by the time Mister Kemal talked to you this morning, Hamet had long since changed his mind back and is quietly playing chess and reading alongside the Nile."

"I had heard he was in Egypt."

The *Philadelphia* Problem

"A ways down from Cairo, I understand. He still has access to a fair amount of money and is living in reasonable comfort."

"More than either of us, most likely. In any event, that answers my first question. I am generally doubtful of what Mister Kemal says, and fortunately I don't have the ability to commit myself to anything, even though, if it were possible, I'd help; anything to give Yusef trouble at this point."

Nelson swirled his wine. "How are your lads holding up?"

Preble coughed, then took a swallow. "The last I heard, they are well, if not being housed in the best of conditions."

"We are doing what we can, Commodore. You have my word on that. 'There but for the Grace of God,' you know."

"It's appreciated, Your Lordship. I will make sure my Government is aware of that. And speaking of my Government, that leads to my next question. Your Lordship . . . " Preble shook his head sadly as he looked out over the island. "Your Lordship, these men back home . . . My God, they have no, absolutely no, conception of what's going on out here. Congress is terrified of spending money. President Jefferson is terrified of offending people who routinely ravage our shipping and deface our flag, and half my men are on the verge of desertion, mutiny, termination of enlistment, or any combination of the three. What in God's name am I expected to do? I have barely enough ships to cover a single port, much less the entire North African coast, and now I am trying to figure out the best way to do all that, short one frigate, twenty marines, and one of my best captains. I can deal with combat; God knows I'd prefer that anytime. But I'm not even allowed to do that. I have rules of engagement that would drive a saint mad. I have to do something, but I'm not allowed to actually do anything."

Nelson reflected on that for a moment, then drank the rest of his wine. "Commodore, it's not an unusual situation on this side of the line either. We do have the advantage of a few more years experience with this sort of thing, and a few more ships, but in the end, you're still dealing with men who have never left the warm, safe parlors of their homes asking us to deal with other men for whom the most enjoyable thing they can imagine is death, violence, and random theft. My country has to deal with it for centuries, and I dare say we're not done with it yet. Frankly, and I say this as a friend who has been through this more times than I care to remember, if this is the worst your nation deals with, consider yourself damned lucky."

"Hardly words to reassure, Your Lordship."

"They weren't meant to be. They were meant to be truthful. I've seen far, far worse, my friend, and that has been a fact of life around here for a very long time. We can't simply go in and clean up what needs to be cleaned, no matter how much the little darlings deserve it. We have to follow the steps of the dance first." Nelson gave a little twirl of his hand, as if calling Lady Hamilton to join him for a quadrille.

"Sorry, Your Lordship," Preble said with a wry grin. "I never could dance worth a damn. However, whatever advice you can provide would be appreciated."

"That's the problem with you Colonials . . . " Preble raised an eyebrow, and Nelson gave a mock bow. "A thousand pardons, Commodore; you Americans don't appreciate the social graces. But, as far as advice goes, let me suggest something. There's a problem with the natives in this part of the world. They hate the people who run their country, but they won't do anything about them. We go in and clean it up, and they hate us for doing it, and all too soon we're right back where we started from. So it seems to me that what you need to do is find someone to go in and do it for you."

Preble thought about this for a moment, then looked at Nelson, his eyes wide in realization. "You mean, support Hamet? Good Lord, you sound like that lunatic consul of ours . . . "

"Eaton? Hate to tell you this, old man, but Eaton's right. The only way this can be dealt with is for your country to put an end to Yusef and get someone in there who's a bit more helpful. Spend it willingly or spend it in tribute and ransom. That's the choice you have. My country spends it willingly. Well, not as willingly as we would like, but willingly enough, and we get left more or less alone."

"Simply burning the damned place to the ground isn't an option?"

"Only if you pay to have someone else do it for you."

Preble sighed, and nodded in resignation. "I see your point, but that doesn't mean I like it any better. In fact, Eaton is back in Washington now trying to make the same argument."

"Any chance they'll listen?"

"Lord knows. I have a feeling that it will depend on how belligerent the Secretary of State is that day, how close Congress is to recess, and how much bad news is in the papers."

"Hmm," Nelson mused, then looked at Preble sympathetically and said, "Commodore, from what I know of your nation, then you are in very deep trouble."

The *Philadelphia* Problem

Preble shook his head. "That's what I was afraid of. In any event, I am going to have to take my leave. Might I ask when you have a ship leaving for Siracusa?"

"You're leaving us, then?"

"All good things, Your Lordship; all good things. If there's any chance at all that Eaton may be able to convince them to back Hamet, then I'd better be back with the Squadron. See what we can do to get ready for it. And if he doesn't . . . Well, we've still got the *Philadelphia* to deal with, and if I'm not there Decatur will take the whole damned Squadron in to get it back."

"Any ideas you'd like to try?"

"None whatsoever."

"Think fast then, my friend. I believe we have a sloop, the *Angel*, leaving day after tomorrow. Say the word, and you'll have the Captain's cabin."

"A simple bunk in the wardroom will be fine, as long as she's fast."

"That she is."

"In that case, Lord Nelson, I shall take my leave." Preble put his hat on and saluted.

Nelson returned it, followed by a handshake. "Fair winds and following seas, Commodore. The *Angel* is at your disposal."

"Thank you, Your Lordship. Until next time." There wouldn't be a next time; Preble would be back home in a few months and Nelson would go on to glory at Trafalgar. But, in the meantime, there was the friendship of two highly skilled professionals. One was one of the best that a new nation could turn out, the other quite probably the supreme achievement of nearly three hundred years of naval tradition and skill, but they were doing the same job for their countries.

Preble headed out of the Governor's Castle, knowing that Nelson was right. Something had to be done, God only knew what, and if he didn't get back to Siracusa someone else was likely to do it for him.

Or worse, they'd tell him what to do.

FOUR: CALL HER *INTREPID*

The trip back to Siracusa was every bit as fast as Nelson had promised it would be, and Preble was back aboard *Constitution* on January 25th. The good news was that Decatur, as the commander who had captured *Mastico*, was about to officially classify her as a blockade-runner. This was important because, at that time, a captured runner had to be quarantined first, in order for her crew and government to make any arguments they might. Fortunately, Yusef Dey wasn't likely to admit that *Mastico* was trying to get past the United States Navy, and there was the little matter of that Tripolitian flag they'd been flying. Before Preble went to sleep his first night back, he had signed the necessary paperwork. On the 28th, Preble officially posted his order that *Mastico* was a lawful prize of the United States Navy. The crew of *Enterprise* was as elated as if they had gotten a delayed Christmas present. After all, this meant that every ounce of *Mastico's* cargo was now contraband, and therefore prize money.

The bad news was that Will Bainbridge and his men were still captive in Tripoli. The worse news was that *Philadelphia* had been dragged off that gravel bar in Tripoli Harbor, intact and afloat, and the Dey's men were busily hauling up every last piece of the ship that Bainbridge and his men had thrown over the side. In a few months, or less, the Dey would have one of the most lethal warships in the world at his disposal. With hindsight, we know that it might not have been all that simple. The Dey's corsairs were considerably less complex than the *Philadelphia*, and his crews certainly wouldn't have had the constant training, training, and more training that men like Preble and Nelson and Bainbridge pounded into their crews. Had it come down to it, Preble's Boys would have been able to deal with *Philadelphia*, but it wouldn't have been easy. Had the Dey let *Philadelphia* go after British shipping, a ship of the line like Nelson's flagship HMS *Victory* would have turned her into matchsticks.

Call Her *Intrepid*

But there was no way for anyone to know that then; no way for anyone to sit down and consider all of this, no communications intercepts and no overhead satellite photos, no regiments of analysts to tell the commanders and politicians what was going on. The politicians were two months away under the best of circumstances; the commanders always a little bit behind events and doing the best they could. And every instinct that the commanders had was telling them they needed to do something, and do it now. Needless to say, this would have led to some spirited discussions among the commanders of the Mediterranean Squadron. These were men accustomed to action; not Hamlet-like debates on what to do or not to do, or whom to consult with and for how long. Their councils of war would have been not to decide what and when, but where and how hard.

They would have argued about it aboard *Constitution*, in Preble's day cabin that spanned her width across her stern. It would have been later in the evening, after the day's duties were finished, with a gentle rain falling across the panes of the stern galleries, turning the reflected lights of the Squadron and the Sicilian hillside into uncountable diamonds. It would have been genuinely cold outside, where the sailors did their best to stay warm and the marine guards could only stand upright and steadfast in the rain; but, in the day cabin, the food would have been good, the wine would have flowed, and the little world bounded by those four bulkheads would have been dry and safe and warm.

The senior commanders present, at least until the reinforcements arrived from the United States, would be Preble, of course, sitting at the chart table against the starboard bulkhead, and John Dent, his Flag Lieutenant aboard *Constitution*, perched on the edge of the big daybed. Daniel Carmick, the marine detachment commander aboard *Constitution*, is leaning against a bulkhead; sometimes in the conversation, sometimes out. Marines are not always welcome visitors in a council of war, but they are necessary ones, because they and they alone have the knowledge and the skill to execute a military operation such as that being contemplated here. Stephen Decatur would certainly be there, his feet up on the oaken cabinet over the rudderpost as he argued his views with Lieutenant Richard Somers, commanding the sloop-of-war *Nautilus*. Somers, perched on the rudderpost cabinet, is almost as young as Decatur, every bit as aggressive as his friend, willing to try anything to get just one more hit in against an enemy and truly believing that a death in combat would be as glorious and honorable as a man could ask for. He'll get his wish a few months down the road in Tripoli Harbor but, sadly, we won't remember him for that. We'll remember the ship named for him, the brig *Somers*, as the scene of the first and only executions in United States Navy history for mutiny, but those are other stories for other times. Right now, the Old Master and the Young Turks are at war, planning their next strike against their enemy. Maps are laid across tables, oil lamps flicker, and the fate of nations is at hand.

To Barbary's Far Shore

Preble puts down his empty goblet, rubbing the bridge of his nose. "I'm not at all sure how long we've been here . . . "

"At least nine or ten hours," Decatur cracked.

"No, no, no, Leftenant," Dent smiled with a sip of wine. "Don't exaggerate. Only five or six. . . "

"At worst," Carmick interjected

"In any event," Preble said, in a tone that brooked no further discussion of matters temporal, "we do need to come to some kind of decision, if we're going to do something about the *Philadelphia*. Whatever time we have is short, and our resources are even shorter."

Dent gave a derisive snort. "Sail in there with every gun blazing and take the damned ship back."

"No, Mister Dent, no. I do not know how many times or how many ways or in how many different tongues I can tell you this. We have exactly one frigate, of which, I might note, you are the Flag Leftenant. We know now through bitter experience that this frigate cannot get safely into Tripoli Harbor, at night and under fire, mind you, and out again without an utterly unacceptable risk of running aground. And with no disrespect to Leftenants Decatur and Somers," the two lieutenants raised their glasses in acknowledgement, "bravery and skill cannot always overcome sheer numbers. One lucky hit on any of our ships could be enough to maroon them there with the *Philadelphia*."

Somers took a swallow of wine and shook his head. "Commodore, I understand the risks; everyone here does. But, damn it, sir! We have a responsibility to the men and the ship. If we can't get the one, for you say that Bainbridge and his men are probably beyond reach in the Fortress, then we have to get the other back. It is a matter of principle, sir; principle!"

Everyone in the day cabin began arguing again, except for Decatur, who was standing behind Preble at the chart table, looking over the Commodore's shoulder at a map of Tripoli Harbor. It was a new one, though not so new that Kaliusa Reef hadn't had to be penciled into place. *Philadelphia's* position when she ran aground was there, and what they believed her new one to be as well: tucked safely under the guns of Tripoli Fortress, but still some distance from the docks.

She's gone, Decatur thought as the others volleyed their views back and forth. *No matter how badly we want to bring her out, just to poke the Dey in the eye. We cannot get her back from where she sits.* He hated himself for the thought, but there was no way to escape it.

Call Her *Intrepid*

We can't get the Squadron in there to get her back, he'd said, and Preble was right. Damn him, Decatur thought, *he was always right. We can't get the Squadron in there.*

Decatur looked ever more closely at the chart, leaning over until he was hunched over like some twisted schoolmaster. *To get her back* . . . His mind absorbed and calculated every contour, every landmark, every sounding.

Get her back. Decatur stood up with a triumphant grin. "We don't have to get the *Philadelphia* back!"

Preble, Dent, Carmick, and Somers went silent at once, and looked at him as if he had suddenly grown six heads. Preble looked behind him in surprise, not realizing that Decatur had been there.

"Mister Decatur," he said past a cough, "what in God's name are you babbling about?"

Decatur was still young enough to look like a schoolboy, but the expression on his face now was that of a ferocious raptor. Spinning the chart around as the others gathered beside the small table, Decatur gestured with his right hand, outlining what he had in mind.

"Gentlemen," Decatur explained, "There is no need at all to get the entire squadron into Tripoli to bring the ship out. As the Commodore has pointed out, it is simply beyond our means to do so."

Preble put on a pair of spectacles and peered at the chart. "Mister Decatur, I have learned that when you agree with me, trouble awaits. Go on."

"Thank you, Commodore. As I said, you are correct. We are outnumbered, outmanned, and outgunned. But we do not have to get *Philadelphia* back; we need only deprive the Dey of her use."

Somers blinked, not understanding. "Stephen, I'm afraid you've lost me. The only way we 'deprive the Dey of her use' is to get in there and either take her back or destroy her. And we cannot get in there with sufficient firepower to do so."

Decatur leaned forward, almost nose-to-nose with Somers, a grin of truly evil proportions crossing his features. He pointed out the stern gallery and quietly said, "We . . . don't . . . have to." Following his gesture, the officers saw, a hundred yards or so behind them, the dim silhouette of the *Mastico*, outlined only by a few safety lights, riding gently at anchor. While they looked at the little ship, they listened to Decatur.

"Gentlemen, we have in our possession a marvelous vessel, one that looks like a pirate, sails like a pirate and is rigged like a pirate, because it *is* a pirate!

To Barbary's Far Shore

We take that runner out there into Tripoli Harbor with a boarding party, set a match to the *Philadelphia,* and get out."

There was silence in the day cabin, broken only by the ticking of a clock on one bulkhead.

"How many men?" Carmick asked.

"What time of day?" Somers asked.

"Who's going to take her in?" Dent asked.

Looking to each of the officers in turn, Decatur answered, "I have no idea, night, obviously, and me, obviously."

The officers started arguing back and forth over the details, while Preble sat silently, looking back and forth between the map and the *Mastico*. Preble was known for solid planning and forethought, and he was bringing all his gifts to bear on this. There was no time to plan, no opportunity to sit down and practice the run until every member of the crew could do it in their sleep. Edward Preble didn't like doing things off the cuff; he never did. But there is one advantage to being a perfectionist, you eventually get a very solid grasp of what will and will not work.

And this, thought Edward Preble, *might just.*

Preble held up a silencing finger, and the argument trailed off. "Mister Decatur," Preble asked, his jaw resting on one hand, "just what would you need to accomplish this?"

Decatur thought about it for a moment, then started pacing back and forth under the low overhead, rattling off a list of points almost as if talking to himself. "We'll need the combustibles, obviously. We'll need to get the replenishment ships back here for that. We'll need about forty to handle her; another forty, maybe fifty, for the boarding party. We'll need to modify her a bit; perhaps a couple of swivel guns as well, just to be on the safe side." Decatur looked up as if he assumed no one had heard a single word and said, "Commodore, I'll need . . . "

"I know what you need," Preble said, looking over his spectacles at Decatur. "What I need to know is how long."

A pause. "Five days?"

Dent asked, "Stephen, are you asking us or telling us?"

Decatur's jaw hardened slightly. "Telling you. We'll be ready by the third of February."

Preble nodded. "No question there. You're not going in by yourself, though."

"But . . . "

"But nothing. I'm willing to accept that you might be able to get in without getting being spotted, but not even the Dey's men will be able to miss a burning warship in their harbor. You'll need some help that can get in fast, start shooting and keep shooting, and cover your escape."

Somers stood up, grinning in triumph. "That would be *Nautilus*, Stephen. We'll be there."

Preble shook his head. "That would be Charles Stewart on the *Syren*. They can get in, cover you, and get out, and they have the shallowest draft in the Squadron."

"Commodore," Somers said pleadingly, "You have to . . . "

Preble raised one eyebrow in warning. "The only thing I have to do is my daily reports, Mister Somers, and if I want I can give those to Mister Dent. *Syren* it shall be."

Somers sulked like a child put in his place by a nanny, but didn't argue the point.

"Any other questions, gentlemen?" Preble asked, but there were none. "Right then. Decatur, get to it at first light. Carmick, whatever the boarding parties need from the armory, get it to them. Dent, get someone over to the replenishment ships, and tell 'em no nonsense. And while you're at it, my compliments to Leftenant Stewart, and I want him here first thing in the morning."

Standing up to indicate the discussion was over, Preble shook each man's hand as they left, but motioned to Decatur to remain behind.

"Yes, Commodore?"

Preble stood with his arms folded as he looked out the stern gallery towards *Mastico*. "You know, Leftenant," Preble said quietly, "you are out of your mind."

Decatur gave a gentle smile. "So I've been told."

Preble nodded in agreement. "Then I shall indeed have you confined to Bedlam itself, Mister Decatur. After you destroy the *Philadelphia*."

Even before the Mediterranean sun began its slow winter ascent, Edward Preble was writing the official orders to do what had to be done. Preble had no doubt at all that Decatur would take care of everything, but it was perhaps best to make sure Decatur knew exactly what was expected of him.

> **UNITED STATES' FRIGATE CONSTITUTION.**
>
> *Syracuse Harbor, January 31, 1804.*
>
> Sir: You are herby ordered to take command of the prize ketch, which I have named the *Intrepid*, and prepare her with all possible despatch for a cruise of thirty days, with full allowance of water, provision, &c., for seventy-five men. I shall send you five midshipmen from the *Constitution*, and you will take seventy men, including officers, from the *Enterprise*, if that number can be found ready to volunteer their services for boarding and burning the *Philadelphia* in the harbor of *Tripoli*; if not, report to me, and I will furnish you with men to complete your complement. It is expected you will be ready to sail to-morrow evening, or some hours sooner, if the signal is made for that purpose.
>
> It is my orders that you proceed to Tripoli, in company with the *Syren*, Lieutenant Stewart, enter that harbor in the night, board the *Philadelphia*, burn her, and make good your retreat, with the *Intrepid*, if possible, unless you can make her the means of destroying the enemy's vessels in the harbor, by converting her into a fireship for that purpose, and retreating in your boats and those of the *Syren*. You must take fixed ammunition and apparatus for the frigate's 18-pounders, and, if you can, without risking too much, you may endeavor to make them the instruments of destruction to the shipping and Bashaw's castle. You will provide all the necessary combustibles for burning and destroying ships.
>
> The destruction of the *Philadelphia* is an object of great importance, and I rely with confidence on your intrepidity and enterprise to affect it. Lieutenant Stewart will support you with the boats of the *Syren*, and cover your retreat with that vessel. Be sure and set fire in the gun-room births, cockpit, store-rooms forward, and births on the birth-deck.
>
> After the ship is well on fire, point two of the 18-pounders, shotted, down the main hatch, and blow her bottom out. I enclose you a memorandum of the articles, arms, ammunition, fire-works, &c., necessary, and which you are to take with you. Return to this place as soon as possible, and report to me your proceedings. On boarding the frigate, it is probable you may meet with resistance. It will be well, in order to prevent alarm, to carry all by the sword. May God prosper and succeed you in this enterprise.
>
> I have the honor to be, Sir, your obedient serv't,
>
>   **EDWARD PREBLE.**
> *Lieut. Commandant Decatur, Intrepid.*

Call Her *Intrepid*

Mastico was on her way pierside to begin her transformation. Decatur went ashore with a team of shipwrights and sailors, and started crawling over every inch of her. It was obvious that her former crew had not kept the same fastidious standards that Commodore Preble insisted upon, but there wasn't time to correct that; her new crew would have to deal with it until they returned. Bulkheads were punched through or taken out entirely, and new ladders put in to enable fast passage from belowdecks to topside. Anything that wasn't absolutely necessary to get *Mastico* to sea was going over the side.

Decatur watched with a gimlet eye as the shipwrights worked around the clock, measuring, cutting, sawing, assembling and improvising when necessary. When he wasn't at her side, he was sailing around the Squadron, asking for volunteers for a mission of 'daring and importance.' Officially, only Preble, Decatur, Stewart, and the other senior commanders knew what was going on, but the sight of the replenishers pulling alongside *Mastico* and transferring kegs of black powder and racks of muskets and flintlock pistols spoke volumes. They were going a-raiding, and though no one knew where, there was bound to be a grand fight somewhere and that was enough for them. Most of *Enterprise's* crew volunteered, so that covered the men who would keep the ship running. *Enterprise's* marines volunteered as one for the job, with Sergeant Wren in the lead, and Dan Carmick had no problems getting enough men from *Constitution's* detachment to round out the boarding party.

By the afternoon of February 2nd, Decatur could stand alongside *Mastico* in a light drizzle, and be pretty satisfied with himself about how things had gone. As he strode aboard, he noticed *Mastico* was riding a little high in the water, but that could be remedied easily enough with a touch of ballast. Saluting the flag as he came aboard, Decatur looked around, before ducking below to the hold. The shipwrights were cleaning up, but they'd done a bang-up job. The belowdecks spaces had been completely reconfigured, and, although it wouldn't be the most comfortable place to spend a few days at sea, the lads really wouldn't have much to do until they arrived off Tripoli Harbor, so they could relax a bit in the meantime.

The sound of a bosun's whistle drifted down from overhead, followed by a bellow of, "Squadron, arriving!" Decatur raced back up the ladder to see Preble wandering about the deck, looking at various details until he saw Decatur racing towards him. Decatur saluted smartly.

"Good morning, Commodore, I hadn't expected to see you; a pleasant surprise, though."

Preble returned the salute with a cough. "I always visit a ship I'm sending into combat, Mister Decatur." Looking around once more, Preble said, "I have to compliment you, sir. You said five days, and you were as good as your word. Do you want to take her out and see how she handles?"

To Barbary's Far Shore

Decatur shook his head. "No time, Commodore. The Dey's shipwrights are almost as good as ours, and they've had more time. We need to go, and we need to go now."

Preble nodded. "Didn't think you'd want to, Mister Decatur, but I wanted to give you the chance. Can you sail with the tide?"

Decatur checked his watch, then looked at the gray woolen sky. "Probably not, but we will. I'll get the word out to the ships and get the men over here at once. I'd like them in civilian clothes; it'll help in getting into the harbor."

"Done. And speaking of getting into the harbor . . . " Preble leaned over the rail and motioned to someone on the pier. Decatur heard footsteps going up the gangway, and a short, stocky Sicilian bounded aboard.

Doffing his hat, he bowed to Preble. "Si, signore?"

Turning to Decatur, Preble said, "Leftenant Stephen Decatur, may I present Signore Salvatore Catalano, ship's pilot, late of Palermo, Siracusa, and points east? Leftenant Dent suggested that we speak to him about guiding you into Tripoli Harbor."

Catalano reached out and took Decatur's hand in a strong, dry grip that Decatur had to fight not to wince at. "Buongiorno, Commandante," Catalano said. "Commodore Preble say I might be able to help you."

Decatur regarded Catalano with mild suspicion. "Signore Catalano, I am very pleased to meet you, and I am sure you would be more than willing to help, but . . . "

"Decatur," Preble interrupted, "Signore Catalano has been a guest of the Dey. He was released the day *Philadelphia* was taken."

"Si, Commandante," Catalano said quietly to Decatur. "I see your friends taken. They fight; they fight hard, but, after while, poof. Too many Barbary, too many ships. I no want to see anyone go do what I do, Commandante. You need help, I give. Just let me help."

Decatur considered this for a moment. "You know the harbor at Tripoli?"

"I know west approach you take, Commandante. I no know the eastern approach, where your friends go aground. Also, I speak Barbary – Ana takallam Berberyi."

Impressed, Decatur's attitude changed. "You get me to the *Philadelphia* and I'll be happy with that, Signore Catalano." Decatur extended his hand again, and the two men shook, sealing the agreement. "Get your things and see the officer of the watch; he'll get you signed on."

Call Her *Intrepid*

"Si, Commandante." Catalano trotted back down the gangway, and Preble started heading that way as well. Falling in beside him, Decatur said, "Well then, sir, I think that about does it."

"Not quite."

They left the ship and went down onto the dock, and Preble led Decatur around to the stern. Pointing to the Arabic characters beneath the tiny stern gallery, Preble shook his head chidingly. "Mister Preble, I consider myself a tolerant man, but I'll be damned if I'm going to allow any ship to sail under my authority to carry a Berber's name."

Decatur looked at the flowing green characters for a moment. "Does the Commodore have any suggestions?"

Preble looked at the stern, then back to Decatur with a grin. "Call her *Intrepid*. It fits." Coming to attention, Preble saluted Decatur. "Mister Decatur, good hunting. Don't leave that harbor without the *Philadelphia* in flames."

Decatur returned the salute. "And if I can't, sir?"

There was no humor at all in Preble's response. "Like I said; don't leave that harbor, Mister Decatur." Preble turned on his heel and headed for the whaleboat that would take him back to the *Constitution*.

It was a little after ten that evening when the newly christened *Intrepid* cast off from the Siracusa dockside. The night was overcast; only a few stars peeked through occasional portholes in the clouds, as if afraid to see her sail. Normally, Decatur would be calling out commands in his best quarterdeck voice, but tonight he was keeping it down. They didn't need to attract any more attention than they absolutely had to. *Intrepid* felt a little strange as he gave the helm a few commands to see how she moved, but he assumed that would be because of her rig. Decatur was used to handling something bigger and heavier, but he had at least a day or two to get the hang of her.

To have stood on the deck of the *Intrepid* that night as she swung into line ahead of *Syren* would have been to see history. Stephen Decatur commanded her, and beside him were lieutenants and midshipmen whose names would eventually ring down into American history. There was James Lawrence; slender and gregarious, a natural on deck. He would eventually lead his own ships into combat, and, one day in 1813, he would die on the deck of the seemingly jinxed frigate *Chesapeake*. His last words would become an unofficial motto of the US Navy: "Don't give up the ship." Joseph Bainbridge stood leaning silently against the rail, built like his brother Will; towering, barrel-chested, and powerful. He must have known that there was no way to get his brother out, at least not yet. But he had happily volunteered for this mission, just to remind the Dey that making the Bainbridges angry was a very serious

To Barbary's Far Shore

mistake. Thomas MacDonough, wiry and strong, stood forward, supervising the line handling. He would go on to find himself commanding US naval forces on Lake Champlain in the War of 1812; slightly more difficult than it sounds, because there were no US naval forces on Lake Champlain. None. Not a single ship. MacDonough had to build his fleet from scratch, and ended up soundly defeating a British fleet built the same way.

A few hundred yards away across the choppy seas, *Syren* was following *Intrepid's* stern lantern, with Lieutenant Charles Stewart on her quarterdeck. Stewart had been born in 1778 and shipped out on a merchantman before the US Navy ever existed. He was known as a hard case, but he was far from the worst. He was also a superb commanding officer, more than willing to put himself and his ship in harm's way, and he never lost a fight. He would become the Navy's first senior flag officer in 1859, and would complain two years later that he was too old to go back to sea against the Confederacy. He was promoted to become one of the Navy's first rear admirals in 1862, and lived long enough to see the Union victorious, before he passed away in 1869 at the age of ninety-one.

The future, though, was far away, holding both tragedy and triumph. For now though, the entire effort of the United States Navy was down to two little ships sailing into an unsteady night, with legends yet untold on their decks.

It took five long days to inch their way to the coast of Tripoli, the lights of the city finally glowing a soft gold on the southern horizon. Unfortunately, the winds that made the trip south take the better part of a week also kept them from entering the harbor under anything resembling favorable conditions. It took nine days for the wind to turn decently enough for them to try and enter. By the time the afternoon of February 16th arrived, dark and cold, *Intrepid's* sailors and marines were quite willing to kill anything to get off the ship, including each other.

Dinner, such as it was, had been served around four. Then *Syren* and *Intrepid* parted company, Stewart taking *Syren* westbound. Decatur had all but those men absolutely necessary below decks, and had turned the helm over to Signore Catalano, who was more than capable of answering any challenges. Decatur had even changed into civilian clothes, something of an effort for a man who loved his uniform as much as he did.

The sun was down below the horizon when *Intrepid* slid into the mouth of the harbor, a thin northeast wind behind them. Decatur and James Lawrence stood behind Catalano, watching the glow of lights brighten, then lift like a curtain to reveal beneath it the whitewashed fortifications of the city. Lawrence had a spyglass lifted, scanning the horizon before him.

"Anything?" Decatur asked, tensely.

"Nothing yet," Lawrence replied before lowering the glass. "I'm hoping we'll actually see the damned ship sometime this evening. We'd look awfully foolish if we have to stop and ask someone where she's at. Where did the Commodore say she was moored?"

"Under the fortress. We'll see just how far under."

Photo # NH 56745 Destruction of USS Philadelphia, 16 Feb. 1804. Chart by C.W. Furlong

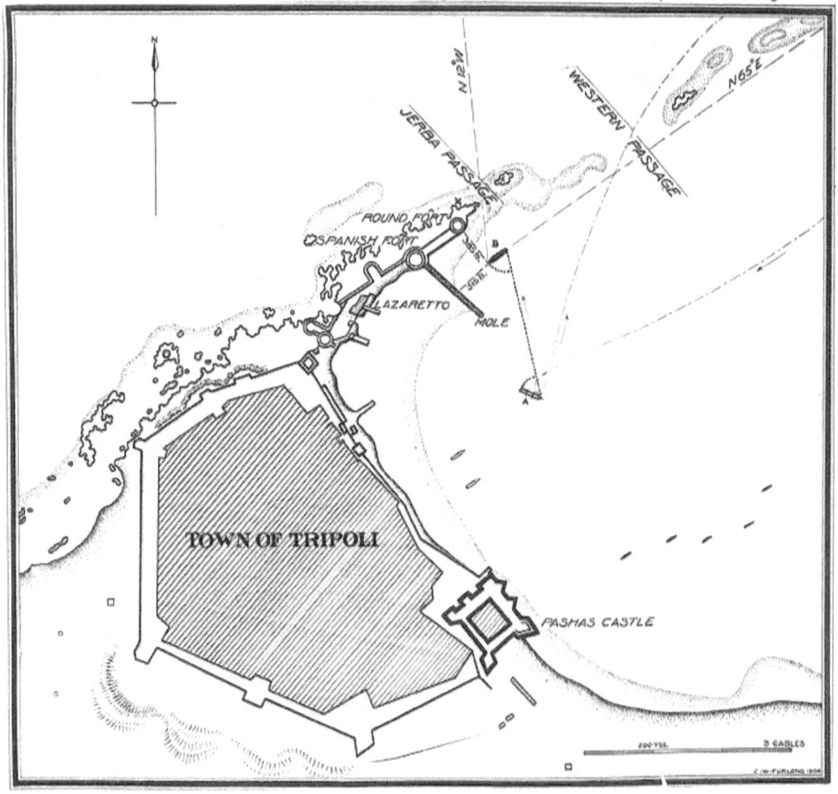

Catalano turned to look over his shoulder. "Quarter mile at most, Commandante. Too shallow to go closer with such a big ship." There was suddenly a low whistle from overhead. Decatur looked up to see one of his lookouts gesturing forward. Taking the spyglass from Lawrence, Decatur shot forward to the bow and peered at the looming Tripoli shoreline.

Not yet ... not yet ... There ...

Her foremast was back up and she still seemed to be tilted slightly to port, but, in the deepening gloom, there was no mistaking United States Frigate *Philadelphia*. She looked to be a bit further away from the shore than they'd

To Barbary's Far Shore

thought, and that was fine with Decatur. Every foot away from the walls of Tripoli was another few seconds they had to get it done and get out of there.

"Mister Catalano," Decatur said as he returned to the quarterdeck, "I believe we've found our destination."

Catalano motioned for Lawrence to take the helm as he looked through the spyglass at the masts that just cleared the horizon now. "Si . . . that her, Commandante." Catalano thought for a moment, then rushed back to a chart table where a chart of the harbor was laid out, Decatur beside him.

"Commandante Decatur, this the western channel, here . . . " Catalano traced the narrow cut that ran through the center of the harbor. "This," he pointed to the immediate east of the channel, "is Kaliusa. We fine, not enough draft. Problem is we come close, very close to English Fort." Catalano touched a spot on the map where the 'English Fort,' a medium fortification built decades ago by English traders, guarded the approach to the harbor proper. "Signore Lawrence good helmsman?"

"Exceptionally," answered Decatur.

"Good, good. Signore Lawrence, I take her in, then you take wheel. Not move an inch, si?"

"Understood completely, Mister Catalano," Lawrence replied.

On the English Fort itself, only a few sentries walked the ramparts. As always, gunners manned the handful of cannon that could easily reach out as far as the northern edge of Kaliusa Reef. Their job was to open fire first, giving the gunners at the Dey's castle enough time to get to their weapons and provide the main defense of the harbor. One of the sentries saw white sails glowing dimly in the Kaliusa Channel, and called for his senior. The senior guard raised his spyglass and saw a ketch with Tripolitian rig inbound, moving steadily closer. Turning to the sentry, he said, "Burn the signal lights to make sure."

On *Intrepid*, they saw a light suddenly flare into life, followed by a second. Decatur looked at Catalano.

"Signore . . . should we be worried?"

"Only if Commandante doesn't have Berber flag."

Decatur had been ready for that. One of his sailors ran up the red-and-gold banners. From the English Fort, the senior guard watched through his glass as the banners slid up the halyards towards the top of the mainmast. *One of ours*, he thought, handing the glass back to the sentry.

"Hail them when they're closer; find out who they are."

81

Call Her *Intrepid*

The fort grew closer and closer, and the few men on *Intrepid's* deck could now see lanterns on the Fort's walls, and worse, they could see the cannon embrasures atop the fort itself. No one was under any illusions of what would happen if anyone decided to try and stop them. Intrepid only had four small cannon aboard, not even enough to dent the walls of the English Fort. But the guns of the Fort each probably threw as much iron as all of Intrepid's guns put together, and one salvo would probably stop her in her tracks. God alone knew what the second one would do.

"Signore Lawrence?"

"Mister Catalano?"

Catalano gave the wheel a slight turn to starboard. "You take wheel now. Do not move unless I tell."

"Yes, sir, Mister Catalano." Turning to Decatur, Catalano said, "They probably hail us soon; I answer, si?"

"You answer, Mister Catalano."

At the English Fort, the commander of the guard was on the rampart, watching the ship come steadily closer. Although there was nothing resembling a schedule, it was still a bit unusual for a ship to come in this late, especially through the Kaliusa channel, without a pilot having been sent out. On the other hand, that probably meant a native Tripolitian crew . . . but, still . . .

Picking up a hailing trumpet, the commander of the Guard called. "Marhaba, hello!"

On *Intrepid*, Catalano spoke tersely. "Trumpet, prego."

Decatur tossed the trumpet to the pilot, who trotted to *Intrepid*'s bow. Lifting the trumpet, Catalano replied. "Marhaba salam alekom, sadiq! I thank Allah we have returned home safely!"

The Guard commander listened to the reply. *It is right . . . but it didn't feel quite right.* "Ismack eh, who are you?"

Decatur and Lawrence stood where they were, trying to look as casual as possible.

Thomas MacDonough poked his head through the hatch and asked, "What's going on? Are we going to be there by daylight?"

Without looking down, Decatur growled through his teeth, "Tom, if you don't get back down there, I swear unto God I'll throw you overboard."

MacDonough knew that Decatur was a man of his word, and he really had no desire to challenge him on this one. Ducking back down the ladder, he turned

To Barbary's Far Shore

to the boarding party, forty-six marines led by Sergeant Wren, Joseph Bainbridge, and himself. Wren shot him a questioning look, and MacDonough shook his head.

"Not quite yet, gentlemen. Please, keep your seats."

Back on the *Intrepid's* bow, Catalano could now very clearly see the guns of the English Fort, and, in the dim light of the torches on the wall, the men manning them. Decatur took a quick, discreet look through his spyglass; not only could he see all that, but the crews manning them as well, their faces faint shadows in the guttering light of their slow matches.

Catalano took a deep breath and called out, "Ana *Mastico*, taa'la hena Tunisya!"

Now the commander of the guard was becoming a bit unsure. *Blockade-runners come in all the time, but there is always, always, some kind of warning from the Castle or the outposts to the north. Perhaps a quick look at this ship would be in order . . .*

"Heave to now and drop your anchor, I wish to inspect you!"

"Dio," Catalano breathed to himself, then turned to Decatur. "They want us stop and let them board."

Decatur was afraid of very little in this world, but a brief chill had to have gone through him at that point. Thinking quickly, Decatur said, "Slow us down, but keep us going on course. Then tell them this . . ."

The commander of the guard watched as the ship continued on course. One of the crew waved at him cheerfully. *They weren't making any effort to slow down*, the commander thought. Turning to one of the sentries, he said, "Tell the gunners to get ready. I want that ship blown out of the water when I give you the word. But, if anyone opens fire before I say, they will answer to me."

"Na'am fahamt, Effendi." The sentry ran off to get the word down to the gunners as the commander raised his trumpet once more. But before he could speak, he heard a call come back from the ship.

"Effendi, we have lost our anchors in the gale, how shall we stop so quickly? Besides, the Dey himself waits on our cargo! Well, the Dey and his hareem!" Catalano laughed heartily, but he didn't feel it. *Intrepid* was now less than a quarter mile away from the guns of the fort; more than close enough to make any salvo quick, accurate and final.

Now it was time for the guard commander to get worried, but for a different reason than just the possibility of someone who wasn't supposed to getting into the harbor. The Dey had standing orders that runners were supposed

Call Her *Intrepid*

to be brought into the Harbor without delay and immediately unloaded, no questions asked. The Dey was most unhappy when his pleasures were delayed, and 'most unhappy' was a euphemism for something almost unspeakable. *But it still doesn't feel right . . .*

Raising the trumpet, the Guard commander called out. "Ayez eh sheif, what do you carry?"

Catalano asked Decatur. "What do we carry, Commandante?"

Decatur rolled his eyes. *The next time,* he thought, *we just shoot our way in and be damned.* "Tell him women and money. Tell him something."

Catalano thought that sounded reasonable enough and called back. "Hareem wa feluus, Effendi, what else?" He was laughing once more.

Time was short, too short; the men on *Intrepid*'s deck could see faces on the fort's walls now. If they didn't make the turn to starboard in a few more seconds . . .

The Guard commander watched the ship inch closer below him. He could stop them and search them . . . But they would also tell the Dey that they'd been delayed . . . and it was just a feeling on his part. He didn't want to meet the bastinado for just a feeling.

"Roah, go!" The commander waved towards the harbor itself, now just one nautical mile away.

Catalano and Decatur tried not to sag too visibly as the smiled and waved at the fort. "Signore Lawrence, just a turn to starboard, grazie."

"With pleasure, sir." Lawrence turned the wheel a few points to the right. *Intrepid* glided past the English Fort. The few men on deck waved happily, still waiting for those wicked black muzzles on the fort's ramparts to speak their mind. *Philadelphia* was now dead ahead. Only a few lights shone on her deck and her stern gallery dark. That was, at least, a good sign; meaning that there was most likely not a senior officer, and his attendant guards, staying aboard.

Catalano moved back to the wheel. "Commandante, I take us in. We tie up alongside *Philadelphia*; you take care from that."

Decatur smiled. "I'm looking forward to it." He ran to the hatch and jumped down the ladder, where MacDonough, Bainbridge, and Wren were standing by. "Right, then," Decatur said. "We're in the harbor . . . "

"About bloody time," Bainbridge shot back. "I was beginning to think we'd have to spend another damned night down here."

To Barbary's Far Shore

"Be of good cheer, Joseph," Decatur said. "In a few minutes, you can spend as much time up on deck as you want, as long as you don't mind someone trying to kill you. Sergeant Wren?"

"Sir?" Wren stepped out of a shadow, his bulk seeming to fill the converted hold. He looked even more dangerous with a few days' beard in addition to the two pistols and dagger tucked in his belt, along with the cutlass on his left hip.

For that matter, Decatur thought, *all of them look positively homicidal, which is just fine with me.* "You and your lads ready?"

Wren nodded, patting the cutlass' hilt. "Point us in the proper direction and it'll be over right quickly."

"I'm holding you to that, Sergeant . . . " Decatur looked up as he heard a whistle from the main deck. Stepping up the ladder, James Lawrence reached down to pull him up, his face grim.

"Stephen, we have a problem . . . ," Lawrence said, gesturing towards the *Philadelphia*.

It was only then that Decatur saw them. Two Tripolitian corsairs, black and silent, moored just a hundred yards or so off *Philadelphia*'s starboard bow. As *Intrepid* continued to glide forward, it got much worse. Moored in line astern behind the corsairs were two gunboats; roughly the same size as *Intrepid,* but with a far heavier punch. Each of the four ships was easily within range of *Philadelphia,* unseen as they approached because the frigate's bulk concealed them from the low decks of the little ketch.

Dear heavens, Decatur thought, *what do I do now*? One of the gunboats alone had more firepower than *Intrepid* did, and each of the corsairs probably had twice as many men. Even if they somehow got aboard the frigate and got her fired, they would still have to sail around those ships to get out.

Now, Stephen Decatur loved combat and he never in his life ran from a fight, but even his warrior's blood had to have run as cold as the wind that night to see those four ships lined up alongside the path he was planning to take. Aggressive, yes; that word could be engraved on his tombstone, but foolhardy would not be. Commodore Preble's words not withstanding, no one could have faulted him at this point if he had decided to just keep going past the ships that outnumbered, outgunned, and outmanned him at four to one as inoffensively as possible until he was back in open water so that he and Stewart could come up with some sort of plan.

Decatur's mind was racing, and he was only dimly aware of Catalano looking at him nervously. Lawrence stood there with his arms folded, waiting for some kind of decision. *Philadelphia* was starting to loom up ahead of them

Call Her *Intrepid*

now, far from the friendly, welcoming ship they had known for so long, the ship he had visited so often when his father commanded her. She was a monster now; enemy territory to be feared and hated instead of treasured and protected. A few lanterns shone dimly on her rail, casting faint cones of light down onto . . .

Oh no, NO . . .

Black, threatening muzzles of cannon, one in each gunport.

She was armed; somehow they'd either raised enough guns off of Kaliusa Reef, or they'd stripped a dozen corsairs to do it, but it didn't matter. She was armed, and she was rigged, and she was ready.

At that moment, when Stephen Decatur realized that the *Philadelphia* was not the partially crippled wreck they had hoped her to be but instead a fully capable weapon, prepared to go out and wipe the sea clean of anything flying an American flag, his choice was made for him.

There would be other moments like that in the history of the US Navy. The one most people remember is Torpedo Squadron Eight at Midway, going after the Japanese fleet with no hope of survival, knowing that their sacrifice would buy time for the dive bombers that would arrive minutes later to change the course of history. But Decatur was probably the first, and he set the example.

"Mister Catalano," he said quietly, but with iron in his voice. "Bring us along *Philadelphia's* starboard side. James, tell the marines there's been a change in plans. Steel only; no pistols or muskets. No sense in waking those fellows up."

Catalano nodded. As Lawrence vaulted down the ladder, he spun the wheel to starboard, sending a silent prayer to the Virgin as the *Intrepid* glided towards her target, leaving a phosphorescent trail behind her as wispy as gossamer.

The day's work aboard the *Philadelphia* had been good. The last of the cannon had been horsed into place. The Americans had been kind enough to leave them a magazine full of shot and powder, and not the poorly made Italian or Spanish powder either, but the beautifully milled and reliable fine-corn from the DuPont works in Delaware, that would save the gunners that much more work tomorrow. The foremast had been reattached and rerigged, though not without considerable difficulty. In the morning, she would be inspected by the swashbuckling figure of Murad Reis, the Dey's senior naval commander. Born Peter Lisle in Glasgow, Reis was a good match for the Dey himself, cruel, avaricious, and amoral, but with exquisite taste. Admittedly, those were not what other navies would consider good qualities in a commander. However, the Dey's fleet was a for-profit operation and, as sad as it is to admit, they were qualities that tended to maximize earnings.

To Barbary's Far Shore

On the other hand, Murad did take care of his men, though it wasn't at all for altruistic reasons. He simply understood that crews who were well fed, well paid, and well cared for were less likely to tell the other side what was going on. And since the work on the *Philadelphia* had gone so well and so quickly, and, incidentally, under budget, Murad had sent over a boat full of food and wine earlier that evening with his thanks. Most of the refit crew was therefore below decks, mostly sleeping the sleep of the just with full bellies and slaked thirsts. There was a fire watch, about ten men doing their best to stay out of the chill breeze that night, but no more than that. One of them was walking slowly to and fro on *Philadelphia's* quarterdeck, stopping every few steps to shiver and clap his arms to stay warm.

"Marhaba! Marhaba salam alekom!"

The sentry looked up to see a small ship, its sails reefing, gliding slowly up behind the frigate.

"Ayez eh, what do you want?" the sentry called.

"Saadni, sadiq, help us, friend! We lost our anchors in that gale, and we carry the Dey's cargo! Throw us a hawser before we run this ugly beast into something!"

Well, at least helping would mean moving, and moving would mean staying warm. Motioning for the ship to continue its course around to the frigate's starboard side, he called for the other sentries on deck to help. They were supposed to be keeping other ships from getting too close to the frigate; but, after all, she was carrying the Dey's cargo.

On *Intrepid*, Decatur stood by as Catalano eased the ship past *Philadelphia*'s darkened stern galleries; windows into the captain's day cabin where he and his brother James had played as children and his mother had smilingly chastised them for getting underfoot. The solid brown hull of the frigate now loomed terrifyingly overhead, like the walls of a fortress. Decatur could see, faintly, the glow of a single lamp inside the starboard gallery.

So with any luck there was either no one there or they were asleep.

As the breezes gusted then died, snippets of laughter and music drifted down from the walls of the city like leaves floating gently down from a tree. Belowdecks, MacDonough had extinguished the few candles that had had glowed softly in the hold. Now, MacDonough, Bainbridge, Wren, and the marines were up against the portside bulkhead, so that no one looking into the hold from above would see anything more than a black pit.

Someone yelled something from above. A hawser the thickness of a man's arm came sailing over the frigate's rail, thudding to the deck of the smaller ship.

Call Her *Intrepid*

Catalano motioned for the crew to get it secured, and Decatur, Lawrence, and some sailors grabbed the rough, abrasive rope and wrapped it around one of the cleats that lined *Intrepid*'s deck. As the hawser began to go taut, Catalano nimbly spun the wheel to slide *Intrepid* to a stop along the larger ship's hull. The boarding parties belowdecks grabbed whatever they could to brace themselves as the ship bumped and shuddered, then came to a halt.

"Come on, come on," Bainbridge whispered nervously.

"Joseph?" MacDonough asked solicitously.

"What?"

"Shut the hell up."

It was probably just as well that MacDonough couldn't see the look Bainbridge shot him, but he did stay quiet.

Catalano and Decatur looked at each other with quick nods. If this was going to happen, it was going to have to happen now. Two rope ladders hung over the frigate's side, pinned there by the smaller ship, for which Decatur said a silent prayer of thanks. A single face leaned over the rail with a questioning expression.

Here goes nothing.

Decatur bounded up the ladder.

The man on *Philadelphia*'s deck asked, "Eh dah, what is this?"

Decatur had a smile on his face and one hand extended as he cleared the rail. The other men who had helped send the hawser down to *Intrepid* had already dispersed, the closest one a good fifty feet down the deck.

The sentry was suddenly feeling a bit disturbed. It was one thing to give the other ship a hand, but another for them to start coming aboard without permission.

Decatur took a step toward the sentry. He leaned in closer to see just who this fool was, just in time for the sentry to realize that this was an infidel . . .

But before the sentry could do anything, Decatur reached out with that extended hand. He yanked the startled sentry to him, and then flipped the dagger out of his belt and up alongside the sentry's throat.

"Not a word, Effendi" Decatur snarled as quietly as he could.

The sentry, of course, couldn't understand a word of it, but stark fear was having the desired effect as Decatur started backing them toward the rail.

To Barbary's Far Shore

On *Intrepid's* deck, Lawrence looked up at the rail that Decatur had disappeared over a few seconds before. *Come on, Stephen,* Lawrence thought, *you don't have to have tea with him. There!*

Decatur's arm whirled frantically over the rail. *Time to go . . .*

Lawrence stomped twice on the deck, the signal for the marines to come up, then raced for the ladder. Halfway up, he looked back to see the boarding parties streaming up out of the hold in remarkable silence, Tom MacDonough and Joe Bainbridge at their head. Clearing the rail, Lawrence turned to see Decatur still holding the sentry hostage. Nodding quickly to Decatur, Lawrence reached back over the rail to start helping the heavily laden boarding party aboard.

Boarding the Philadelphia.

Call Her *Intrepid*

A few feet down the deck, one of the sentries who had helped throw the hawser over the side heard something behind him. More by reflex than by any real interest, he turned to see what was going on. He saw men pouring over the deck aft.

Now that isn't right, he thought, wondering why all these men were coming up from that ship. Then, squinting into the darkness, he realized that something was very wrong.

"Hey!!" he called.

As Decatur looked up at the sound, the sentry elbowed him in the gut and broke away. He had to make it to the hatch; he had to make it to the hatch.

But Solomon Wren, six feet of angry marine as solid as a tree trunk, was in front of him, swinging his cutlass down in a hissing arc that mirrored the lights of the city in the last thing the sentry ever saw.

Damn!, Decatur thought, *so much for surprise!*

"Go, go, go!" MacDonough called as the marines thudded across the deck, heading for their goals, the storerooms and the gunroom cockpit forward. The sentry who had realized that they were being boarded was now trying to scramble over his fellows' heads as they all tried to get down the narrow hatch first, followed by the rampaging marines.

Wren led the first group forward. MacDonough, Bainbridge, and Decatur took a second to the storerooms. Lawrence stayed at the ladder with a handful of marines to keep a look out and make sure no one came back the other way who wasn't supposed to.

Decatur himself was never sure just how many Tripolitans were aboard *Philadelphia,* and he honestly said so in his after-action report. However many there were, they didn't seem to make the kind of effort that had been expected to defend her. About twenty Tripolitans died at the hands of the marines and Decatur's officers, every one of them by cold steel. Decatur's orders for no firearms was obeyed to the letter.

The marines never actually did secure the ship; in the event, it wasn't necessary. The surviving Tripolitans went over the side as best they could, to one of *Philadelphia*'s boats or into the water. The boarding parties were able to get about their work unimpeded. Although they brought enough of the charmingly named 'consumables' with them to turn the frigate into the most glorious of bonfires, there was more than enough already aboard her to do the trick, including the very fabric of the ship herself. Her storerooms had been full of the tar, pitch and other flammable materials that were needed to keep the ship in good repair, and her magazines were literally overflowing with thousands of pounds of Mister DuPont's finest products. Somehow, poor Will Bainbridge had

To Barbary's Far Shore

never gotten around to throwing that over the side. In addition, every one of those cannon Decatur had seen was loaded and ready to fire, making it indeed likely that *Philadelphia* was within days, or less, of going out after American merchant traffic. Had that happened, with Preble and *Constitution* days away at best, it might have been the end of the American presence in the Mediterranean for a very long, long time.

Now, to this point Decatur had somehow pulled off two of the four miracles he needed to accomplish his mission; he'd actually gotten into Tripoli Harbor and then aboard the *Philadelphia,* and done it in magnificent style. However, the mission was only half finished. He now had to ensure the destruction of the frigate and then get out again.

The first part was easy. The marines set matches not only to the consumables, but everywhere; the storerooms, the gunrooms, the magazines, the sail lockers and even the rigging. This had the dual effect of not only making sure that the *Philadelphia* was going to die that night no matter what, but also to insure that even the Tripolitans would not fail to notice, finally, that something very wrong was happening a few hundred yards from the castle of Yusef Karamanli.

As the refit crew went over the side, the crews of the corsairs finally stirred themselves. They were easily within gun range of the *Intrepid*. Decatur's report says that they were within a 'cable and a half's length,' or about a thousand feet. That is point-blank for today's weapons, but, as a practical matter, approaching long range for a corsair's cannon. The corsairs were never known for their gunnery; they tended to use their large crews to overwhelm the lightly manned merchant ships that were their prey. But even that close to the *Intrepid,* they should have easily been able to cripple or sink her and still not have done serious damage to the *Philadelphia*; or, failing that, have sent their crews over to board *Intrepid* and take her back.

They did neither. The corsairs did send some of their crews out on boats, but they never left the shelter of their ships' sides, simply sitting there seemingly unable to believe what they were seeing. None of the corsairs or gunboats ever opened fire, although the gunners of the Dey's castle and the surrounding shore batteries did. Their shooting was less than effective; although they opened up a heavy fire, only a single round even came close, and that one sailed harmlessly through one of *Intrepid's* sails. There has never been any satisfactory explanation for the utter inability of the Dey's men to make much impression on the *Intrepid*. Surprise was probably an important part of it, as well as confusion over what exactly they would have been shooting at. After all, as Decatur quite rightly pointed out, she looked like a pirate because she was a pirate.

But there is also the fact that we have seen this kind of behavior in more recent Middle Eastern madhouses. When faced with well-organized opposition

Call Her *Intrepid*

from well-trained men with guns who aren't afraid of them, well-armed bullies have shown a tendency to keep their heads down and remain as harmless as possible.

By the time the boarding parties got back onto *Philadelphia's* deck, the flames had taken fatal, terminal hold. Two large, well-fed fires were burning below decks, waiting to ignite the magazines, and tongues of flame were crawling up the lacework of rigging that soared upwards from the sides of the ship's hull. Now, it was surprisingly tough to kill a large sail warship. They were always strongly built from the best oak that could be found in either Empire or Republic, and, if they were well maintained, they were even stronger. Magazine accidents or explosions, though common, were no guarantee of a sinking. A full magazine detonation was often the only sure way to send a large warship to the bottom.

We see in our collective memories that film of the battleship *Arizona* exploding at Pearl Harbor, the tombstone black feather rising above the grave of a thirty-thousand-ton warship being blown in two as a Japanese bomb finds her forward magazine. But just a few months before Decatur dashed into Tripoli Harbor, damaged powder aboard the frigate USS *New York* was accidentally detonated by the foolish actions of a gunner's mate, who lit a candle in its vicinity. The explosion was a powerful one, killing fourteen men, but *New York* still sailed under her own power into Valetta a week later. Large multiple fires, if dealt with smartly, could be brought under control, albeit frequently crippling a ship beyond the capability to fight. Even impressive holes in a ship's side, if handled quickly, could be patched and flooding brought under control.

But *Philadelphia* had none of the advantages that might have kept her afloat a little longer. Her Tripolitian crew had no incentive to stay aboard and fight things. The normal rules and routine that did so much to ensure safety had ended when Will Bainbridge struck her colors. It was going to end there and then for that lovely ship, and it was going to end quickly.

Wren was sending his men back over the side as quickly as he could, with the occasional boot in the backside to those who seemed to be lagging behind. MacDonough, Bainbridge, Lawrence and Decatur made sure of the last details needed to finish her off. Hacking through the thick ropes that held the gun carriages in place against weather and recoil, two cannon were manhandled, more quickly than anyone had probably thought possible, to the open hatch covers midships, then levered downwards so that their muzzles were pointed into the open hatches. With that, the officers went over the side to *Intrepid*.

Decatur went down the rope ladder so fast his hands burnt. As he shot past the open gunports, he could see orange flame inside, devouring *Philadelphia*. As his feet hit the deck, he ran for the quarterdeck.

To Barbary's Far Shore

"Mister Catalano, get us out of here, avanti!"

"Si, Commandante!," Catalano replied with gusto. James Lawrence had been a bit ahead of things. When Decatur hit the deck, the crew was already letting the sails out. Others were loosing the hawser that *Philadelphia's* new crew had so kindly sent down to them. Decatur looked around as the sails snapped full with a sudden breeze. Flames suddenly snaked out over his head from the starboard side gunports and the stern gallery with a whoomp of combustion and venting gases. The men on the quarterdeck slammed themselves flat. Fire and superheated gases shot out over their heads, leaving flaming embers and sparks in their wake all over *Intrepid's* deck.

Intrepid was inching forward, but only just. Shot fell all around her now from Tripoli's shore batteries. The distance between her and the dying frigate was opening so slowly that a fit sailor probably could have swam faster. *Intrepid*'s crew and marines were using anything they could find to push themselves further away. Catalano was spinning the wheel like a virtuoso conductor at La Scala, doing his best to keep the *Intrepid* in the wind and keep her from scraping back to a stop along the *Philadelphia's* smoldering flanks.

Suddenly, a good solid breeze filled her sails. The lightly loaded ketch jerked forward, just as a hollow thud emanated from inside the frigate, followed by a soul-tearing crash as part of the after magazine began to go, blowing off gunport covers and blasting huge divots of teak from *Philadelphia*'s deck into the cold night sky over Tripoli.

About ten miles offshore, Charles Stewart was pacing the quarterdeck of the darkened *Syren*, trying to keep his aggressiveness and temper in check at the same time and failing miserably at both. He had every spyglass on the little warship in the rigging, with his best lookouts behind them. All they had seen so far was the flashes of gunfire reflecting off the low clouds over the Tripolitian coast. With no radio to tell them what is happening, no radar to show moving and stationary ships, all Stewart could do was berate every crew member he saw until he saw something that gave him some sure idea of what the Hell Stephen Decatur was doing in there.

"Captain Stewart!"

Stewart looked up. A lookout was gesturing at a spreading glow in Tripoli Harbor. Racing to the rail, Stewart snarled, "Give me that!" at one of his lieutenants, yanking a spyglass out of his hands. Leaping as far into the rigging as he could with one hand, he trained the glass on the horizon. Whatever was happening, it was spectacular. Another flash lit the sky; the explosion that was lancing out over the heads of *Intrepid*'s crew.

Enough of this nonsense, Stewart thought. "First Leftenant, make full sail, helm; get us in there!"

93

Call Her *Intrepid*

The first lieutenant gave the commands, but looked at Stewart as if he'd gone insane.

"Wipe that expression off your face, First, or I'll lighten ship and start with you!"

"Captain, we don't even know . . . "

Stewart fixed the lieutenant to the spot with a glare that could send the most hardened tar into fear of his life. "We know there's a fight in there, and we'll sail to the sound of it! And besides . . . " Stewart looked southward again as *Syren* started to glide forward with a northerly wind behind her. "I'll be damned if I'm going to let Stephen Decatur pull this off on his own!"

There was also a far more practical aspect to Stewart's efforts. The northerly winds that are now propelling him towards the gunfight to the south will also make it difficult at best for *Intrepid* to get out of the harbor again. *The Intrepid*, Stewart thought, *may have accomplished her mission, only to find herself trapped in the harbor by the fickle winds.*

Intrepid is slowly crawling away from *Philadelphia*, which is now aflame along her entire starboard side. Catalano is keeping *Intrepid* moving, but just barely. It is terrifyingly slow going. The men aboard her decks are brave, but no fools. They are taking cover anywhere they can, especially as cannon rounds from the castle are now hissing just above the deck.

There is the sound of a cannon letting go almost directly above them. The loaded and shotted guns on the frigate's deck, heated to a cherry red, are letting go on their own, like the demented, mindless screams of a dying monster. Fortunately, *Intrepid* is too close to be damaged. The balls are sailing overhead, menacing but harmless. On the other hand, the portside cannon are letting go directly at the walls of the Dey's castle, random balls crunching into the fortress' masonry.

Philadelphia is almost gone now. The two cannon that Decatur had moved into position around the main hatch, remember them, are about to deliver the *coup d'grace*. As they start to glow bright red, their heavy oaken carriages now in flames beneath them, they both cook off with colossal roars, sending their shots down through the inferno that now rules the ship. The rounds punch downwards at a shallow angle, through the gun and berth decks, through the holds and bilges, and out from the copper-sheathed hull. Their journey came to an abrupt stop just a few feet later in the muddy bottom of Tripoli Harbor. That harbor starts pouring into the *Philadelphia*. From here it will be fast as she starts to settle onto the bottom of Tripoli Harbor; no longer a state of the art warship, but merely a bonfire that has put to sea.

MacDonough looked at the corsairs, still blacked out to their right, with boats full of screaming Berbers still tethered safely to their sides. Turning to Joe Bainbridge, he said, "Why the hell are they just sitting there?"

Bainbridge shrugged as he ducked below the rail to avoid a splash of silty water from a cannonball. "They probably can't believe this is happening." Another round sailed overhead and into the water. "God knows, I certainly don't!"

On the walls of the Dey's castle, the master gunner was up now, and was marching behind the gun crews, motivating them with profanity, kicks, and the occasional stroke of the heavy wooden staff he carried. It didn't seem to be making much difference. And as *Intrepid* moved further away, the glow from the frigate's funeral pyre started to wreck what aim they had. But, apparently working on the theory that it was better to do something, no matter how ineffective, than nothing at all, they kept firing.

The western approach to the harbor was in sight now, but *Intrepid* was still only crawling. There were now signs of movement on some of the ships in the harbor. Lawrence looked to Decatur and said, with masterful understatement, "Stephen, this isn't good."

"Pleased you noticed. Any suggestions?"

Lawrence took a quick look around at the situation and swallowed hard. "Well, you're in charge, Stephen. I was hoping you had some idea."

They both ducked as two cannonballs came uncomfortably close to *Intrepid's* stern. "Sorry, James," Decatur said apologetically. "Fresh out."

No one on *Intrepid* wanted to admit that the whole thing might go under within sight of escape, but no one on that ship would have been human if the thought hadn't started to cross their minds. Without a doubt, everyone had considered the possibility before hand, and it is unlikely that *Intrepid* would have gone back to her former owners without a very sharp fight.

Syren was closing the distance fast now. The topmost lookouts could clearly see *Philadelphia*'s burning silhouette; they could even see the guns going off aboard her. But there was no sign of . . . *wait* . . .

"Captain Stewart, there's a ship moving in the harbor, outbound!"

Stewart could barely make it out, but by God there it was. A dim shape moving directly for the Western Approach, lit from behind by the burning frigate. But she was moving slowly, far too slowly for safety. The wind was still against them, and, by all appearances, it looked to stay that way.

Call Her *Intrepid*

Do something Stewart, and do it damned quickly, he thought, *but what?* It was then that the idea struck him. *If they can't get out on their own, give 'em a hand.*

"Helm, continue making for the *Intrepid*. First Leftenant, get every boat crew we can standing by to go over the side when I give the word!"

The first lieutenant complied with alacrity, but he was now more than convinced that Captain Stewart was as mad as everyone thought.

Intrepid cleared the western light, but they were still less than a mile from the raging fire they'd left behind them. Lights were coming on aboard every ship in the harbor now. There was no mistaking the signs of their crews making preparations to get underway. Catalano held her steady, making sure she stayed in the narrow channel that led out of the harbor. A grounding now would have been extremely bad form.

"Ship ahead, Captain!"

At that, even Decatur's heart froze. An unexpected corsair or even a gunboat now would be the end of it, because they would have the wind and the firepower. *All right then*, Decatur thought, *let's show them how Americans can really fight.* Decatur was about to give the order to get everyone to battle stations, when the lookout called again.

"She's flying the Stars and Stripes, Cap'n; she's one of ours!"

Stewart had run the colors up so that there would be no mistaking who he was tonight. A brief break in the clouds had illuminated it just long enough for *Intrepid's* lookout to spot it. And sure enough, from *Syren*, Stewart watched as the outbound ship changed course to head directly for them. The distance was now roughly three nautical miles to *Intrepid* and about five to Tripoli Harbor, where ships were starting to move. The outer batteries were opening fire, with shots falling across their path and that of *Intrepid*.

"Helm, keep her steady," Stewart called, pacing the deck confidently. "They couldn't hit themselves in the ass with a hatch cover."

One gunner looked to his comrade across a cannon barrel and whispered with a frown, "Now if only they were shooting hatch covers at us."

The frown brightened considerably as Stewart came tromping past. The Captain was known to be extremely unhappy when his crew wasn't cheerful. Stewart took another look at *Intrepid*, then thought for a moment about what to do next. It didn't take long; Stewart just needed to get a sense of the timing necessary.

"First Leftenant!"

To Barbary's Far Shore

"Sir!"

"Bring us about, due west!"

"Aye aye, sir, due . . . west?"

"Are you going deaf, First?"

"No, sir!"

"Good, then see if you can hear this. After you get us about, I want this ship brought to a full stop, and get the boat crews over the starboard side!"

"Aye aye, sir!" The first lieutenant went about his work, more sure than he had ever been in his life that, when they got back home, he was going to resign his commission and go to work in his father's dry goods store. At least his father occasionally made sense.

"Captain Decatur!"

Decatur was starting to feel a bit safer as the outer batteries' gunfire started to drop astern, but he wouldn't feel safe, period, until he was well away from the hell he'd raised in Tripoli Harbor. Looking up at the lookout's call, he saw the sailor pointing ahead and lifted his spyglass. Even in the darkness, he could see it was *Syren* all right, but she was turning away from them and slowing to a stop.

What in God's name was Stewart doing?

He got his answer as he saw two whaleboats go over the side, fill with sailors, then start rowing. It was clear now that what Stewart had in mind was to send the boats out to *Intrepid,* to actually tow her out of harm's way. It was a surprisingly easy trick, especially if you had some wind behind you. Mind you, it was a bit rough on the men in the whaleboat, but, if a captain had a choice between a lost ship and some sore backs, it was an easy choice every time. Decatur had Catalano steer for the approaching boats.

In a few minutes, the *Intrepid's* crew had lines going over the side to the boats, and the crews were pulling for all they were worth. It took some time, and some truly impressive motivational work on the part of the coxswains, to get *Intrepid* moving. Once she did, she started gaining speed; within half an hour, she was alongside *Syren* and casting off the boats.

The US Navy has been privy to some magnificent celebrations in its history, but its unlikely that any was ever as heartfelt as the ones that erupted aboard *Intrepid* and *Syren* as they headed southwest, opening the distance between them and the Tripolitians with every second. Backs were slapped, hands were shook, and even marines and sailors forgot for a few minutes how much they disliked one another. Best and most wonderful of all, aboard *Syren,* Lieutenant Charles Stewart was actually seen to smile. It was brief, and there

Call Her *Intrepid*

was heated discussion afterwards as to whether or not it really was a smile and not indigestion, but all agreed that Mister Stewart was, at least, pleased.

Which was more than could be said for the Dey of Tripoli and his admiral, Murad Reis.

Although there is no proof that this specific incident took place on Philadelphia, the pictute captures the atmosphere of a boarding action fought with cold steel.

To Barbary's Far Shore

As Decatur and Stewart guided their ships back to Siracusa, *Philadelphia* continued her slow descent into oblivion. She burned until just before dawn, when the last flaming part of her hull collapsed into the turbid waters of Tripoli Harbor. She was a total loss this time; she sank up to the gun deck and everything above that burned down to the water. Her charred bones rotted where they sat. Presumably, what's left of her still rests there today as supertankers and Libyan patrol boats pass by.

The smoke from her pyre still hung heavy over the Dey's castle as dawn broke, thin and gray, over the North African shore. It might as well have been a shroud for the commander of the English Fort, as he knelt in the Dey's castle, waiting for his fate to be decided. In the West, his mistake, grievous though it was, would have been an understandable and perhaps even a forgivable one.

Not so in Tripoli.

The first inkling that he had done something terribly wrong would have come when he saw *Philadelphia* erupt like a volcano, and remembered the ship that had slipped into the harbor, past him, and headed into the harbor. He probably thought, very briefly, about just keeping silent about the whole episode, but that was no help. Even if he did, someone would have said something, if for no other reason than to make sure they didn't catch the blame.

He knelt on the floor of the throne room, sobbing and shivering and hiccupping in fear, held down by the weight of two massive soldiers on either side of him, his head bent down so that he could not behold the glory of the Dey as he entered. The doors to the throne room slammed open, and he entered.

The Dey was a big man, well over six feet in a world where most men in his nation were only about five feet tall. He was built like one of the fortresses that studded the harbor around his capital, with a silken black beard, a raptorial nose, and cruel, heavily lidded eyes. He was widely regarded as extremely intelligent, rapacious, and utterly ruthless. William Eaton described him as 'bear-like,' but he tended to say that about most of the Berber rulers he dealt with. Behold, therefore, His Majesty Yusef Karamanli, Bashaw and Dey of Tripoli. A most unhappy man this morning, as the ship that was about to become the pride of his fleet was now just so many ashes floating beneath his castle walls.

The commander of the English Fort saw only two feet come slowly into his field of view. There was silence for a few long heartbeats, followed by a slow, disappointed sigh from above.

"I give you my trust, sadiq. My . . . trust. Such a thing is not easily given, not in such a cruel world."

The commander looked up pleadingly. "Effendi, I had no . . . "

Call Her *Intrepid*

One of the guards slammed the commander in the back of the head, sprawling him out on the floor. The Dey placed one foot on the side of his head, putting his weight down on it. The commander yowled in pain. Another foot crashed into his stomach, and the Dey spoke again, in tones of sadness and disappointment.

" . . . And then, when I try to make my feelings known . . . you interrupt me. Is that any way to treat your Dey?"

"No, Majesty," came the groaning reply.

"And then, you allow infidels to enter my harbor and destroy my property. If I did not know better, sadiq, I would say you wished me ill."

"I do not," the commander sobbed, "I live only to serve OOMPH . . . "

Another kick in the ribs silenced him as the Dey squatted beside him, shaking his head sadly. "You have been a good friend. I should hate to have to find a new commander for my fortress."

Gasping for breath, the commander wheezed, "Effendi . . . please . . . I beg . . . beg . . . "

The Dey stood again, his arms folded, looking down at the commander.

"Do I understand that you are sorry for your failings, sadiq?"

"Oh . . . yes . . . yes, Majesty . . . "

The Dey smiled, the wolfish grin of a predator who had made up his mind to play with his dinner first. "Well, then . . . if you are truly sorry, you would not mind accepting a mild . . . punishment . . . as part of your regrets."

"Oh no . . . of . . . course . . . Effendi . . ."

"Good." Turning to the guards, the Dey spoke with solicitude and kindness. "Please take the commander to the lower galleries and give him five hundred with the bastinado.

"Then walk him back to the English Fort."

It took an effort, but the commander was still able to scream.

It only took a few days to get back to Siracusa, and Preble was waiting. No word had gotten back yet, so he genuinely had no idea what to expect when *Syren* and *Intrepid* came racing back into the harbor. Preble ran up a quick signal:

BUSINESS OR ENTERPRISE HAVE YOU COMPLETED

To Barbary's Far Shore

Decatur must have been grinning from ear to ear when he sent back a single word:

ENTERPRISE

Not a single life was lost, and there was only one minor injury, a sailor who hadn't been able to avoid a sword thrust quickly enough, and was probably never able to live it down.

And so Philadelphia *died. This painting belonged to President Franklin D Roosevelt and was reputedly one of his favorites. Source: U.S. Navy History and Heritage Command*

When all was said and done, the situation was still in an odd sort of limbo. *Philadelphia* was well and truly gone, and the Dey would never be able to use her to threaten shipping in the Mediterranean. It had been a brave, brilliant plan that had succeeded beyond anyone's wildest dreams. Horatio Nelson himself called it "the most bold and daring act of the age;" coming from him, that was high praise indeed. But on the other hand, Will Bainbridge and his men were still in their dank cells in Tripoli. Despite the force and skill Decatur had applied, the Dey seemed not to have quite gotten the message. He notified Commodore Preble that he was willing to back off somewhat on the ransom for

101

Call Her *Intrepid*

the *Philadelphia's* crew, from $3,000,000 down to $500,000. But, on the other hand, restitution would have to be made for the lives of his loyal subjects who were lost during the raid. The impression one gets from this is that the Dey felt that this might have been the best the Americans could do, and that he could afford to be generous.

That attitude was guaranteed to send Preble right over the edge, and it did. Having finally had enough, Preble decided to show the Dey what the United States Navy was really made of. Calling his captains together once more, the Commodore made it clear that this time they were going to get the Dey's attention.

Preble had an exceptionally fine grasp of what he was up against. Tripoli harbor was a fortress in every sense of the word, with heavy walls and strongpoints mounting about one hundred and fifty cannon. With the ships that were in the harbor as well, Preble reckoned he was up against two hundred cannon compared to his 156, almost a third of which were aboard *Constitution* herself. The rest were divided through the fleet on the US ships and a handful of single-weapon gunboats sent over to help by the Kingdom of the Two Sicilies, which was also one of the Dey's favorite targets. On the other hand, Preble had an edge in training and technology, not to mention a pretty solid understanding of his enemy. He therefore made up his mind to go in shooting and do as much damage as he could. The diplomats would have probably gone into shock upon hearing of the plan, which could be why we find no record that Preble ever told them what he was going to do.

The attack was brilliantly planned and a textbook example of how to use one's resources to the fullest. Stephen Decatur was in the lead, taking the gunboats in first, in order to prove a theory he and Preble shared about their adversaries. The Berbers in general and the Tripolitians in particular tended not to take their enemies by maneuver and tactics but by sheer physical force. The corsairs tended to be overmanned, with the surplus crews creating huge boarding parties that usually outnumbered their prey by several to one. Added to the ferocity and viciousness that they tended to display, many ships simply surrendered when challenged.

But Preble thought that if the Tripolitians were faced with a trained adversary ready, willing, and able to fight back, it might be another story. Decatur's job was to justify that theory, and he did so magnificently.

Taking the gunboats in first, Decatur ordered his own ships to close, grapple, and board. The pirate crews, used to having things their own way, had no idea what to do next. Decatur boarded the largest ship he could find, and led every crewmember he could over, even though they were outnumbered by the pirates at somewhere around two to one. Carrying swords, axes, flintlocks, marlinspikes and tomahawks, the sailors were in no mood to grant any quarter at

To Barbary's Far Shore

all. They literally chopped the crew of the pirate to shreds in a few short minutes. Only five survivors out of thirty-six survived to be led back to the US gunboat. The same grim drama was being played out on two other pirate ships. No Americans were lost in the opening gambit, and only a handful wounded.

Preble, leading the rest of the squadron from aboard *Constitution*, plowed right through the opening Decatur had cut for him and brought the fleet to within eighteen hundred feet of the walls of Tripoli Harbor. From where he was at, Preble could see thousands of Tripolitians crowding the roofs and battlements of the city. They had been expecting an American defeat, but that wasn't the show they were going to get. The Dey was among them and seems to have suddenly decided that things might be just a little dicey in the open, so he led his advisors and ministers into a shelter far beneath his palace.

The Dey missed a magnificent show. *Constitution's* gunners gave a two and a half hour demonstration of American firepower. Preble ordered the squadron back out to reorganize, and four days later they came back with a vengeance. This time, the corsairs didn't even try to come out, and the rooftops of Tripoli were noticeably empty. Having gotten his range the last time, Preble brought *Constitution* in as close as he dared and opened fire directly on the Dey's palace. He probably came close to emptying *Constitution's* magazines. Records show over five hundred solid shot and fifty explosive shells expended before he ordered her brought about and sailed out.

Preble wasn't done yet, but he waited just long enough to make the Dey think he was. Seventeen days later, Preble sailed *Constitution* in just before two o'clock in the morning, opened fire, and stayed that way until sunrise before he sailed out, his ship's sails catching the rays of the rising sun. The commodore repeated his nocturnal symphony twice more in the next two weeks, generally raising hell and terrifying the citizens of Tripoli. By this point, some of the Dey's gunners had found enough of their courage to open fire on the departing frigate, but they don't seem to have had their hearts in it. Most of their shots sailed over Preble's head, giving a good idea of just how close to the walls the commodore really was. In his final report Preble would say only that they were 'wretched' at their tasks.

Not long after this, a message from the Dey arrived for Preble. After considerable thought, and in the interests of peace and the welfare of the poor souls whom the Dey was looking after, His Majesty has decided that the entire unfortunate matter could be cleared up for a single lump-sum payment of $150,000. Edward Preble must have exhausted his vocabulary of maledicta at that, but his orders were clear. He was required to give the Dey a counter offer in the amount of $120,000. The Dey found that laughable, and said so in his refusal. And at that, the matter of the officers and crew of the USS *Philadelphia* settled into an ugly deadlock, neither side giving in any further.

Call Her *Intrepid*

Preble had tried, tried as hard as anyone possibly could. But at the end of the day, he was the victim of the dual nature of a senior field commander two hundred years ago. He was trained as a warrior, and he was a fine one. But he was also expected to turn it off on demand from his civilian superiors and become a diplomat. But the nature of the problem called for a coherent military and diplomatic policy, neither of which was going to be forthcoming from Washington any time soon. We've seen this ourselves, sadly more than once, in policies that have swung wildly from cruise missile strikes to 'diplomatic engagement,' a term that seems to stand for begging thugs not to hit you again. And when it doesn't work, so many people who should know better are surprised.

Actually, a diplomatic policy with military overtones, more incoherent than not, was on its way to the Mediterranean, under the care of a diplomat who believed himself more warrior chieftain than ambassador. He turned out to be not very good at either, but had a monumental self-image that convinced him that if only he could be in charge of his own Bold and Daring Act, he could change the course of events. He would do so, but only with the aid of a warship, three hundred mercenaries, a handful of light cannon . . . and eight marines.

To Barbary's Far Shore

FIVE: THE PRACTICES OF POLITICS

The war, however, was very far away from Washington, DC, in June of 1804. *Constellation* glided almost soundlessly to the pier at the Washington Navy Yard, the only sounds heard being the echoed commands of her skipper as he brought her alongside and the sounds of hammer, saw and plane a few yards away as older ships were rebuilt and new ones arose from bleached oak skeletons. Other ships were at the dock ahead of her, dismasted and stripped, festooned with ropes and cables and men swarming over them to get them into shape. The men who had fought for and against the Armada two hundred and sixteen years before would have recognized the scene instantly. The only thing that had changed was the scale. Beyond the docks and sheds and masts, Anacostia lay tucked up alongside the Yard's long brick wall, an uneven jumble of houses, shops and a few small farms.

"Always a welcome sight, isn't it?"

Will Eaton turned to see Captain Alexander Murray standing beside him, keeping an eye on all the myriad activities that were needed to get the ship secured after her long voyage.

"That it is, Captain," Eaton agreed. "Is your family there?"

Murray shook his head. "Norfolk. I'll have to double back to get there, but it's no worry, just another day or so. We have a little house that looks right out towards Craney Island and the docks. My wife could spit nails at the view sometimes, but I love it. Sit there with my boys and just watch the ships go in and out. My oldest one, he's ten, can't wait to get out there himself."

Eaton smiled. "Wife can't stand that either?"

William Eaton. Source: U.S. Navy History and Heritage Command

The Practices of Politics

Murray grinned. "Drives her mad. And when the day comes that he ships out . . . " He leaned over the rail and shouted forward, "Mind those lines, you deckapes!"

Turning back to Eaton, he said, "Damned hard to get men who know what they're doing these days. Italian, French, Dutch, and God only knows where else. I suppose we're lucky they understand English, if only after a fashion. Where's your family, Mister Eaton?"

Eaton tilted his head towards the gentle hills that unwound towards the Capitol. "Georgetown. We moved down here when I was appointed Consul." Eaton's eyes focused on the horizon now, a pensive but determined look on his face. "A bit difficult to pay for a house when one doesn't have a job, but I'll think of something."

Murray thought about this for a moment, swaying slightly as *Constellation* bumped to a stop against the dock, her journey finally finished. There were no cheers, no celebrations, just a collective sigh of relief of being home at last.

"No doubt, sir. You're nothing if not inventive in the face of adversity." Murray paused for a moment, trying to frame his next words. "Mister Eaton, if it's any consolation, a great many people believed in you and what you did, even if it didn't work. At least you got the Commodore to get off his arse and do something. We were by ourselves out there . . . in a ship that shouldn't have been that close to shore in any event. As unpleasant as it must have been, I would have given anything to have seen his face in that Algerian prison once he realized what kind of trouble he was in."

"Afraid it wasn't all that bad," Eaton replied. "They were under house arrest. Actually, a damned sight better than what he had here on the *Constellation*, though the food wasn't as good."

Murray winced. "Food worse than what we have to put up with? Dear Lord, that qualifies as torture. How did the Commodore handle it?" The deck crews were now spilling topside and prying the hatch covers open to unload the myriad number of items that *Constellation* had brought back after nearly a year. Eaton's baggage would be near the top of the pile, and it would be off in just a few short minutes.

"Badly," Eaton replied. "Kept muttering about how those heathens needed a good Boston cook."

Murray shook his head in both amusement and exasperation. "You'd think he'd have listened to you. After all, you were the diplomat in charge."

"Sadly, Captain, many people have not learned to listen to my advice until it was entirely too late." A dock crew was manhandling a gangway to

To Barbary's Far Shore

Constellation's side, and now the crewmen were starting to cheer and good-naturedly jostle each other as they lined up to leave the ship.

"Well, Mister Eaton," Murray said, extending his hand. "I can say it has been a pleasure having you aboard. Wish you could go back with us. It would be nice to have someone there who wants to get something done."

Eaton shook Murray's hand firmly. "Well, one never knows. Once I get the Secretary of State to listen to me, I will be back over there with the entire fleet."

Murray smiled politely, having heard the details of Eaton's intended interview with James Madison more than a few times on the trip over. He thought it was a wonderful idea, but he was a professional. He knew how politicians thought, how they acted, how they looked at a situation. Eaton's plan had absolutely no chance of ever getting past the present Administration, no matter how persuasive the Consul might be about it. But the man was so dead set on the thing. Around the wardroom table, on more than one evening, his officers had patiently explained every flaw, outlined every possible disaster, pointed out every hurdle that would be in its path. Not once did he listen, or, for that matter, even acknowledge the conversations. Oh well . . .

"Mister Eaton," Murray said in farewell, guiding him to the gangplank, "I am not sure if your adventure will receive official backing. If, for no other reason, than it requires us to use serious, actual force against one of those damned heathens. But I rest secure in the knowledge that if it is not approved, it will not be for lack of trying on your part." Murray came to attention and saluted.

Eaton snapped to as well and returned it with a smile. "Keep your bags packed then, Captain, for I intend to succeed. Thank you for getting me home safely, sir, and my best to your family." Eaton marched down the gangplank, whistling a cheerful little ditty.

Constellation's first officer stepped up beside Murray as they watched Eaton go ashore. "What do you think, sir? Can he browbeat those people," he tilted his head towards Capitol Hill, "into actually approving it?"

Murray shook his head. "First, I don't know what possibility bothers me more. That they'll deny him the chance to even try, or, worse, they'll let him."

Spring in Washington gets warm and stays that way with only the occasional deluge to provide a bit of variety. In 1804, with no air conditioning or electric fans, and most suits made out of wool or fairly heavy cloth, it was miserable. James Madison could only endure it as he walked from his boarding house towards his offices in the District. At just about eight o'clock in the morning, the heat was already approaching an intolerable level, but it was important that a Cabinet secretary carry himself in such a manner that he appear

The Practices of Politics

to be utterly unruffled by the temperature instead of on the verge of heatstroke. He was recognized by more than a few of the populace, who tipped their hats in salute and respect. Madison, being a man of manners and culture, returned the courtesy. It is difficult today to picture the Secretary of State simply walking down a dirt track to his office, but Washington itself was barely more than a concept then, and America was a very different place.

Madison, like most Cabinet secretaries and legislators, would have probably maintained a room in a boarding house near to wherever offices could be found for the growing young government. Madison would have been in one of the nicer boarding houses while his wife, the redoubtable Dolley, kept an eye on things at his home in Virginia. On the other hand, the offices the Government paid for tended to be provided by, then as now, the lowest bidder. So, men whose names appeared on the Constitution itself, who had fought the King's Men to ultimate defeat, often made the decisions that would affect the young nation in dark, cramped, and occasionally vermin-infested rooms that no self-respecting drunkard would sleep off a bender in. But the work had to be done somewhere, and, until the grandiose plans of Pierre L'Enfant could be translated from parchment and ink into stone and marble, they would have to do. And besides, the awesome monuments of Whitehall and the graceful stonework of the Quai D'Orsay were for kingdoms and empires, not republics. There was something rightly plain and simple about history, about the future, being made in such humble surroundings.

James Madison. Source: U.S. Navy History and Heritage Command

Madison would have walked up the stairs to his office, which would almost certainly have been the best of the available rooms; big enough for a secretary to scribble away in one corner and keep track of appointments and callers, while Madison tried to make sense of what was going on in the world. Most different from modern Washington, it would have been quiet, oh, so very quiet. With the windows open, Madison would have heard birds singing and horses clip-clopping past, the conversations of people passing by on the street. No

To Barbary's Far Shore

telephones, radios, cars, buses, sirens, fax machines, televisions or protesters. Just sounds more appropriate for a farm than a nation's capital.

Madison's secretary stood as the Secretary of State entered his office. "Good morning, Mister Secretary," he said, bowing slightly.

Madison nodded to the secretary as he hung his hat on an upright rack that stood beside the door. "Good morning, Edward. Any dispatches from last night?"

Edward shook his head. "No disasters, emergencies, or insults that I am aware of, Mister Secretary. Shall I put some tea on?"

"Please." Madison walked over to Edward's desk and spun the daily ledger around to face him. "What do we have on the agenda for today?"

Edward turned slightly from where he was assembling a pot of strong orange pekoe. "Most importantly, you have a dinner engagement with the President at his residence this evening at six o'clock, but until then the day is essentially clear; with the exception of an interview at ten with Consul Eaton."

Madison gave Edward a puzzled look. "Consul Eaton? Who the devil . . . wait, Tunisia?"

Edward closed his eyes briefly, bringing the necessary facts up to consciousness. "Appointed to Tunisia by one of your predecessors, sir. Was declared *persona non grata* by the Dey after the most unfortunate episode with Commodore Morris and Consul Cathcart, returned here last week and immediately contacted us regarding an opportunity to review his report and make arrangements for him to return."

Madison blinked, trying to remember the salient points. State was still an infant as bureaucracies went, but it was growing rapidly and Madison was responsible for many administrative details that today would be handled by a junior-grade civil servant. "Why on Earth would he want to return to that God-forsaken place? If the good Lord convinces the Dey to send you home, be happy with it."

"Given the brief conversation I had with him, sir, he seems to think he has some unfinished business there."

"Damned if I know what it could be. Did you send for his reports?"

"Yes, sir, after you left for the evening yesterday. They are on your desk."

"Fine, fine. Have to admit I haven't paid as much attention to Tunisia as perhaps I should have. Been relying mostly on Cathcart and the Navy to keep us informed, but I suppose that's not much of an option any more. All right then,

The Practices of Politics

let me look at a few of his reports so I have some vague idea of what he's talking about."

"Very good, sir," Edward replied as Madison opened the door and went into the inner office.

Edward was placing cups and saucers onto a tray when he heard Madison say through the open door, "Good Lord," in a tone of bemused amazement. Edward left the tray to step quickly over to Madison's office. The Secretary was standing, arms crossed, in front of his desk. The desk, made of beautifully carved and richly polished cherry wood, was buried beneath several large piles of books, binders, ledgers, envelopes, maps, and other items, with Madison's pens and ink vanished under it all.

"Yes, Mister Secretary?"

"Edward . . . " Madison made a gesture that encompassed the desk and the piles atop it. "What is this?"

Edward smiled like a teacher enlightening a pupil. "Consul Eaton's reports, Mister Secretary."

Madison's jaw dropped, but he caught it before Edward saw him gaping like a landed fish. "Edward, how long was he there?"

"Just over four years, Mister Secretary."

Madison slowly walked around the desk, examining the mountain of information and correspondence from every angle before he spoke again. "He only had to send one report a month, for goodness' sake . . . you're sure it wasn't forty years?"

"Quite positive, Mister Secretary."

"Mmm." Madison sat down at his chair, and promptly vanished behind Eaton's missives. Edward stood silently for a moment, waiting for some kind of message from beyond the paper colossus.

"Edward?" came an almost plaintive voice from behind one particularly large and ominous looking stack of ledgers.

"Sir?"

"Bring me my tea. I fear this could be a long morning."

"Very good, sir."

It was close to nine-thirty when Secretary Madison came out again, while Edward was scratching industriously away at one of the dozens of letters he had to slog through every day.

110

To Barbary's Far Shore

"Edward."

Edward looked up from his writing. "Sir?"

Madison looked at the multi-page letter he was holding, and then down at Edward through the spectacles perched on the bridge of his nose. "Have you ever met Consul Eaton?"

"Yes, Mister Secretary, when he was first appointed. He was assigned here briefly before proceeding to Tunisia. Why do you ask?" Madison shot one more glance at the letter he held. "Does he speak the same way he writes?"

Edward placed his pen down, and then clasped his hands, a thoughtful expression on his face before he looked up to Madison. "Actually, Mister Secretary, he is far more voluble in person, and infinitely more opinionated and forceful."

Madison closed his eyes in resignation. "That, Edward, is what I was afraid of. He'll be here soon?"

"Yes, Mister Secretary, about half an hour."

"Send him straight in when he gets here. Perhaps I can end this before luncheon today."

"Very good, sir."

Precisely at ten, Eaton strode into Madison's outer office, clad in a new suit and shaved and barbered in a manner befitting a diplomat.

"William Eaton, Consul to the Dey of Tunisia, to see Secretary Madison, and be quick about it."

Edward raised an eyebrow, but said nothing besides a politely murmured "Very good, sir"

He rose and knocked on Madison's door. Edward opened it and said something discreetly, then opened the door wide, stood tall, and announced.

"The Honorable William Eaton, United States Consul to the Dey of Tunisia and His Associated States."

Eaton swept past Edward, with his hand outstretched to Madison. Madison rose with all the gravitas he could muster as he stepped from behind the desk and shook Eaton's hand.

"My dear Eaton," Madison smiled, "it is a true pleasure to finally meet you at last."

111

The Practices of Politics

"The pleasure is all mine, Mister Secretary. I have been waiting for this opportunity for so very long now, to finally explain to you how we shall overcome the depredations of the Tripolitans . . . "

Madison smiled and held up a silencing hand. "In due time, Eaton, in due time. How was your voyage back? Did the Navy take good care of you?" Madison gestured Eaton to a chair beside the desk as he took his place again.

"The captain of the *Constellation* was a most gracious host, Mister Secretary; I could have asked for nothing better. His staff was kind enough to review my proposal to you and give me most wonderful advice."

Madison looked politely confused. "Your proposal?"

Now it was Eaton's turn to look just a bit befuddled. "Yes, Mister Secretary, my proposal for dealing with the Dey of Tripoli and rescuing our imprisoned countrymen from the *Philadelphia*. You did read it, did you not?"

"Ah . . . that proposal," Madison lied smoothly, having no honest idea what Eaton was referring to. "I must confess to not being as familiar with it as I should be, Eaton. After all, " Madison waved vaguely towards the towering piles of correspondence, "you were most . . . ah . . . conscientious about keeping us informed of events there, and I am afraid I may have skimmed over it far too fast to truly appreciate all its significance . . . "

Eaton suddenly brightened and sprang up out of the chair. "I understand completely, Mister Secretary. Allow me!" With one sweeping move of his arm, Eaton cleared half of the desk onto the floor, simultaneously snatching a map from the pile that was cascading to the worn wooden floor with a series of thuds that brought Edward rushing in.

"Is everything all . . . "

"I did not send for you, sir!" Eaton said to Edward with wounded dignity. "Kindly remain outside unless called for. These are most sensitive matters!"

Madison waved Edward off with an understanding smile, and the secretary slowly stepped back into the outer office with a suspicious expression, closing the door behind him. Eaton unrolled the map, a beautifully drawn and detailed depiction of the North African coast. Hunched over the map, Eaton turned to Madison with all the intensity of a scientist within reach of a lifelong quest.

"Mister Secretary, behold North Africa! He who controls it controls the Mediterranean . . . "

Madison looked at Eaton skeptically. "The British at Gibraltar may have something to say about that."

To Barbary's Far Shore

"They will say that they control who gets in and who gets out; but, once someone is in, the pirates, Morocco, Algeria, Tunisia, and Tripoli, control who is permitted to continue onto their destination. Warships are unmolested, sir, but merchant ships, especially our merchant ships, must run a gauntlet of desperadoes who seek only plunder!"

"We are aware of that, Eaton. It's been our primary concern here now for some time . . . "

"And I propose to relieve you of that concern, sir!" Eaton traced the coast with one finger, his eyes alight. "Now, we are on good terms with the Sultan of Morocco, so we need not concern ourselves with him. Algeria raids because Tunisia raids, and Tunisia raids because . . . " With this, Eaton stabbed his finger onto the map at Tripoli Harbor. "Tripoli does! If we could take Tripoli out of the equation, Algeria and Tunisia would cease their raids almost immediately!"

Madison studied the map, but looked doubtful. "Eaton, that is a grand leap of faith . . . "

"But an accurate one. If Tripoli were to stop raiding, or, better yet, permit American warships to operate from there, the Algerians and Tunisians would be boxed in between us and the British at Gibraltar. They would have nowhere to go, sir; no safe harbor at Tripoli to hide in and resupply! More than half of the Mediterranean would be cut off from their assaults!"

Eaton does have a point, Madison thought, *even if it is stretched to the limit. With Tripoli out of the business or under American influence, a huge stretch of the Mediterranean would be out of Berber control. But there is no way to do that.*

"Eaton," Madison said carefully, "your concept certainly has merit, but I think it is at least possible you underestimate the difficulty of such a task . . . "

"No sir, I do not!" Eaton proclaimed triumphantly. Looking over the desk, he suddenly started burrowing through what was left of the piles of paper, throwing it into the air as if it had been flung there by a tornado. Suddenly, he held a sheet of paper aloft over his head, as if it were some magnificent reward. As notes and reports fluttered to the ground about them, Eaton thrust his prize into Madison's face. The Secretary staggered backwards a step so his eyes could focus on the spidery writing.

Madison scanned the first few lines, then paused. He looked up at Eaton for a moment, then carefully took the paper from him and then sat down to finish reading it. Eaton hovered over him, grinning in victory; waiting to hear the Secretary's opinion. He would go back, he would go back in triumph, and he would read the terms of surrender to the Dey on the very walls of his own castle.

The Practices of Politics

When Madison was finished, he carefully placed the paper back on his desk. Then he looked up at Eaton with a mixture of mild amazement and surprise. Eaton stood there with a smile that literally went from ear to ear, almost unable to contain himself. Finally, Madison spoke, quietly but firmly.

"Eaton . . ." He paused for a second, then looked at Eaton as gently as he could. "William . . . please . . ." Madison gestured towards the other chair. Eaton sat down, still looking like a child expecting a wonderful present.

"William . . . I understand your desire to get the *Philadelphia's* crew out. I certainly understand your desire to put a stop to the damned Tripolitians once and for all. But it is simply not possible to do it as you suggest . . ."

Eaton looked at Madison in puzzlement. "I am sorry, Mister Secretary, I thought I had made clear the adventure's requirements." Eaton reached for the paper, but Madison blocked him. It was only then that a look of realization began to dawn on Eaton's face.

"William," Madison said softly, "You outline, very professionally and very clearly, what would be needed to overthrow the Dey of Tripoli and establish his brother back on his rightful throne. If the desire and resources were available to do it, I would certainly entertain the thought, but . . ."

"But they are, Mister Secretary."

Madison held up a warning finger. Though his voice remained quiet, there was more steel in it now.

"They are not, Consul Eaton. They. Are. Not. You ask for the entire Mediterranean Squadron."

"But is that not their purpose – to bring war to an enemy?"

"Yes, it is, and it is also to protect others from the enemy. Your plan would strip every ship from its station and have them cruise at your command until you accomplish your mission. That would be utterly unacceptable, Eaton. Every American ship merchant ship from the Straits of Gibraltar to the Dardanelles would be fair game. Then you ask for thousands of men; where are we to find them?"

Eaton leaned forward to make his point, his voice rising slightly now. "The regular Army and the militia, sir; the marines who are already there."

Madison's jaw dropped. "The Army and the militia? Eaton, have you lost your mind? Besides the fact that the marines have their own duties, and are undermanned to begin with, how do you think we could ever convince Congress to authorize that many men to be called up for duty in Tripoli, for God's sake? Not to mention the problem of getting the state legislatures to recruit and help pay for what they will almost certainly condemn as 'foreign adventurism?'

To Barbary's Far Shore

"Most of Congress believes that any use of military force outside of purely self-defensive purposes is a violation of the very principles of this nation. And while we are on the subject of legislative approval, where do you expect the money to come from?"

Eaton opened his mouth to reply, but Madison cut him off. "I shall tell you where it will not come from, sir: the United States Treasury! I do not know what funding you had under Secretary Pickering, but I assure you it was a damned sight better than we get under Mister Gallatin. The man had butchered the Federal budget by more than fifty percent, sir; across the board, no one was safe. Only the fact that those lunatic Berbers declared war on us saved us from having to sell the very desks we work from for firewood! Even with what we have to accomplish now, Mister Gallatin guards every penny as if it were his own, and President Jefferson is not at all inclined to overrule him. Though he can find fifteen million dollars to buy that godforsaken wilderness from Napoleon!"

Eaton challenged was Eaton enraged, and it showed. Springing up, he leaned forward over Madison. "Sir, I am giving you a chance to get our fellow countrymen back and destroy a threat to us at one blow! You would throw barricades at it?"

Madison rose, and, although he was a good head, or more, shorter than Eaton, he feared no man and showed it, leaning right back and thrusting his jaw forward. "Do you not think that I have tried to remove them instead, sir? Do you not think that I have tried every stratagem at my disposal to get them out?"

"Then cease playing these idiotic diplomatic games with the pirates and declare war!"

Madison turned bright red, and his voice went dangerously low. "Mister Eaton, I would remind you of a minor detail known as the Constitution. You may not be entirely familiar with it, sir, but I am; I helped write it! I assure you sir, there is nothing, nothing, in that document that permits a Secretary of State or a President, or anyone else except Congress assembled, to declare war on their own! And one other thing, Mister Eaton. The next time you suggest, imply, or state that my efforts on behalf of those poor devils are 'idiotic,' you and I shall step outside and settle this in the traditional manner. Am I understood, sir?"

Eaton had been in the military, where for decades, deaths from dueling were often more a cause of officers' deaths than combat, long enough to know that he had just stepped across a very thin and dangerous line. If he didn't step back now, he could very well find himself facing down James Madison at ten paces some morning just at dawn. Men went at each other with pistols and swords over far less than what he'd just suggested. Madison may have been

The Practices of Politics

physically small, but Eaton was not foolish enough to confuse that with an inability to put a ball or blade through him in a most thorough and final fashion.

Taking a pace backwards, Eaton bowed deeply with a sweeping gesture of his hand. "Mister Secretary," he said with as much regret as he could muster, "I most humbly and deeply apologize if I have caused you any hurt. I meant no insult to your work or your efforts on the prisoners' behalf. Please, sir, I beg your pardon."

Madison gritted his teeth for a second; his color dropped from lobster red to a mild pink. "Your apology is accepted, sir," he replied, "but, in the future, have . . . a . . . care."

Madison sat down in his chair, keeping his eyes locked on Eaton's. "I am going to say this but once, Mister Eaton, and I wish you to make very sure you understand it: There is no support here, either in the President's House or in the Legislature, for any formal war against Tripoli. With that indeed the case, sir, how in God's name do you expect anyone to convince Congress to appropriate twenty thousand dollars, a dozen ships and two thousand men to invade a nation whose state of war with us is barely acknowledged?"

Eaton said nothing, simply looking at the seams between the boards in the floor.

"Mister Eaton? Please sir, answer me."

Eaton's voice was subdued. "I have none to give, sir."

Madison was silent for a moment, then leaned back in his chair. "Eaton, you have been gone for four years. This is not the government that posted you, sir. It is one that counts every penny and values peace above all else. And unfortunately, pennies and peace are more important to some than three hundred and eight of our sailors. Now, in the end, they will come home. I do not know when or by what bribe or tribute or whatever else you wish to call it, but they will come home. It sickens me, sickens me, sir, to know that they rot at the Dey's pleasure. But I am one voice and one voice only, and I cannot act alone."

Eaton was silent for a moment, then he lifted his face defiantly. "And if I could find another voice?"

Madison gave an exasperated sigh and shook his head sadly. "Eaton, I believe we just discussed that . . . "

"Sir," Eaton pointed out reasonably, "all we did was discuss why it could not be done. Now that I know the limitations that you must deal with, it is my job to overcome them."

Madison lowered his face into his hands, quite unable to deal with Eaton's persistence. Slowly, as if lecturing a particularly dim student, Madison looked

To Barbary's Far Shore

up and spoke. "Eaton, without Congressional support and pressure on the President, there will be no chance of military action at all and certainly not ever at the scale you suggest."

Eaton stood tall and grinned. "Then, sir, it is my job to obtain that Congressional support!" Madison shook his head in disbelief, but Eaton forged onwards. "Mister Secretary, I have been there. I have seen what we face, I have seen what is needed, and more importantly, sir . . . " At this, Eaton thrust his chest out as if he were posing for a statue. " . . . I know what must be done and can lead it! Mister Secretary, I beg of you, please, give me the chance." There was silence for a moment, and Eaton's voice dropped to a whisper.

"Please. They are our fellow citizens. They are Americans. We owe them a debt, sir. Let me try and pay it."

Madison rolled his eyes at the ceiling. "Eaton," he said in a stunned tone, "you really don't take no for an answer, do you?"

Eaton's smile was joy itself. "No, sir!"

Madison sighed. "I didn't think so." The Secretary thought about the situation for a moment, drumming his fingers on the desk top before looking up at Eaton again. "You know, you're the only other person I've met here in Washington who wants those men out as badly as I do . . . and is willing to try anything to get them back."

"You are wrong, sir. No one in Washington wants them out as badly as I do."

Madison gave a snort of laughter and shook his head. He was not normally an impulsive man, but he could be when opportunity arose, and this was an opportunity that might never arise again. Up to now, the plight of the *Philadelphia*'s crew and the other assorted disasters were things that had to be dealt with, but abstract things that happened thousands of miles and months away on the other side of the world. Eaton had been there; he had 'seen the elephant,' as the old saying went.

President Jefferson and Congress might be writing off Will Bainbridge and his men until everybody arrived at the right amount, but wouldn't it be worth it, just once more, to try and make the argument? And Lord knew William Eaton could make an argument.

"Mister Eaton," Madison said, "I may regret this for the rest of my life. I know I will certainly regret it for whatever remains of my career in government service. Do you think you could turn this," Madison waved a hand at the piles of documents that surrounded him, "into a coherent briefing by four o'clock?"

"Absolutely, Mister Secretary! May I inquire for whom?"

117

The Practices of Politics

"No, you may not. This will be difficult enough without you getting any ideas about sticking your nose farther into it. And, while I'm thinking about it, take any mention of yourself out of the plan. No one here knows you, and I'll be damned if I'm going to keep explaining who you are. One more task for you. Do you know Senator Sumter?"

Eaton shook his head. "I know of him, sir, but we've never met."

"You will, as of tomorrow. I am going to arrange an appointment with him for you. You are to give him the same argument you gave me, but I strongly suggest that you avoid angering him; he is nowhere near as forgiving as I am. In the meantime, get that briefing finished by four."

Madison give him a dismissive wave, then looked at his desk to try and figure out some way to clean it off so he could get some work done. Madison was thinking about having Edward go get a barrel for him to dump all of it in when he heard a semi-discreet clearing of the throat. He looked up to see Eaton still standing there, an expectant look on his face.

Madison simply raised an eyebrow. "Yes?"

"A thousand pardons, Mister Secretary, but . . . about my title?"

All Madison could do was blink. "Your title?"

Eaton's tone and expression were divine logic. "Well, of course, Mister Secretary. If I am going to assist in this effort, I shall have to have a title. I certainly cannot make these efforts as a private citizen."

James Madison had helped to found a nation, and he had helped fight for it on the battlefield, but he had never in his travels met anyone quite like Will Eaton. Leaning back against his desk, Madison thought for a moment while he rubbed the back of his neck.

"I shall tell you what, sir."

Stepping up to Eaton with his arms folded, Madison spoke. "William Eaton, by the authority vested in me by the Constitution, Congress of the United States Assembled, and anyone else who may be an interested party, I hereby appoint you Special Naval Consul to the Several Barbary States. Congratulations, please see Edward about your credentials, and no, you may not ask me about your pay, for I have no idea what a Special Naval Consul makes. Now . . . " Madison spun Eaton around and half guided, half pushed him out the door.

A few minutes later, there was a gentle knock on Madison's door. Edward poked his head carefully in. "Mister Secretary, are you all right?"

To Barbary's Far Shore

"I'm not at all sure, Edward. Did you take care of Consul Eaton's credentials?"

Edward's jaw dropped nearly to the floor. "Dear God . . . "

"What?"

"He was serious, then. He said something about you making him a Consul again!"

Madison nodded in affable agreement. "I did. Special Naval Consul to the Several Barbary States, if I remember correctly. I shall need his credentials as soon as possible, and let us see if we have any spare operating funds we can divert for his pay and expenses."

Edward looked at Madison in a combination of utter horror and disbelief. "With respect, Mister Secretary, I have served with every Secretary of State since Mister Jefferson, and I have never . . . ever . . . seen anyone like that carry our nation's credentials abroad."

Madison nodded as he sat back down. "Neither have I, Edward, and I must carry the responsibility of letting someone like him run around loose on my watch. But he knows the score, Edward. He has been there and seen how bad things are, unlike these armchair diplomats we have running around up on Capitol Hill. If he cannot convince them that we need to go in there and clean it up, he might at least be able to persuade them how bad the situation really is. Heaven knows I don't seem to be able to. Now, if you would please do two things for me."

"Certainly, Mister Secretary."

"First, please send Senator Sumter my compliments, and ask if I may send a representative to see him first thing in the morning."

"Consider it done, sir. And the second?"

Madison looked around his office with a look of utter defeat. "Clean this up . . . "

William Eaton strode confidently up a residential street in Washington the next morning, politely tipping his hat and giving a gentle bow to all who passed. His energies had been renewed by his conversation with Secretary Madison, and now he knew that he had an ally in a high place. All he needed to do was convince one more person, and then President Jefferson and Congress would have to listen to him.

The second briefing for Secretary Madison last night had gone much more smoothly. Madison had carefully helped Eaton tweak the outline and the plan itself to the point where it now sounded not only coherent, but actually doable.

The Practices of Politics

Madison promised that he would bring the matter up with President Jefferson that evening, but, and at that, Madison had held up a stern warning finger, Eaton needed to speak about it only when directed to, and only to those people who he was told to speak to. That would have been a major effort for William Eaton, but he was determined to give it his best.

Coming around a corner, Eaton looked for the address Edward had given him yesterday. There it was, two houses down, a medium sized two-story home of elegant design. Stepping through the gate, Eaton walked up the steps and tapped on the door with the head of his cane. He heard footsteps inside, and a tall, stately Negro in dark blue servant's livery opened the door.

"G'Day, suh. May I help you?"

"Ambassador Eaton to see Senator Sumter."

"V'good, suh. Please come in."

Eaton stood in the foyer as the servant trotted upstairs to fetch Sumter and took a quick look around. Thomas Sumter was a wealthy man by 1803 standards, and it showed in his taste. This was his second home, and it was far grander than anything Eaton could ever aspire to. There was beautifully carved woodwork, with potted and carefully trimmed magnolias and palmettos around the house. A tasteful, well-decorated sitting room with a massive fireplace was off to his right. The room was dominated by a huge painting of Thomas Sumter, in the uniform of a general in the Continental Army, astride a rearing horse on a battlefield.

"Y'all shouldn't be too impressed by that," Thomas Sumter drawled. "The ol' boy who painted it just had a good imagination."

Eaton turned to face Sumter, who was approaching with his hand extended. Tom Sumter was a solidly built, stocky, barrel-chested man, who looked every bit the legend that he was in South Carolina. The man known as the Gamecock of the Revolution had lost almost as many battles as he had won, but his tenaciousness and ferocity in combat made him a mortal threat to His Majesty's forces in the Carolinas. At the

Thomas Sumter: Source: South Carolina State Government

age of sixty-nine, he still looked every inch a man to be reckoned with. That was one of the reasons South Carolina had sent him to the Senate in 1801. He would live another twenty-nine years, the last surviving general officer of the Revolution.

Eaton smiled and shook hands firmly. "General Sumter, I am William Eaton, Special Nav . . . "

Sumter grinned as he shook hands. "Never been much one for titles, Eaton. Jemmy Madison sent y'all here; that's enough for me. Please, have a sit." Sumter motioned him into the sitting room to take a seat. "Can I offer you something cool to drink? People here think that it's warm here already, but I guarantee they've never spent any time in South Carolina."

"Thank you, General, I would be greatly obliged."

Sumter motioned for the servant, who was passing by the door. "Abraham, please fetch some iced water for me and our guest." Abraham nodded silently and headed for the kitchen.

"Now, Mister Eaton," Sumter began, "Jemmy Madison tells me you'd like to propose a remedy to our problems in North Africa."

"That I would, General. I explained it to Secretary Madison yesterday, and he feels it has some merit."

"Does he now?"

Abraham glided in and gave each man a tall glass of water, diamond-like chunks of ice glittering in the glass. Eaton took a long drink, savoring each swallow.

"General, that is wonderful," Eaton said, smacking his lips. "I haven't seen ice in, good heavens, almost five years now."

"Enjoy then," Sumter smiled. "There's plenty more where that came from. But, about your thoughts?"

"Ah, yes. I have assembled a proposal for a military expedition to Tripoli that would have two aims. First, liberating the crew of the frigate *Philadelphia*, and secondly, remove the usurper Yusef Karamanli as Dey and replace him with his brother, the rightful Dey."

Sumter gave Eaton a skeptical look. "You don't think that's a mite optimistic?"

"No, sir. We have the ships, we have the men, and we have the money."

The Practices of Politics

"Now I know you haven't been here in the States for some time. Mister Secretary Gallatin tends to be a bit stingy with the Treasury's funds. How much do you propose to ask for?"

"Twenty thousand dollars."

Sumter laughed as he took a long swallow of water. "Mister Eaton, we in the Senate, the Senate, mind you, must justify every piece of paper we use now. The President is a close personal friend of Mister Secretary Gallatin, and even he has a difficult time getting all but the most basic expenses out of him. But, back to you, what do you project in terms of men and ships?"

"Twenty vessels, including all of our frigates, and two thousand men."

"In other words, all of the Navy's major fighting vessels, and more men than we actually have in service."

"The ships are there to serve as needed, General, whether it be one or all of them. Likewise our fighting men, Army and Marine. As far as the extra men we shall need, there are a great many soldiers at liberty in that part of the world. I should think we would easily be able to hire enough to round out our force."

"And in the meantime, the frontiers are unguarded, and the red Indian runs loose."

Eaton looked straight into Sumter's eyes. "Such a risk has to be run."

Sumter looked straight back, and in his eyes Eaton could see the General who had ridden roughshod over Lord Cornwallis' troops in the Carolinas. "Mister Eaton, that 'risk' you refer to has killed a great many of your fellow citizens, and a few of my family. If I were to even suggest pulling the Army off the frontier, people might very well rise up against the government. Your idea runs out of practicality right there."

But before Eaton could answer, Sumter continued, "But I don't think we're done discussing things yet. Assuming we could somehow do everything you suggest, what would happen once they get there?"

In reply, Eaton stood up and strode to a beautifully inscribed map of the world that hung on one wall. "We would land here." Eaton's finger stabbed the map at Tripoli. "We would go ashore just northeast of the city, then attack it. While the soldiers stormed it from the unguarded west, the Navy would enter the harbor and bombard from there. Once we have secured the palace and disposed of Yusef, we would place Hamet on the throne, eliminating at one effort an unfriendly ruler, securing Tripoli as an American ally, and depriving the Barbary States of more than a third of their sphere of control. In short, a *coup d'etat*, a stroke of state!"

To Barbary's Far Shore

Sumter thought about this for a moment. "Never had to execute a siege when I was in. How long do you think this would take?"

"No more than a few days, a week at most."

A pause. "You ever been in uniform, Mister Eaton?"

"Yes, Senator." Eaton stood just a bit taller as he continued. "I was a sergeant in the Twelfth Connecticut during the last two years of the Revolution. I then had the honor of serving under General Wayne in Ohio, commanding an infantry company."

"Mmm. A real soldier, then, and under Mad Anthony to boot. Not one of those staff pansies. That always registers well with me, Mister Eaton, always very well. There's just one thing with never having done staff work, though. Sometimes you don't develop a suitable appreciation for what somethin' like this can require. For instance, you've taken into account the Tripolitian fleet?"

"The Navy will be more than capable of dealing with them."

"And if you lose even just one or two ships, you're in trouble. What will you do about attrition among your troops from disease or wounds?"

"The campaign will be over long before . . ."

"If everything goes right, Mister Eaton. It never, ever does in combat, down on the lines where you served it should be easy to see that. And as I see it, you also have the possible threat of troops and ships from Tunisia and Algeria. Did you take that into account?"

Eaton was happy to field that one. "Yes, General, I have. Again, the campaign would be over before they could get there."

"But once they do, where do you get your reinforcements?"

Eaton thought for a moment before replying, "Locals and soldiers of fortune."

"All of whom are notoriously fickle with their loyalties, and horribly expensive to keep on our side in any event."

"General, I realize there may be some things needed to refine the plan, but it is a viable one."

Sumter took a long swallow and then shook his head. "It's been my experience, Mister Eaton, that any plan that needs 'refinement' is one that's going to have a great deal of problems." Looking at Eaton, Sumter asked, "Mister Eaton, why do you think Jemmy Madison sent you to see me?"

The Practices of Politics

Eaton looked as if he wasn't quite sure of the question. "General, I assumed that it was because you are an influential voice in Congress and could help us win backing for such an effort."

Sumter shook his head. "Jemmy Madison has three very important qualities. First, he's honest. Second, he's smart. And third, he's smart enough to know when to be honest. Did he tell you it was because I'm influential?"

Eaton had to be honest. "No sir, he did not."

Sumter nodded. "The reason he sent you to see me is that President Jefferson owes me a very great favor, because I supported him when it was a draw between him and Mister Burr, and if I hadn't he might never have become President. Now, I have no compunction whatsoever about using that leverage on the President at all; that's the nature of politics, Mister Eaton. But I must be very honest with you. I can not see using that leverage for this plan."

Eaton wanted to jump and scream as loudly as he could to try and get his point through. It was maddening that no one, no one, would understand that what he had here was the only possible plan. But after his experience with Madison, Eaton understood that he dared not anger anyone in a position to help him. It took an immense effort on his part, but instead of roaring in protest, Eaton simply spoke. "I understand, General. I thank you for your time." He made ready to depart.

Sumter gave him a confused look. "Now where might you be going, Mister Eaton?"

"General, you have made your point clear. I must try and find some other way to pursue this."

Sumter's response was to laugh heartily, and Eaton's expression went from dejection to indignation.

"General Sumter," he said heatedly, "I see no reason to laugh at my efforts here! I am trying to free my fellow Americans!"

Sumter didn't stop, and said between laughs, "Mister Eaton, you have a lot to learn about how business gets done here. I said, 'that plan.' Everything else I've seen so far says we need to blockade them, and then pay them off. I've been asking for a real military plan, but no one wants to offer one. You've got one; it's just that there is simply no way it is going to succeed. On the other hand, we might be able to come up with something else, and it is at least worth the effort to try. So, let us sit down and see if we can come up with a plan to take up the Hill. Unless, of course, you'd rather quit?"

To Barbary's Far Shore

Eaton's eyes widened as he realized what Sumter was saying. "No! I mean, yes. That is, no, I don't want to quit! Please, let us see what we can accomplish!"

Sumter grinned. "Thought you might feel that way. Let us see what other options are open to us."

Eventually, changes were made to the plan that was finally sent, quietly and discreetly, up the ladder for argument and consideration. Sumter and Madison were certainly correct in their assessment; suggesting Eaton's original plan as a realistic option would probably had about as much chance as one of the contingency plans for invading North Vietnam would have had in 1968. The problem with getting any plan approved was that there was no real 'discomfort' back home. The economy was in good shape, the nation was expanding, and there was no realistic chance that the Tripolitians were going to come to America and invade us. In the early nineteenth century, piracy was considered an occupational hazard of seaborne commerce. As bad as the pirates' raids were, they still had not reached a 'discomfort level' that would bring down our wrath on them. The pirates were extremely careful to avoid that. Raiding and ransom were extremely lucrative sources of revenue, and suddenly starting to kill Americans and sink their ships would have put that at risk.

The situation is very comparable, disturbingly so, to the attitude of more than a few governments towards terrorist attacks from the late 1990s onward. All was well at home, no one considered any of these groups realistic threats to the world's only superpower, and certainly no one had any desire to go to war about it, because wars upset people. It was unfortunate when marines died in Lebanon, diplomats died in Tanzania, or sailors died aboard USS *Cole* in Yemen, but it was an 'occupational hazard,' wasn't it? Until, of course, four airliners suddenly veered off course one beautiful Tuesday morning towards Washington and New York, and then the game was very, very different. In the end, it probably would have taken an event equivalent to our 9/11, an actual physical invasion of US territory, or the death of several hundred Americans, to ignite a full-dress military campaign against Tripoli, and even then it might not have been enough.

The plan, whatever it was, was actually moving pretty quickly. It's a reasonable guess that whatever backing it was receiving was quiet and discreet, because there still was no great public outcry for a war. But, on the other hand, there was a perceived need to "do something;" no matter what it was, or whether or not it was the right thing to do. It could not have been widely distributed throughout Washington, but at least a few senators and congressmen had to have been in on it, as well as Madison, Secretary of War Henry Dearborn, and Secretary of the Navy Robert Smith. Given the amount of funding that the plan was finally given, it is possible that Treasury Secretary Gallatin never did know

The Practices of Politics

what was going on or that the funds were taken out of other discretionary spending.

There is a temptation to compare this effort with other covert efforts to hit back at America's enemies, but it isn't quite the same thing. First, there was legislative involvement in the project; whereas in Vietnam and especially in the 'Iran-Contra' affair, every effort was made to keep Congress out of the loop. Secondly, there was at least some public and media backing for doing something about the Tripolitians; though, as has been mentioned previously, there was little or no support for a full-blown war. Third, and most importantly, there was Executive Branch knowledge and approval throughout the entire affair. President Jefferson knew about the plan, as did Secretaries Madison, Dearborn, and Smith.

The question that does come up is just how much all concerned actually believed that what they were trying to do would have any realistic effect at all on the situation. These were not flighty or trivial, far from it; they were remarkably well grounded, hardheaded men who had built an entirely new form of government from the ground up, then fought to keep it from being strangled at birth. More than a few of them had defended it on the battlefield. Secretary of War Dearborn, for example, was a POW during the Revolution and later rose to the rank of lieutenant colonel. James Madison fought alongside George Washington, and quite a few of the senior members of both houses of Congress were veterans as well. These men understood the concept of 'minimum force,' in that they knew how to assess what level of military strength would be needed to obtain a given result. It appears, however, that whatever decisions were made were made with political considerations first and foremost, instead of military ones.

The final decision would have been made around the middle of May, at the latest. In keeping with the need for secrecy, it probably would have been a very small meeting, with just enough people in authority to let Eaton known what he had finally won. Eaton would have shown up dressed to the nines, confident he had defeated every foe, won over every doubter, and not incidentally convinced them that he was the best man to lead it.

He was wrong on all but the last count. The final version of his plan authorized him two thousand dollars, one ship, and eight marines, not quite one half of one percent of the manpower he'd originally requested. With that he was authorized to contact, just contact, Hamet Karamanli and assist him in raising an army to overthrow the Dey of Tripoli. There was apparently no mention of any American, or, for that matter, Eaton-led, effort to directly attack Tripoli with an all or mostly American force. The plan had been, quite literally, hacked to pieces and none of it had survived in any recognizable form.

To Barbary's Far Shore

If this was not the most humiliating moment in William Eaton's life, it had to be close to the top of the list. He had worked and planned diligently for months in order to give his nation a realistic, useful option for dealing with a military and political threat and this is what they had done to it. One has to wonder if this was not meant as a backhanded way of getting Eaton to simply drop the entire idea. That, if he was presented with the gutted remains of the only plan that had made any sense, even he might finally throw up his hands in frustration and walk away, allowing the diplomats and politicians to avoid the messy realities of war.

The explanations and justifications would have been couched in the usual 'we regret,' 'unfortunately' and 'circumstances do not permit' of political language. Eaton knew better than to lose his temper there and then in what would have been the most spectacular, and the last, explosion of his public life. Instead, I prefer to think he would have sat quietly, taking notes and waiting for the assembled solons to say their piece. Then, he would have stood, tugging at the hem of his coat to straighten it, stretching his neck, and clearing his throat before he spoke.

"Gentlemen," Eaton said quietly, "I mean no disrespect when I say I am indeed disappointed that given this nation's resources and its natural genius, that no other way can be found to try and liberate our captive citizens."

Someone would have thought that he was taking this pretty well, and it was a shame he had to be slapped down like this, but it was for everyone's own good.

"I have been in service to my country long enough to know that when such a group of distinguished gentlemen state that these are the only options open to us, then I must accept them at their word."

Someone would have thought that they should have all been embarrassed that this was the only option open to the United States.

"With that in mind, then, our course is clear."

"We understand, Mister Eaton, and we hope that you understand that sometimes, we cannot . . . "

"I gladly and proudly accept your charge. If these are all the resources that can be spared, I shall do what I can with what I have."

Some of the most august mouths in Washington opened, intending to say "Thank you, Mister Eaton." instead, they flapped in sheer disbelief. This was most definitely not supposed to have been the way things turned out. "But . . . but . . . "

The Practices of Politics

"But nothing, gentlemen!" Eaton said with a wide smile. "You have entrusted me with our resources and I shall treasure them as if they were my own." Bowing slightly, Eaton picked up his hat and left the room. As he reached the door, he turned with a smile that spoke of far greater determination than the rest of them had ever fathomed.

"And I shall be successful. Good day, gentlemen."

Eaton strode out into a beautiful summer day in Washington; the sound of people going about their business and horse-drawn carts rumbling through the streets, all beneath a gloriously blue sky.

And as he put on his hat and pulled on his gloves, Will Eaton would simply not have been human if even with his vast self-confidence and self-assurance, he hadn't wondered just what in God's name he was going to do next.

President John F. Kennedy once hosted a dinner for American Nobel Prize winners, and opened his remarks with, "I think this is the most extraordinary collection of talent, of human knowledge, that has ever been gathered together at the White House, with the possible exception of when Thomas Jefferson dined alone." Jefferson's reputation for intelligence was well deserved, but, for all that, he was still no more than an intelligent politician; an exceptionally intelligent one, but a politician nevertheless. Jefferson had agreed with Eaton's plan and endorsed it, however reluctantly, but that didn't mean he thought the military option was the best, or only option. There had to be a diplomatic option in case everything else went to hell in a hand basket; there had to be a plan to sit down and talk if Eaton failed, and, given his reputation and actions so far, that was at the very least a possibility. So Jefferson quietly sat down and considered who Plan B was going to be . . . and sent for Tobias Lear.

Lear looks forward to us from a picture painted approximately about this time. The confident and mildly cocky smiles one sees in similar pictures of Eaton, Hull, and Rodgers is not here. Rather, Lear wears a supercilious smirk, the look of a salesman who has just spotted a sucker approaching from some distance away and has money falling from his pockets.

Lear was actually something of a public figure by this point, but for reasons he probably would have preferred to have avoided. He was born in New Hampshire in 1762, to a fairly well-off family that was able to send him to college in preparation for a teaching career. He managed to avoid service in the Revolution, instead graduating from Harvard in 1783. He taught for a while before his uncle, General Benjamin Lincoln, the officer who took the British surrender at Yorktown, recommended him to Martha Washington as a tutor for her grandchildren. He ended up not only as a tutor but as General Washington's personal secretary, a position which gave him constant access to the General,

To Barbary's Far Shore

almost complete control of the General's personal papers and the General's complete trust.

Tobias Lear. His character was such that he contributed his name to the American language as the adjective "Leary" applied to questionable or shady dealings

Which, as others have learned since, is not always a good thing. Lear appears to have first convinced the General, now President, to invest in a land scheme, followed by a canal scheme, which appears to have been the 'green energy' scam of those long-ago days. Neither one succeeded and the President ended up bailing Lear out on both occasions. But it appears possible that Lear never completely leveled with the President as to the extent of his liability, or, just as likely, kept making questionable investments.

The reason for this possibility is that President Washington discovered that rent owed to him by his tenants had been given to Mister Lear . . . and never quite made it to the President. Washington, whose temper was well-controlled, but when it appeared could be most impressive, is said to have been furious when he realized what was going on, but eventually forgave his secretary, even obtaining a commission as a colonel for him. Lear stayed on as Washington's secretary until the President's death in 1799, taking on more and more of Washington's business work, as well as some irregular real estate deals involving friends that required him to pay back even more money. He was even apparently the last person to see Washington alive, passing on the Great Man's final words: "it is well."

When Washington died, Colonel Lear, as he preferred to be called, though he never served a moment in the field, had possession of all of Washington's business and personal papers, a treasure trove that has present-day historians salivating over what may have been in there. We say 'may,' because it appears that Colonel Lear hid or destroyed quite a bit of that collection. The reasons are unclear, but boil down to some letters being less than respectful to other great figures in American history, in particular Alexander Hamilton, John Marshall and Thomas Jefferson. Jefferson specifically seems to have benefited from the disappearance of these letters, and as Lear kept getting lucrative diplomatic jobs from Jefferson, it's difficult to believe there was no quid pro quo.

The Practices of Politics

Jefferson, of course, was of mixed feelings regarding the war against the pirates. On the one hand, he knew full well what orders Eaton had been given, as well as what restrictions he was under. On the other hand, public opinion, while not exactly demanding war, was starting to be a bit unhappy with the way things were being handled, and no President likes to hear, or, for that matter, think, that his policies are being met with less than approval and adoration. So Jefferson decided that Eaton, who, as one will recall, had left North Africa in less than good standing with the region's leaders, would be replaced by Tobias Lear as a new Consul General, with orders to try and settle things down in the wake of Eaton's departure.

That plan very nearly blew up in the Squire of Monticello's face, for Lear and his family were originally ordered to depart aboard the *Philadelphia* with Will Bainbridge. Scheduling problems kept the Lears ashore when Bainbridge sailed towards his own appointment with Tripoli, so they sailed aboard *Constitution*. It's not clear whether or not Eaton knew about Lear's appointment, or whether or not he would have even cared. But the bottom line was that there were now two separate initiatives underway to try and free the crew of the *Philadelphia*. Only one could succeed.

To Barbary's Far Shore

SIX: UNDER ORDERS

Aboard Frigate Constitution, *Seventeen Nautical Miles North of Tripoli Harbor, September 9th, 1804*

It is said that a happy captain makes a happy ship. Edward Preble could not in any way be described as happy, so it is easy to imagine the state of mind aboard *Constitution* that breezy September morning.

The Mediterranean Squadron was hanging grimly on its station north of Tripoli, now down to only three ships keeping an eye on the Dey: *Constitution* herself, along with *Vixen* and the new *Argus*, just off the ways a few yards down the Charles River from where *Constitution* was born. The rest of the squadron, *John Adams, Syren, Enterprise* and *Nautilus*, along with a handful of Jefferson's gunboats, had trudged back to Sicily three days before, after a quiet, earnest conference aboard *Constitution* that morning, about which more in a moment. *Intrepid*, tiny, gutsy little *Intrepid*, was gone, along with her entire crew in a badly misfired attempt to take out the Dey's fortress and a good chunk of the pirate fleet by using the same trick that had gotten them in to destroy *Philadelphia*.

Intrepid was too small to really stay with the fleet, though she could give excellent service in shallow water. But, without enough size or room to carry decent armament, she was actually more of a hindrance than a help, not to mention that she was using up precious manpower as well. Preble reflected on this long and hard, and decided to send her to glory in a manner befitting a warship of the United States Navy.

Preble decided to use her as a fire ship; essentially, turning her into a floating bomb by stuffing every square inch of her hull with explosives. An all-volunteer crew would take her into Tripoli Harbor, light the slow fuse, then get into the boats and get out. Tricky, yes; suicidally dangerous, without question, but possible. The British had tried something similar, but on a much larger scale, against the Armada in 1588. It hadn't had quite the desired effect, but it did further rattle those already nervous souls who made up *El Invencible*. Preble

Under Orders

himself would most certainly have been aware of a little known fire ship attack on August 16th, 1776, against a British squadron moored in the Hudson River off northern Manhattan Island. The British lost one of their tenders, HMS *Charlotte*, and very nearly lost their flagship, the 44-gun frigate HMS *Phoenix*. Given the potential for heavy, and possibly war-winning, damage, Preble would have thought the game was worth the candle.

Richard Somers, who had wanted to follow Stephen Decatur into Tripoli Harbor, wasn't going to let this chance get past him. He browbeat Preble into letting him command *Intrepid's* last ride. Two others, Midshipman Henry Wadsworth and Acting Lieutenant Joseph Israel, stepped forward as well, along with ten enlisted men. A surprisingly large number of volunteers had to be turned down, which says something for the fighting spirit of the US sailors at that point.

On the night of September 2nd, after an unsuccessful attempt the night before, Somers carefully guided *Intrepid* into the harbor without alerting a soul. So far, so good. Somers smoothly guided her through the crowded anchorage and under the walls of the Dey's fortress, moving her like a prima ballerina on her greatest night. Two guns, likely warning shots from the fortress itself, were heard, then nothing for ten minutes. What happened next will never be known with any certainty, but the night was torn apart without warning by a nightmarish flower of orange, red, and yellow that tore Intrepid apart, killing every member of her crew instantly.

There are a couple of possible explanations, none of them comforting. The Dey's gunners always claimed afterwards that their phenomenal shooting blew them out of the water, but these claims should perhaps be regarded with the same skepticism as those of elderly farmers who have supposedly shot down heavily armored attack helicopters with a single round. More likely, the inexperienced volunteer crew may have made some fatal error that set off the seven tons of explosives that were crammed into *Intrepid's* tiny hull. It's also possible that the time fuse may have been the culprit. Time fuses of that period were an iffy proposition and one could never be completely sure if it would burn the required amount of time or suddenly just flash along its entire length.

There is, however, a last, and far grimmer, possibility. Since no one outside the harbor was able to see what happened after the two guns sounded, it is possible that *Intrepid* may have reached her intended target, only to find herself surrounded by newly awakened, and thoroughly angry, Tripolitian pirates, none of whom wanted to risk the Dey's wrath by letting another boatload of infidel raiders get away. If that happened, it would have taken far less than ten minutes for the pirate crews to get to their boats and head for *Intrepid*, now lying still with her sails slack. There would have been no possibility of escape; given the

To Barbary's Far Shore

large pirate crews, they would have been able to overwhelm Somers and his men in mere seconds.

That left three options for Somers. First, he could surrender, and end up in the same rathole with Will Bainbridge and no idea when or if they would ever get out. Second, they could fight it out to the end. Given the mood of the pirate crews, that might have been quick, but it would have been remarkably unpleasant, and it wouldn't make much of a dent in their numbers; not to mention, it would also have handed the *Intrepid* back to the Dey with a hold full of munitions that would be quickly turned back against their former owners. But the last option, the one Bainbridge turned down as the pirates came over *Philadelphia's* rail. It was an option that would let the Tripolitans know that some victories could be more costly than they were worth.

If that was the option Dick Somers chose, he would have had to take it quickly, before he lost his nerve. He would have jumped down into the hold, where he was surrounded by fourteen thousand pounds of explosives, and quickly cut the fuse to just a few short inches. With a shaking hand and a whispered prayer, he would have struck the quickmatch and held his breath for the few short heartbeats until the world turned glaring white and vanished. The blast would have devastated any ships nearby, inflicted gruesome casualties on the pirates, and thrown flaming wreckage for hundreds of yards in every direction. It may very well have put a good dent in the wall of the Dey's fortress, though this is unclear. It certainly would have awakened that dear soul and let him know, in no uncertain terms, that Dick Somers had been there.

The loss of Somers and *Intrepid* had devastated the squadron, all the more so because there was no apparent effect on the Dey. Squadron morale was not one hundred percent to begin with; the weather had been miserable, which had either kept the Squadron away from its stations or kept them in port completely, but the men were still eager to do their duty. Preble, however, was about to rotate back to the United States. In any unit where a well-liked and respected commander is about to leave, there is always a brief drop in morale. When added to the disaster with *Intrepid* and the perception that, for all their effort and loss, they really weren't accomplishing anything, this sent morale into the bilges. In addition, Preble himself was not at all up to speed. His tuberculosis had flared up again. As the weather got cooler, it was physically tougher on him to even keep up with his paperwork, much less the physical demands of leading the squadron.

So, on the morning of September 6th, Preble and his senior commanders met in the day cabin aboard *Constitution*. We don't know exactly what was said, but from the events that followed it is easy to figure out. Preble himself was tired and ill, the Dey hadn't budged, the weather was going to start closing in, Will Bainbridge and his men were still captive, and a new commodore was on

Under Orders

his way out with God only knew what new guidance from Washington. No one had any ideas of what to do next, and Preble simply wasn't up to thinking of any on his own. Bring in most of the fleet for provisioning and repair, keep a few ships on station and let the new man handle it. Accordingly, the squadron split up and most of it headed back to Sicily.

Argus, Lt. Isaac Hull, commanding, was at the western end of the line off Tripoli the morning of the 9th. Hull was thirty-one that morning, had already been at sea since he was fourteen and an officer in the USN since he was sixteen. Hull was a big man, tall and built like a barrel, and had a reputation for being a skilled, capable officer who was also devoted to his crews. We tend to expect that today, in an era when skippers of the most lethal warships in history are required to take sensitivity training to make sure they treat their crews with the proper respect. But two hundred years ago, the captain was quite literally God Incarnate on that narrow stretch of oak planking, and his word was the Alpha and Omega of life at sea.

USN skippers were, as a rule, far more tolerant and beneficent towards their men even then, but some could be what we would consider today brutal. Flogging was still legal, and practiced, in the USN for some years yet. Edward Preble in particular was known for bending regulations into near pretzels so as to permit the maximum possible number of lashes. As late as 1842, the captain of the USS *Somers*, named for poor lost Richard, had three men hanged at the yardarm for merely talking about mutiny, and was cleared by two formal investigations.

Without question, Hull resorted to the lash on occasion; but, by all accounts he disliked it and preferred to just get rid of troublemakers. Like many captains of the time, he often raided his own purse to pay for sufficient, and healthy, food for his crew. He believed that the first three principles of naval combat were training, training, and training, especially gunnery training. Hull's ships were known for their gunnery skills, and he'd show how much the training had been worth ten years later against HMS *Guerriere*. But that is another story.

Pacing the deck behind him would have been *Argus'* first officer. The first officer, known today as the executive officer, has a very simple job: handle the care and feeding of the ship and crew while the captain takes care of the serious work. It would have been an officer like Lieutenant Joshua Blake, USN. What we know about Blake says that he was a tough man to work for, demanding and unwilling to grant a great deal of leeway in anything. Younger officers like Blake tended to be that way, until they'd gotten a bit of time under their belts. On the other hand, some never did, but most eventually came to an accommodation with those they commanded, realizing that they were not perfect and never would be.

To Barbary's Far Shore

The winds were good and steady that morning, and although the sky was intermittently covered over with a gray blanket, enough sunlight broke through every now and then to give everyone a brief taste of warmth. Hull was in his element as *Argus* sliced through the waves, *Constitution* a few ship lengths behind, and visible behind her the *Vixen*'s sails. For a few minutes at least, he could forget about the eternal curse of the paperwork and everything else that kept him chained to his desk as *Argus* cut effortlessly through the water. Hull could take a great deal of pleasure in the fact that the shipwrights in Boston, the same ones who had built *Constitution*, had done another excellent job. Of course, she had her share, possibly more so, of minor quirks that had to be ironed out, but that was to be expected in a day where each ship was built by hand and no two were exactly alike.

"SAIL HO! SAIL HO!!"

Everyone on deck looked up towards the mainmast, where the one of the lookouts was gesturing towards the horizon. Hull reflexively checked his watch, just before noon, and then strode forward with his own spyglass. Blake stepped to where Hull had been beside the helm, and took a quick glance over the deck to make sure everything was in its place. The odds were good that it wasn't a runner. That sort of thing had fallen off rather precipitously lately, but one always had to be ready.

Hull looked back to Blake. "Josh, signal to *Constitution* that we have company and are closing to investigate."

"Aye aye, sir!" Blake turned to the signalman and snapped his fingers.

That was one thing about Hull's crews. They knew their jobs almost reflexively, which was one reason he never lost a battle. The colored flags went racing up the halyards as Blake gave the orders for the helm to go over just a couple points to starboard. *Argus* responded like the thoroughbred she was and raced forward towards the nearly indistinguishable dot on the horizon that marked the stranger's location.

It took some time for the two ships to close, but soon enough her masts were above the horizon. Hull was able to get a good look at her. It was a frigate, all right, and she was flying a massive American flag and, below that, a commodore's pennant.

The sight was enough to make Hull blink for a second, and take a quick glance over his shoulder to make sure there was still one flying from *Constitution*'s foretop. Then, it clicked; Commodore Barron was here. *Oh, dear*, thought Hull as he lowered his spyglass. *Suddenly the day doesn't look quite as pleasant as it had.* Motioning Blake forward, Hull leaned on the rail, watching the ship grow closer with every second.

Under Orders

Blake came to attention beside Hull. "Yes, sir?"

Hull simply nodded towards the other ship and handed his glass to Blake, who extended it with a practiced snap of his wrist. Blake looked at it for a moment, then folded it and gave it back to Hull.

"Have to admit, Captain. I was hoping for a few more days with Commodore Preble. He's a good man."

Hull nodded in understanding. "He's also a sick one, Josh. He should have been relieved months ago. That's unfortunately one of the reasons we're sailing around in circles out here, no disrespect to the Commodore. He just needs a rest, and it shows."

"Yes, sir. Shall I notify *Constitution*?"

Hull nodded. "By all means. Better we get the formalities over and done so we can get back to the business of trying to figure out what to do next."

One of *Constitution's* lieutenants caught the message first, turned to take it back to the quarterdeck and almost ran headlong into Edward Preble, now thinner than he had been that morning when they had run across the ship that would become *Intrepid*, and stooped with stress and illness.

The lieutenant pulled up short, stepped back and saluted. "B . . . begging the Commodore's pardon, I . . . I . . . "

Preble dismissed it with a weak smile, a wave and a cough. "Quite all right, Leftenant. I saw it too."

Patting the young officer on the shoulder, he turned to slowly half-walk, half-shuffle back to the day cabin. Stopping as he passed *Constitution's* flag captain, Preble spoke.

"Please see to it that we have a proper welcome for Commodore Barron. He is most deserving of the honor."

Then Commodore Edward Preble, USN, walked off his quarterdeck, knowing that his career had just ended in what he considered failure, because he had failed to defeat the Tripolitians. It wasn't failure, of course; he had simply run out of that most priceless of all commodities, time.

The inbound frigate turned out to be *Constitution's* sister *President*. She took almost another hour to come up alongside *Constitution*, which had come to a stop while *Argus* and *Vixen* circled protectively nearby. A boat went over the side and six brawny sailors pulled for *Constitution* carrying three passengers. Dan Carmick had his marines lined up in three bolt-upright blue rows at shoulder arms, while *Constitution's* sailors manned the rails at parade rest, ready to greet the new commodore upon his arrival.

To Barbary's Far Shore

Everyone had noticed three passengers coming over, and, as with anything unexpected in a military situation, it started the rumors flying. Barron had brought his family with him, like the unlamented Commodore Morris. No, it was some inspector general that had come out with him to see how they were running the ship. Whoever it was, they were being treated with the utmost respect.

As the boat pulled closer, it could be seen that the passengers were two officers and a civilian, which put paid to the suggestion that Commodore Barron might have been making this a family cruise. Of course, this made the mystery even more intriguing. After all, an unexpected officer suddenly showing up always means something is in the air.

The boat bumped alongside *Constitution's* hull. The handling crew reached down to help secure it. The oars swung upright out of the water, sending a fine spray into the air, followed by a snarling bellow from below the rail.

"Be careful, ye damned fools! I'll find who taught you jackasses how to row and give the lot of you hell!"

As the words blistered the paint along *Constitution's* rail, more than a few faces went white as the mainsail. They suddenly realized that they knew from whence it came.

Constitution's flag captain roared. "Attention on deck!"

The bosun's mate expertly trilled his whistle in that three-note signal that commands attention in the US Navy to this day. The sailors snapped to. The marines, of course, were already there. The first figure moved slowly, almost painfully, up the ladder.

Another bosun called, "Commodore, Mediterranean Squadron, arriving!"

Samuel Barron moved with halting, hesitant steps over the rail as Preble and his flag captain saluted smartly. Barron, a stocky, solidly built man, brought his hand up with a smile that wasn't quite able to belie his physical illness. Preble smiled and shook his old friend's hand. He thought to himself that Barron looked worse than he did.

Barron was seriously ill with a liver ailment of some kind, and it showed. The record is unclear as to when he developed it, but given that soon after September 13th he was extremely ill, he had probably first started showing symptoms before he left the United States, and he never should have been there in the first place. But Barron was as stubborn as Preble was when it came to denying illness, and he went anyways.

The bosun called out once again. "Flag Captain, Mediterranean Squadron, arriving!"

137

Under Orders

A thick head of curly black hair came over the rail. Most of *Constitution's* crew rolled their eyes heavenwards in hopeless supplication to a merciless God who had clearly abandoned them. Stepping onto the deck was Captain John Rodgers, United States Navy, and the source of that raging snarl a few moments before. And, as anyone who had served with him could tell you, from Johnny Rodgers that had been a mild reproof. Rodgers' eyes swept the deck of *Constitution* with a glare that would have frozen hellfire. Those closest to him had the good sense to keep their eyes locked forward, without so much as a twitch. Rodgers stepped away from the ladder, surveying the deck that was soon to be his, as he reached forward to salute, then shake hands coolly but respectfully with Preble and his flag captain.

For every sailor and more than a few officers, John Rodgers was the most unpleasant officer they had ever sailed under. He was considered a prickly, vicious, spiteful, belligerent and arrogant martinet who believed that most people were utterly inferior to him, and that the tiny remainder were only somewhat so. And, in an era when the vast majority of warship commanders in the world were notably difficult men, such an opinion was indeed a grim one. Even if that was not a one hundred percent accurate description of him, it is without question that Rodgers was widely disliked by subordinate and peer alike.

John Rodgers. Source: U.S. Navy History and Heritage Command

However, he had a reputation, and a well-deserved one, as being one of, if not the most, aggressive and combative skippers in the young Navy. In those days that sort of thing enabled one's superiors to overlook more difficult personal flaws. The bottom line was that he had never lost a fight and never would. There is the distinct possibility that his crews were too terrified of him to even think of doing so, but never mind.

Included among those whose day was ruined by Rodgers' appearance was Edward Preble. The two men were, to put it gently, not fond of one another. This dated back to not long before *Philadelphia*'s capture. Commodore Morris was still in the Mediterranean, and was getting ready to go home aboard the frigate *Adams*. Morris then left Rodgers as Commodore of a two-ship squadron, comprised of *New York* and *John Adams*. It seems that Rodgers got it into his

To Barbary's Far Shore

head that this was the precursor to an appointment as Commodore of the entire Squadron when Morris departed, Preble's appointment notwithstanding. There was at least some reason behind this, insofar as Rodgers actually was senior to Preble, and felt he had a justifiable claim to the job.

Needless to say, the acting Commodore was most irate when he pulled into Gibraltar and saw the commodore's pennant waving proudly from *Constitution's* mainmast. Upon meeting Will Bainbridge later that day, Rodgers let go with both barrels about which officer he believed should rightfully have the job. Bainbridge in turn was so concerned about the potential for conflict between the two that he reported the conversation directly to Preble.

Now Preble had his orders in writing, many of the officers he had brought up through the Navy with him were under his command, and he was also as touchy about rank and privilege as any officer. There was no way he would even consider discussing the situation with Captain Rodgers. On the other hand, the tactical situation was such that Preble really couldn't afford to alienate anyone under his command.

According to Dr. William M. Fowler, Jr., in Jack Tars and Commodores, Preble wrote a conciliatory note to Rodgers telling him of the esteem in which he was held, and how there was no intent on anyone's part to insult or anger him. Professor Fowler states that Rodgers apparently swallowed his pride and asked for permission to meet with Preble so they could form 'a common policy.' It should, however, be pointed out that subordinates do not negotiate 'common policy' with their superiors. Preble, by the book, had every right to tell Rodgers to do what he was told or go home. Rodgers was apparently not foolish enough to go off on Preble, but this author finds it difficult to believe that Rodgers took the whole thing with good grace and cheer. But now, he was out here as Barron's flag captain, assuming command of *Constitution* and effectively second-in-command of the Squadron, and Preble was not at all happy about that.

The bosun stood to once more, and took a deep breath before announcing. "His Excellency, the Ambassador Plenipotentiary and Special Naval Agent To The Several Barbary States!"

Will Eaton bounded over the rail with all the joy of a child on a picnic. He leapt forward to shake Preble's hand warmly.

"Commodore, it is a pleasure to see you again. We have so much to catch up on!"

Preble smiled politely back and replied, "Good to see you again as well, Ambassador. Congratulations on your promotion."

John Rodgers watched the exchange with ill-disguised impatience. Rodgers, who had gotten away from the Dey of Tunis through Eaton's efforts,

Under Orders

probably tolerated him more than he did most people, but it wasn't any great improvement, especially as Eaton had most likely already let it be known that he considered himself in command of the upcoming enterprise. That was accurate as far as it went. Eaton's actual orders, as will be seen, were a bit on the vague side.

The round of introductions and greetings were brought to a close and the assorted dignitaries made their way to the day cabin. As they moved aft, *Constitution*'s captain motioned to Dan Carmick, who quickly dismissed his marines and trotted over.

"Sir?"

"Might want to come on in with us, Dan."

"Right. Honor guard?"

"Afraid not," the flag captain said with a distasteful expression. "More likely to keep Preble and Rodgers from each other's throats. This won't be at all a happy reunion."

Most everyone took a seat once they were inside the day cabin; Preble at his desk, Barron beside him. Rodgers and *Constitution*'s captain performed a stiffly polite 'after-you' gavotte before sitting down at opposite ends of a bench seat, while Carmick stood beside the door, trying to look relaxed, and Eaton paced slowly across the deck.

"Well, gentlemen," Preble said with a smile he didn't feel, "I suppose we should get on with the formalities."

With that, Barron pulled a folded letter from his inside coat pocket. He opened it with the crackle of fine stationery. Standing with a noticeable effort, he cleared his throat and began to read.

"July the first, eighteen hundred and four, from the Honorable Robert Smith, Secretary of the Navy, to Commodore Edward Preble USN, commanding the Mediterranean Squadron. Sir, effective upon receipt of these orders, you will relinquish command of the Mediterranean Squadron to Commodore Samuel Barron," Barron gave a brief nod and polite smile "with your flag captain to relinquish command of frigate *Constitution* to Captain John Rodgers." Rodgers, in turn, simply gave a self-satisfied smirk.

Barron paused for a moment with a doubtful look on his face as he scanned the next part of the letter, and then began to read once more. "I also request that Commodore Preble remain in the area to serve under and assist Commodore Barron and Captain Rodgers so as to bring their natural talents to bear against the Bashaw of Tripoli and return our imprisoned countrymen. Please extend to Commodore Preble my deep appreciation for his service and assistance. He has

To Barbary's Far Shore

consistently shown himself an example to subordinate and superior alike and his reputation shall be a hallowed one. Signed, Robert Smith."

Barron looked up to see a look of grim anger frozen on Preble's face. *Constitution's* flag captain caught it as well, and hurriedly stood up and looked at Eaton. "Ambassador Eaton, would you like a walk around the ship? I'm sure there's a great deal you would be fascinated by."

He practically pushed a protesting Eaton out the door followed by Carmick, who pulled it shut behind him. When the sound of the bolt snapping shut finished echoing off the walls of the day cabin, it left Barron, Preble, and Rodgers sitting in an uncomfortable silence.

Barron spoke first, looking to Preble. "Edward, may I ask . . ."

"Yes. After Captain Rodgers has excused himself."

Rodgers had to have known what was coming, but couldn't resist a chance at the dig. "Edward," he said with an unctuous grin, "My responsibility is here now, surely . . ."

Preble stood with an explosive cough, and he looked angry enough to chew through a mast in one bite. Glaring down at Rodgers, he snarled through clenched teeth. "Captain . . . if you call me 'Edward' once more, I shall break your neck and do it with pleasure. Now excuse yourself."

Rodgers stood, ready to start swinging, but then thought better of it. Thin and tubercular Preble might have been, but even John Rodgers knew better than to physically attack a superior. Gritting his teeth, Rodgers bowed slightly. "A thousand pardons, Commodore. Please excuse me while I join the others on deck." Rodgers spun on one heel, and then strode out, slamming the door behind him.

Before Barron could say anything, Preble turned to face him. "Do not say a word, Sam Barron, not a word!"

"Edward, this was not my decision . . ."

"And did you do anything to protest it? You know, you *know,* how Rodgers treated me when I arrived here! The man was at best disrespectful, and at worst mutinous! He fought me at every turn, ignored me when he could, and insulted me behind my back and sowed dissention when he couldn't! You know that, Smith knows that, and yet, he writes orders asking me to serve beneath him?? Samuel Barron, I have thought of you as my brother, damn it! But you acquiesced to this? Not now, not ever!"

Barron tried a conciliatory approach. "Edward, I know of the problems between you and Rodgers, but I need you and your knowledge! You have been

Under Orders

more successful than anyone else so far, and everyone back home knows it! Please, stay here for just a while longer! Surely we can work something out; you can move over to *President*, she'll be all yours . . . "

"NO!" Preble collapsed in a coughing fit, almost falling into a chair. Barron moved to help him, but Preble waved him off, and Barron stood there with his arms at his side as Preble got back under control, wheezing and hacking. Finally, with labored breathing, Preble looked up at Barron and said, "I . . . will . . . NOT. If it were just you, Sam . . . with all we have served through . . .I . . . could find it in my heart to stay." Shaking his head sadly, Preble said quietly, "Not under Rodgers, Sam. Not for you, not for Will Bainbridge and his men . . . and not even for my country."

There was silence for a long, unhappy moment, as Barron looked away from his old friend, then spoke quietly. "I understand, Commodore. I shall, of course, accede to your wishes." Moving slowly to the door, Barron turned as he placed his hand on the doorknob. "Captain Rodgers, Ambassador Eaton, and I shall return to the *President*. I hope that we may arrange to meet at your earliest convenience to discuss the political and strategic situation here. Ambassador Eaton will also need to be briefed."

Preble looked up, saddened anger on his face, but averted his gaze from Barron. "Of course, Commodore. We can begin at first opportunity tomorrow."

Barron nodded, eyes downcast. "Until tomorrow, then."

As he left, the day cabin that had heard the roars of broadsides and the cries of men in battle was torn apart by the quick, sharp and so often unnoticed sound of a door bolt clicking home.

The next four days would have been awkward ones at best. It's not known if Barron made any other attempts to talk Preble into staying, but, even if he did, it was probably without a great deal of heart. Preble had made up his mind to leave and it doesn't appear as if Secretary Smith actually ordered Preble to stay, though he probably expected him to. The four ships cruised in long, slow racetrack patterns while Preble and his captains briefed Barron, Rodgers and Eaton on the situation. They probably already knew the particulars: *Philadelphia*'s crew still hostage, *Intrepid* gone, the Bashaw still defiant.

One thing they probably didn't know until that briefing was that manpower was becoming a problem again. The loss of *Intrepid* and her men had been bad enough, but illness was starting to rear its ugly head. It was particularly bad aboard *Constitution*, who was nearly a hundred men short from her complement. At that point, she might have been capable of going to sea, but taking her into battle would have been an unacceptable risk. Rodgers could not have been happy to hear that, but for the time being he seems to have stayed silent.

To Barbary's Far Shore

When Preble, Barron, and Rodgers met privately at some point with Eaton, Eaton laid out the plan that had been approved back in Washington. Barron and Rodgers had probably already heard it, repeatedly, but Preble had only heard Eaton's original idea and didn't know what had actually been authorized. Preble was asked for his input, and he probably questioned Eaton within an inch of his life. But, in the end, Preble gave his blessing, though one has to wonder if Preble, as ill as he was, just finally gave in to Eaton's constant harping on the subject.

This raises an interesting question, why did Barron and Rodgers feel it was necessary to not only bring Preble in on the plan, but also, as it turns out, actively seek his support and approval? There would seem to have been no need for it; Barron, Rodgers and Eaton were under orders from higher authority. Barron had all the authority he needed by virtue of his position and rank. Rodgers never felt he needed anyone's permission for anything, and, even if Preble had disapproved, it wouldn't have mattered to him in the slightest. And as far as Eaton was concerned, he had his authorization from the President, Congress, and the Department of State, and no mere outbound commodore would have been able to stop him short of physically subduing him, lashing him to a mast, and taking him back to the States. Finally, Preble was on his way home and by the book no longer had the authority to close the plan down even if he'd wanted to.

So, why?

Well, a couple of reasons present themselves. First, Preble was on his way back home, and if he could be persuaded to back the plan, that would be one more supporter on Eaton's side. Secondly, it is not entirely clear as to what extent Barron and Rodgers believed in Eaton's plan. Even to men like Rodgers, who wanted to hit back against the Bashaw somehow, Eaton's plan had to have borne a whiff of desperation, and not being under direct military control would have increased their doubts even further. Whatever their personal feelings would have been towards Preble, his reputation would have insured that if he backed the plan, and it went wrong, the blame would have been spread around a bit wider, and Barron and Rodgers wouldn't have taken the blame alone. In modern terms, the appropriate phrase for such a maneuver is 'cover your ass.' In 1804, that wasn't the phrase those officers would have used, but they would have certainly understood the concept.

The meetings and briefings went on until September 12th, the night before Preble was to shove off for home. There was, however, one more meeting that needed to take place, and it would probably have taken place over dinner aboard *Constitution*, a solid, hearty affair for all concerned. Preble and Barron might not have been eating all that well due to illness, but Rodgers was known to

Under Orders

attack meals with gusto, and Eaton certainly never turned down a healthy plate; for that matter, neither did the guests that evening, Isaac Hull and Will Allen.

For whatever had gone before, its reasonable to assume that the evening would have been a fairly pleasant affair, with toasts to President Jefferson and the United States, and stories told and retold and memories of those who could not be there and those who never could be again. Finally, after the plates had been cleared, the brandy poured, and the pipes lit, it was time to get down to business. Barron, obviously unwell but in a relaxed mood, started the festivities.

"Leftenant Hull," Barron said, "as pleasant as the company has been this evening, we do have some military business to get down to, and you will be playing a most important role in it."

Hull smiled and lifted his brandy in salute. "My ship and I are ready to do our duty any time the Commodore asks."

Rodgers gave him a curious smile. "Glad to hear that, Leftenant. You may find this duty most . . . interesting."

Hull was still smiling, but in the back of his mind there was that little tickle that most military men feel when they hear the word 'interesting' used in conjunction with an upcoming assignment.

"Well," Barron interjected, "let's allow Leftenant Hull to decide that." Reaching back to his desk, Barron picked up a sheet of paper, then passed it over to Hull. "By all means, review them at your leisure, but, to summarize, you will take *Argus* to Malta and refit. Shouldn't take you more than about forty-five days; the usual arrangements with the British will apply. Once you're done there, replenish and make for Smyrna, where you will gather whatever American vessels are waiting to transit outbound into a convoy, then escort them out to Gibraltar."

Hull had been scanning the orders, and when he was finished he handed them to Allen. Looking back to Barron, Hull gave a nod. "Commodore, it looks all straightforward enough. I think the only interesting part will be what kind of chicanery we meet with from the fine merchants of Valetta."

"True enough, Leftenant." Barron replied. "Nothing to be concerned about, however. I want you to ignore everything I've just told you. Those were for the official record, and, in the event you're asked anything by anyone, you refer back to those. Understood?"

Several very loud alarm bells were ringing in Hull's head right now, but he kept his concerns under control. "Understood, of course, Commodore. But . . . if we're not going to refit at Valetta then perform escort duty, what will we be doing?"

To Barbary's Far Shore

Barron leaned back in his chair, and tilted his head towards Eaton. "I believe the Ambassador can explain it far better than I can."

Rodgers gave Eaton a sour look. "Heaven knows he's had enough practice," the Captain muttered.

If Eaton heard, he gave no sign. Instead, Eaton stood, and paced a few steps, before turning to Hull with his arms folded and asking, "Leftenant, how would you like to not only help liberate your imprisoned comrades from the *Philadelphia*, but an entire nation as well?"

Eaton's reputation hadn't quite made it to Isaac Hull just yet, so all Hull could do was look around the room, then back to Barron and Eaton and reply, "Mister Ambassador, I am not at all sure I understand what you have in mind."

Eaton grinned from ear to ear. "Let me explain. The written orders you were just given are intended to deceive any prying eyes. Your actual orders are as follows." Striding to the huge map of the Mediterranean and North Africa that was posted on the forward bulkhead of the day cabin, Eaton pointed to their present position.

"Tomorrow morning, I shall transfer to your fine vessel, and we shall make for Alexandria. Once there, I shall proceed up the Nile to locate Hamet Karamanli, the true Bashaw of Tripoli. Once we have found him, I am going to bring him back and help him raise an army to overthrow his brother, the present Bashaw and an usurper. Once he has ejected the usurper, he shall liberate the crew of the *Philadelphia*, assuming that has not already happened."

"Give him this," Rodgers muttered to no one in particular while taking a sip of brandy. "The explanation gets more coherent with every retelling."

Hull stood and walked to the map, examining it closely. "Mister Ambassador, just how far up the Nile is Good King Hamet?"

Eaton said, "Ah, he is . . . " then paused, looking at the map, trying to find something. After a few moments of searching the sinuous, pale-blue line that marked the Nile, Eaton stepped back and cleared his throat. "Well, Leftenant, he is very well hidden; after all, his brother has made several attempts on his life. But I am sure we will be able to find someone in Alexandria who can assist us."

Hull looked skeptically at Eaton, but, before he could say anything, Rodgers spoke up, his eyes wide in angry surprise. "Are you trying to tell us you do not know where Hamet is?"

Eaton looked at Rodgers with a hurt expression. "That isn't at all true, Captain. We know he is in Egypt. We know he is on the Nile, we are just not sure exactly where."

Under Orders

"Dear Lord," Rodgers said, shaking his head in near shock. "We are risking a ship and crew on –"

Preble raised a hand, and Rodgers was so surprised by that he fell silent. "Captain Rodgers, leave that be for the time being. The Ambassador does have a point; there are no secrets in Alexandria. With a little discreet questioning and some baksheesh you should be able to find someone who knows and can be trusted." Turning to Eaton, Preble gave a polite smile and said, "Pray continue, sir."

"Thank you, Commodore." Eaton gave Rodgers a genuinely sincere smile, having absolutely no idea what he was dealing with and blissfully ignoring a look on the Captain's face that implied brimstone was afire somewhere in his soul. "As I was saying," Eaton went on, "we will find a guide who can assist us once we arrive in Alexandria. We will also need one or two officers in the delegation and a few marines for security."

Hull thought about this for a second before speaking. "Commodore, I feel I should mention that we are somewhat undermanned aboard *Argus* right now. I only have three Leftenants to begin with – Mister Allen, Mister Cable, my surgeon, and my navigator . . ."

"Any midshipmen?" Rodgers asked abruptly.

"Yes, sir, we have three . . ."

"Then send 'em!" Rodgers snarled. "They're there to learn and serve!"

Pointing out that naval cadets may not be the best possible choice for a delicate diplomatic mission seemed to be a bad idea at that moment, so Hull merely gave a respectful nod. "Yes, Captain. We also, however . . ."

Rodgers' brow furrowed and his eyes narrowed. "You also have what?"

"We are short of marines, Captain. In particular, we do not have a detachment officer right now. . . "

"Ye gods!" Rodgers exploded, bursting upwards out of his chair. "What kind of damned playhouse have you been running over there, Hull?"

That was a bit much for Hull, who'd been operating shorthanded from the minute he'd weighed anchor in Boston earlier that summer. He stood to face Rodgers, and drew himself up to every inch of his considerable height. Hull was a stout man as well; now he seemed to fill the day cabin and tower over Rodgers. "Captain Rodgers," Hull said firmly, but politely, "I have labored under the most difficult of handicaps since my vessel was commissioned and have done everything in my power . . ."

To Barbary's Far Shore

"The hell you have!" Rodgers shot back. "It sounds like all you've been doing is playing at sailor over there..."

"Captain Rodgers, I must protest!" Hull's booming voice all but rattled the windows in the stern gallery. "Commodore Preble can show you my reports and requests for manpower! I have tried as hard..."

"But not hard enough, Mister Hull! Maybe you need me to come over there for a few days and show you how a man-of-war should be run, ye..."

"Gentlemen!" Barron snapped. Both Hull and Rodgers fell silent while continuing to eye each other warily. "Captain Rodgers, I have no doubt Leftenant Hull has done everything humanly possible to keep his vessel in fighting trim."

"But..."

"Captain Rodgers." Barron was frowning, and that was usually a sign that he was becoming irate. Rodgers decided that it wasn't worth pushing any further, for now, and sat back down. "Leftenant Hull," Barron continued, "Exactly how shorthanded are you?"

"Six hands, one leftenant, a marine officer and five enlisted marines."

Barron considered this for a moment. "Right, then. Get with Mister Carmick and see if he can find some marines. Tell him he has my authority to comb the Squadron and make any necessary transfers. I realize he will probably not be able to do anything this evening, but he should be able to send you some marines in a few days."

"Yes, sir."

Barron looked at Hull with the utmost seriousness. "Isaac," he said quietly, "There is a great deal riding on this assignment; more, perhaps, than you believe. This is not a matter of simply transporting a passenger to Alexandria so he may speak with someone. This is a matter that, if it succeeds, will make this entire region safe for the trade of all nations... and bring our friends home."

Hull stood at attention again, this time in pride, not in anger. "Commodore, you may rest assured that the *Argus* and I will complete our assignment in keeping with the finest traditions of the United States Navy."

Rodgers looked like he was about to say something fairly rude, but stayed silent while Barron beamed and Eaton vigorously shook Hull's hand.

Barron rose, indicating the evening was at an end. "Leftenant Hull, I look forward to hearing your report when this is over. Please keep me informed of Ambassador Eaton's progress."

Under Orders

"I shall, sir. In the meantime, I had best return to *Argus*. Ambassador Eaton, I shall come alongside *President* at first light to transfer your baggage, if that is acceptable."

"Perfectly so, sir; perfectly so. I cannot wait for us to get underway, we have so much to discuss!"

At that, Preble, Rodgers and Barron all gave pained little smiles, but said nothing. Hull caught them, but wasn't quite sure what to make of them, so he merely said his goodnights, as did Lieutenant Allen.

As they stepped out onto the deck, Hull said to Allen, "Will, I'm going to go see that marine captain and let him know what he's in for . . . " At that moment, Edward Preble stepped out on deck, and Hull said, "Will, strike that. Would you see Captain Carmick, please? I'd like to speak to the Commodore for a moment."

Allen nodded. "Gladly, sir. Meet you back at the boat." Allen stepped off forward for officers' country as Hull walked back to Preble. Even in the dim light of a September moon, Preble was pale and wan as he leaned against the rail, simply looking out at the calm, softly billowing waves.

"Commodore?" Hull asked quietly.

Preble turned, coughed once, and smiled as he saw Hull. "Isaac, lad, ready for your adventure?"

Hull grinned. "I'd prefer adventure at sea instead of serving as a glorified packet boat, but orders are orders. I just wanted to stop by for a moment and wish you a good voyage home."

Preble nodded. "Thank you, Isaac. You've done well, very well out here, considering the handicaps under which you had to labor."

Hull grimaced. "I keep telling myself that even a commodore has to start somewhere, but I had been hoping my first command would be a bit smoother than this."

Preble laughed at that. Hull had gone through more than the normal share of problems on *Argus*' first cruise. When added to that the usual first-command jitters that any new captain went through, it had been a challenging experience. But, on the other hand, Hull had handled it all quite skillfully; better in fact, Preble thought, than some of his more experienced captains would have. Barron knew that, and that's probably why they gave Hull the job of taking Eaton to Alexandria. Probably just as well, though, that Hull was broken in under Preble than Rodgers.

Preble clapped Hull gently on the shoulder. "Isaac, there's only one way to avoid that sort of thing, and that's to never leave the dock. Heaven knows you

To Barbary's Far Shore

don't want that. You've been given a most important assignment, due in large part to the skills and intelligence you've shown as master of *Argus*. You must consider it a gesture of confidence on the part of Commodore Barron."

Hull nodded and smiled in agreement. "I do indeed, sir. And I won't let the Commodore down, rest assured. In any event, sir, I must be back to *Argus*. Please take care, and I hope to see you upon our return." With that, Hull stood at attention and gave Preble a salute so sharp it fairly whistled.

Preble returned it; slower and more deliberately, but with respect and pride just the same. "Fair seas, Isaac."

"And to you, sir." With that, Hull headed for the boat.

While Hull and Preble were parting company topside, Josh Blake was ducking through the confined rabbit warren forward that made up officers' country aboard *Constitution*. Most of the doors were closed, and a comforting blend of warm light, conversations, and pipe smoke glowing softly from beneath each door. One was open, and a tall, red-haired man sat patiently writing in a ledger. Blake stepped forward and knocked on the doorframe and the man looked up, squinting slightly to make out Blake's face in the dimly lit passageway.

"Yes?"

"Evening. Can you tell me where I can find Mister Carmick of the marines?"

The man stood up and extended his hand, minding so that he didn't bang his head into the low ceiling. "You found him, Leftenant . . . ?"

"Joshua Blake, *Argus*."

The two shook hands as Carmick asked, "What can I do for you, Leftenant? None of my men have insulted you, have they? I'll have them thrashed senseless if . . . "

"No, no," Blake said, wondering to himself just how many times that sort of thing happened. "Actually, Commodore Barron and Captain Rodgers told us, myself and Captain Hull, to see you about borrowing a few marines."

Carmick smiled, rubbing his forehead. "They did now, did they? Well, perhaps you'd better come in so we can discuss this." Carmick's tone seemed to suggest this wouldn't be all that easy, so it was with wariness that Blake stepped into the tiny cabin. Carmick motioned to the chair he had been sitting in while he himself sat down on the edge of his bunk. As Blake sat down, Carmick asked, "Now, old Sam and Johnny Rodgers sent you down here to 'borrow' some of my marines?"

Under Orders

Blake nodded. "We've just been tasked with . . . " Allen paused for a second, debating whether or not to tell Carmick what was actually going on, and decided to err on the side of caution. "Let us just say a most sensitive mission, that requires us to have our full complement of marines."

Carmick grinned. "Consul Eaton's cunning plan?"

Blake was taken a little aback. "How did you know?"

Carmick shook his head. "Leftenant, Consul Eaton's delusions of warrior godhood have been the talk of this squadron for days, and far longer than that among those of us who have been here long enough to remember his tour in Tunisia. Dear Lord, I am still amazed that his head wasn't sent back on a pike. He actually got that fool idea of his approved?"

Blake shrugged slightly. "Well, the plan he did get is simply us taking him to Alexandria, finding Hamet Karamanli, helping him raise an army, and then pointing him at Tripoli."

At that, Carmick laughed. Blake looked at him in mystification. "Not quite sure I see the humor in that, Captain."

"You'll have to pardon me, Leftenant. It's just that you must believe me when I tell you that Consul Eaton is most unlikely to let it stop there. Mark my words; he won't rest until he himself is sitting on Yusuf's throne, garbed in the finest silks and insisting that we call him Pasha Will the First while the crew of the *Philadelphia* strew rose petals about him and chant the Te Deum."

All Blake could do was look at him dumbfounded for a moment, before speaking in awe. "You're serious, aren't you?"

Carmick raised his right hand as if taking an oath. "Keep an eye on him, Leftenant. The word 'no' is not in his lexicon. However, you came here to find more marines for your grand affair of nations. Wish I could help you, I don't have any."

Blake blinked once at the abruptness of Carmick's answer, then said, "Captain, perhaps I didn't fully explain myself. Commodore Barron . . . "

"And that monster in human form he calls a flag captain authorized you to get whatever you need. Now I know *Argus* needs a detachment commander, I know you need some marines as well. Unfortunately, I'm short myself a leftenant and half a dozen men. There is simply no possible way that I am going to tell Captain Rodgers that his marine detachment is even more undermanned because I gave them to you. 'tis nothing personal, Leftenant, truly, but this is one request I have to deny. I know what life is going to be like around here with Captain Rodgers at the helm, and believe me, whether it's for raiding parties or honor guards or just to rock him to sleep with a lullaby each night, I'm going to

To Barbary's Far Shore

need every leatherneck I can get my hands on. As far as the other vessels, name one and I'll tell you how badly under they are. The only ship here right now with a full detachment is *President*, and I intend to raid her over the next few weeks to try and even things out."

Blake was about to say something, but then thought better of it; Carmick's sincerity was evident. In quiet frustration, Blake held out his hands in a pleading motion and asked. "Well, can you tell me exactly what it is I'm supposed to do? Consul Eaton has this grand voyage up the Nile planned to find Hamet Bey, and we're going to be expected to provide a marine escort. We can't go meet a king with a sergeant in charge of the honor guard."

Carmick reflected on that for a moment. "Leftenant Blake, I'll grant you that you have a point there. The only thing that would annoy Rodgers worse than inconvenience is embarrassment. I could just go ahead and get Sergeant Bancroft a commission. He's been in long enough; certainly has earned one."

Blake's response was to look at Carmick as if he'd lost his mind. "I hardly think Bancroft is officer material, Captain."

Carmick looked down at Blake and replied quietly, "Mister Blake, a few of the best officers I've ever known in the Marines were once enlisted men. They tend to have an exceptional grasp of what is important and what isn't, unlike some of the commissioned idiots I've had to work with. I'll thank you, sir, not to denigrate the abilities of my men."

Blake was momentarily taken aback by Carmick's quiet defense of his men. In fairness, Blake was the product of a system that believed that the average enlisted man was a sly, lazy, and worthless soul who, without the threat of the lash or the rope, would do no more than he absolutely had to, and certainly didn't have the social graces necessary to be an officer and a gentleman. Coming up from the deck was pretty much unheard of in those days for a sailor, and although it wasn't completely unknown for marines of that era, it was rare to say the least.

Blake blushed slightly before replying, "I beg your pardon, Captain. I mean no disrespect to your men."

Carmick looked dubious at this, but nodded in acceptance. "Fair enough, Mister Blake. About the only thing I can do for you is try and get you an officer, and I can't guarantee when or if, just that I'll try. I've got two new ones coming aboard from *President* tomorrow and I'm going to have to get them sorted out, but I promise I'll do my best. Fair enough?"

Blake nodded in mute surrender. Captain Hull was going to have to handle this one himself.

Under Orders

A few minutes later, Will Eaton stood on *Constitution's* deck, watching Hull and Blake return to their ship. The tiny cutter stood out against the dark sea, but, as it moved away, it became more and more indistinct, until finally the only sign that it was there was the cheery yellow glow of its lantern, bobbing rhythmically.

"Why, if it isn't Consul Eaton." Eaton turned to see John Rodgers stroll up beside him, just a little unsteadily; *the brandy*, Eaton thought, *had been exceptional*. Rodgers leaned against the rail, swaying a bit as he did so.

"Good evening, Captain. Looking forward to getting underway tomorrow?"

Rodgers gave a twisted grin, then spat overboard. "Looking forward to getting this mob a taste of genuine discipline. Discipline, Consul Eaton! Preble let his captains go to seed while he was here, but there's a new . . . burp . . . new hand holding the whip." Rodgers leaned in close, and Eaton stiffened as he felt Rodgers poke him in the chest with a finger. "And it is the whip they shall feel, sir!"

Eaton tried to smile. He was certainly no slacker when it came to making sure men toed the line, but from what he'd seen so far, at least *Constitution's* crew was fairly ship-shape and certainly didn't seem deserving of what Rodgers seemed to be promising. *At least, not yet.* "I have no doubt, Captain, that you will have the most capable crew in the Squadron."

"That they shall be," Rodgers grinned almost sadistically, "or they'll be the saddest!"

Rodgers peered out at *Argus'* whaleboat, now almost back to its home. "And there go Messers Hull and Blake, Esquires, prepared to take you up the Nile so that you may commune with the shades of the Pharaohs. Ready to join them on that miserable little tub of theirs?"

Eaton nodded, grateful for a chance to change the subject. "Oh, very much so. I must admit, I had hoped for a somewhat more . . . impressive ship. Perhaps *President,* or *Constitution* herself."

Rodgers gave a derisive snort of laughter. "Your Excellency, Consul Eaton . . . why exactly do you think *Argus* was assigned to this mission?"

Eaton thought for a moment. "I would assume it was because Captain Hull was considered highly skilled enough to accomplish it."

Rodgers shook his head, still laughing. "In my book, the only one that counts, by the by, Isaac Hull doesn't have the experience to navigate a toy boat in a millpond. He got the job because he's the junior captain out here; everybody else turned the job down."

To Barbary's Far Shore

Rodgers straightened up, and tugged on the hem of his coat, and said with mock cheer, "And with that, Your Most Serene Excellency Mister Consul Eaton, I bid you a fond adieu." Rodgers headed for the quarterdeck once more, weaving slightly and taking an opportunity to kick at a sailor who he thought was loafing. *Rodgers must be drunk,* Eaton thought sourly, *he missed the sailor's behind by a mile.*

The next day, the Squadron went its separate ways. Preble boarded the ship that took him home and sailed quietly away; no ceremony, no fanfare. Sam Barron, with a mildly hung over John Rodgers lurking at his side, was now in charge, and that was that. Isaac Hull pulled *Argus* alongside *President* to load Will Eaton's baggage. As a rule, it is generally assumed that civilians will have twice as much baggage as any given military member. Hull was prepared for that, but not for the trunks, cases, and other assorted impedimenta that *President*'s deck crew gratefully cleared out of the hold. Will Eaton was a diplomat representing the United States of America, and by God he was going to look the part. For his part, Joshua Blake, supervising the transfer, could only look heavenward for guidance, and then try to make a little more space in *Argus'* already crowded holds.

In the meantime, things were not going well with the Squadron itself. Sam Barron's health had quickly worsened, to the point where he was bedridden and unable to function as commodore. He was taken to Syracuse, the nearest friendly harbor with competent medical help, and sent ashore. He would stay there for nearly seven months. To the utter dismay of all concerned, Barron designated Captain John Rodgers as acting commodore. For his part, Rodgers was ecstatic about his sudden good fortune, until Barron laid down a few provisos. First, the winter weather was starting to set in, and there was little chance of, and great risk in, establishing a close blockade of Tripoli again. Accordingly, Rodgers was ordered to leave the line with *Constitution* in late October and start rounding up replacements for his crew, which was now reaching a dangerously thin number. Rodgers would end up in Lisbon, where he made enemies of the US Consul and more than a few unfortunate Portuguese; proving that, if nothing else, Rodgers' moods were consistent with ally and fellow citizen alike. In fact, Rodgers managed to make himself so thoroughly unpopular that it took him four months to get his replacements, not departing Lisbon until February 9th, 1805. One may safely assume that his absence was not mourned.

Argus headed for Syracuse with the rest of the Squadron, but Hull intended to replenish there and then head for Malta-but-really-Alexandria. It took a while to get there, running against the northerly winds. By that time, Hull was going to need more than just a simple replenishment; some simple but time-consuming repairs were in order. One week became two and two became three, and by this point it was Friday, October 26th. Isaac Hull was working frantically to get *Argus* underway, as much to begin his mission as to get out from under the

Under Orders

gimlet eye of Acting Commodore Rodgers. Over the last few days, the final details had been put into place. Now it was just a matter of getting the victuals aboard and getting underway. Unfortunately, that required one last briefing to Rodgers, and that meant talking to him face to face. So, back into the *Argus'* cutter Hull went, across the calm waters of Syracuse Harbor with the necessary paperwork. As he'd expected, *Constitution* was as shipshape as always; the deck holystoned and swabbed to perfection, though Hull was fully aware that perfection wasn't enough for Acting Commodore Rodgers.

The bosun's mate piped him aboard with absolute precision. Saluting the quarterdeck, Hull said, "Leftenant Hull to see the Acting Commodore."

"He's in his cabin, sir. Shall I announce you?"

Hull smiled. "No, no sense in feeding any more Christians to the lions than absolutely necessary."

Striding aft to the nearest hatch, Hull stepped down the short ladder to the gun deck and headed aft again towards the Captain's quarters. Now, on a frigate like *Constitution*, the gun deck separated the two sets of officers' quarters – the Captain's quarters all the way aft, and the rest of the officers, both Navy and Marine, billeted forward. But before one got to the Captain's quarters, one found the marine billeting area. The idea was that, in the event of a mutiny or other disturbance, the marines would be between the miscreants and the captain.

Before Hull got to the bottom of the ladder, he heard the sounds of someone being, well, counseled in the marine spaces. Perhaps counseled is not the correct word, because the tone being used by the counselor was far from gentle.

It is always an awkward moment when one stumbles across someone being disciplined in public. It is sometimes a necessary thing, and when it does happen, one hates to walk through someone's public embarrassment, but military business trumps all else. When that happens, one must simply look straight ahead and march through as quickly and discreetly as possible. That is what Hull did, locking his eyes on the door to the captain's quarters.

However, it was impossible to ignore what was being said. Dan Carmick was in some poor soul's face, a lieutenant from the looks of it, and truly giving him what for. The lieutenant, for his part, was a tall, slim young man who looked to be in his late twenties. For all the effort Carmick was putting into things, the lieutenant didn't seem all that upset, other than the slightly wounded expression on his face; as if he considered himself the victim of some simple yet horrible misunderstanding.

Carmick's voice was a low, harsh snarl as he tore into the lad. " . . . you have succeeded in being on this ship for three, exactly three, days, sir; and, in

To Barbary's Far Shore

that time, you have succeeded in showing yourself incapable of accomplishing the simplest administrative tasks, at best untrained at drill, and unable to meet the most basic standards of this service! What possible explanation can you have?"

The lieutenant started to say, "Sir, I . . . "

"Quiet!" Carmick snapped. "Whatever you have to say is irrelevant! I do not know where you hail from . . . "

"Fauquier County, Virginia, sir!"

Carmick's faced darkened even further at the interruption. "And you traveled all the way from there just to make me angry?"

"Sir, no . . . "

"QUIET!"

At that moment, Hull brushed past as best he could and thought he'd navigated the problem quite well, until he heard Carmick ask, "Leftenant Hull?"

Drat, Hull thought. He turned to face Carmick with as pleasant a smile as possible. Carmick was holding a silencing finger in the lieutenant's face as he awaited a reply. Hull cleared his throat and said, "Good day, Captain. My apologies for the interruption, but I have most urgent business with Acting Commodore Rodgers . . . "

"Oh, quite all right," Carmick said with a thoroughly unpleasant grin. "Just don't let him hear you say 'Acting Commodore;' he'll have you hanging from the bowsprit. Before you go in there, however, may I have a moment of your time?" Hull looked suspiciously at the young lieutenant, still locked in a position of iron attention, but probably glad for the pause. Carmick noticed it and said, "Fear not, Leftenant. He'll keep."

Hull smiled nervously. "Certainly, Captain. What may I do for you. . . ?"

Carmick looked back at the lieutenant for a moment with predator's eyes, then back to Hull. "If I remember correctly, your vessel needs a commander for her marine detachment, is that not so?"

Hull nodded. "Quite so, and we will probably be sailing tomorrow morning . . . "

Hull paused, realizing what was about to happen. He looked at the lieutenant, whose expression was now not so much hurt as seriously concerned. Carmick grinned slowly; his eyes pivoting back to lock with those of the lieutenant.

155

Under Orders

"Leftenant Hull, I think I have just the man for you. Meet Leftenant Presley Neville O'Bannon, United States Marines."

"... whose best qualities were brawling, wenching, and fiddle-playing." Lieutenant Presley Neville O'Bannon, United States Marines. (Photo Courtesy United States Naval Historical Center)

O'Bannon, for his part, gave a confident grin. He extended his hand to Hull. "A pleasure, I'm sure . . . "

Carmick roared. "Who the bloody hell told you to say anything? Stand to attention, you idiot!"

O'Bannon snapped back to attention; the look of hurt feelings passed across his face once more. Carmick turned to Hull, his face contorted in anger.

"Mister Leftenant O'Bannon has come to us from *President*, and Washington before that. According to his efficiency reports, his primary talents are, and I quote, 'brawling, wenching, and . . . '" Carmick paused for a moment, composing himself to say what must be said. ". . . 'fiddle playing.'"

Hull looked mildly taken aback at that. After all, brawling and wenching were traditional pastimes in the Navy and Marines. But the fiddle . . . There was nothing intrinsically wrong with fiddle playing, you understand, it is just that it was something officers and gentlemen did not do.

For his part, O'Bannon merely said, "Begging the Captain's pardon, but my parents always felt that the mark of a well-rounded gentleman was the ability to play a musical instrument well . . . "

With that, Carmick gave a strangled bellow and grabbed O'Bannon by the lapels, shaking him like a leaf in a thunderstorm. Risking his own safety, Hull reached in to get them separated.

"Mister Carmick, please! This man is certainly not worth the trouble you would find yourself in were you to assault him. Remember the regulations!"

At that, Carmick paused. Although it wasn't unknown for one officer to discipline another in a physical manner, if O'Bannon wanted to make trouble for

To Barbary's Far Shore

Carmick, he certainly could. Breathing heavily, his hands balled into fists, Carmick growled, "You are right, Leftenant Hull . . . he isn't worth it."

Looking straight at O'Bannon with murder in his eyes, Carmick said, "You have exactly ten minutes to get your gear collected and be ready to transfer over to *Argus*. If you aren't, I swear unto God Almighty I'll throw you over there!" With that, Carmick stomped off forward, muttering to himself the whole way, with the occasional violent outburst as he went.

Hull could only look in amazement as Carmick disappeared into the maze of ropes, overhead and cannon that made up *Constitution's* gun deck. Almost to himself, he asked in wonder, "What could you have possibly done to Dan Carmick to get him that angry?"

"I'm afraid I'm quite at a loss to explain it." Hull turned to see O'Bannon standing there, arms folded with an utterly puzzled look on his face, shaking his head sadly. Seeing Hull looking at him, O'Bannon smiled and shrugged. "In any event, he did seem rather definite about my being off the ship in short order."

"Well, that's . . . "

"Quite all right, sir. It'll only take me a few minutes to pack. See you on deck." With that, O'Bannon waved cheerily and went forward. Hull still wasn't quite sure what had happened. He knew he had a new marine detachment commander, but he wasn't at all sure what kind. Shaking his head, Hull strode the last few steps to the Captain's quarters and, gathering his courage, rapped sharply once on the door.

The muffled voice from inside snarled. "Come!"

Opening the door, Hull stepped in, came to attention before Rodgers and saluted briskly. "Leftenant Commandant Hull to see the Commodore."

Rodgers continued to scribble away at something for a few seconds before looking up and giving a dismissive wave that Hull took to be a salute. Turning back to his paperwork, Rodgers merely growled. "What?"

Hull placed his reports on Rodgers' desk. "We've finished replenishing, and the Sailing Master is securing the rigging now."

"About damned time. When will you be underway?"

"At first light, Commodore."

Rodgers picked up the papers Hull had brought him and quickly leafed through them. "Looks to be in order for a change. I gather you finally have a new marine commander."

Under Orders

Hull was nonplussed for a moment, then he realized that Rodgers could hardly have failed to hear the entire episode outside his quarters. "Yes, sir," Hull replied, "a Leftenant O'Bannon . . . "

"The man's a complete ass," Rodgers said as he began signing the paperwork. "Amazes me he's survived long enough to make it this far."

"Yes, sir."

"Amazes me even more that Carmick didn't simply shoot him and throw him over the side. I would have."

"Yes, sir."

Rodgers handed the paperwork back to Hull. "Do not let me see that tub of yours in this harbor past first light. And if O'Bannon does anything, anything, to jeopardize your mission, put him in irons and bring him back to me personally. That's an order."

"Yes, sir."

"Get out."

"Yes, sir." Hull saluted, about-faced, and marched out of the Captain's quarters. Closing the door behind him, he thought to himself, *well, that wasn't terribly painful. Now back to* Argus.

A quick trot topside to the main deck brought him out under an overcast Mediterranean sky once more. The first thing that caught his eye was a small pile of baggage that was being handed down the boat ladder, where Presley O'Bannon was standing, securely guarded by two marines, both at port arms. Hull warily walked up to the marines, who both snapped to attention with a rattle of gear and boots.

"Is there some sort of problem, gentlemen?"

One of the marines asked firmly, "Sar, would ye be Leftenant Hull?"

Hull nodded. "Yes, I be . . . that is, I am."

The marine nodded. "Cap'n Carmick's compliments and orders, Sar. We're to make sure Leftenant O'Bannon gets on th' wee boat wi' ye."

Hull smiled, thinking to himself that perhaps Carmick was laying it on a bit thick. With a friendly lilt in his voice, Hull asked, "And if the Leftenant refuses to go?"

The marine never missed a beat. "We're to shoot him where he stands, Sar. Cap'n Carmick also said that once we sure he was dead, we were to personally ferry his, an' I'm quotin' directly here, Sar, 'disgraceful carcase' to yer ship."

To Barbary's Far Shore

Hull blinked for a moment, trying to think of the last time one of his subordinates had been ordered to his ship on pain of death. Unable to recall any other precedent, Hull said, "Well, in that case, I see no reason to waste ammunition. Leftenant O'Bannon?"

Hull gestured politely. O'Bannon gave an equally polite bow, then stepped over the side, easily maneuvering down the rope ladder to *Argus'* boat. Hull followed him down a few seconds later. With a quick, smooth series of commands, the bosun's mate in charge had the boat sliding easily over the waves back for *Argus*. Upon their arrival, word spread rapidly throughout the ship that there was a new Marine officer aboard. On a ship as small as *Argus*, rumor moved faster than a racehorse, and it took no more than a few minutes for the marines themselves to get the word.

Marine quarters on any warship tend to be comfortable or not more as a function of the ship's size than anything else. That meant in *Argus* it was a cramped space just ahead of the captain's quarters, where twenty men lived, worked, and slept. The overhead, what us poor sods ashore call the 'ceiling,' was probably no more than about six feet above the gun deck. Claustrophobic for us, merely an annoyance for the average sailor or marine of the early nineteenth century, who was usually right at about five feet high. But sitting down, on a gun carriage, a crate, or a low-slung hammock, it was more than tolerable, and that is where Sergeants Edmund Bradshaw and William Williams were, carefully cleaning their muzzle-loading flintlocks as if their lives depended on it. Which, of course, they did.

We know little about Bradshaw and Williams. We do know it was unusual for a ship as small as *Argus* to have two sergeants aboard, but that was probably a result of being without a detachment commander for so long. The one roster I was able to locate does not even give their ages, simply their status one day in August, 1805, almost a year later. It is reasonable to assume however that the were still probably fairly young men in their mid-forties. They would certainly have been exceptionally experienced and competent in the many duties they would be required to accomplish, because one thing has not changed to this day, the fact that the United States Marines will not give anyone the title of sergeant unless they are very, very good at their job.

The two didn't even look up to see Corporal Arthur Campbell come tromping down the ladder from the main deck. Campbell, a little banty rooster of a man, was almost breathless as he charged down past the hulking black shapes of *Argus*' cannon.

Bradshaw never even took his eyes off his rifle as he asked, "And what would th' problem be now, Arthur?"

Under Orders

Campbell came to attention as best he could. "We've got ourselves a new leftenant!"

Bancroft and Williams simply continued working on the rifles, acknowledging Campbell's bulletin with only a quick glance at one another. Williams held up the barrel of his rifle, sighting along its length to make sure it was true. "I know, Arthur," he said. "Be a good lad and get the men together down here at once."

Campbell had no idea how Williams knew what was going on, but one does not question the prescience of one's sergeants lest you find out what else they know about you. "Aye, aye, Sar!" Campbell barked, then headed back up the ladder, bellowing his fool head off.

Bancroft started reassembling his piece and, without looking up, asked quietly, "How did ye know that we had a new leftenant?"

"I didn't," Williams answered equitably. "But doing things like that drives Campbell utterly mad."

"Mmm," Bancroft replied noncommittally as he stood up and put his jacket back on, straightening it to hang with precision off his lanky frame. By this time, the gun deck echoed with the pounding boots of twenty marines, lining up into a precise formation, tallest in front, shortest in the rear, each one dressing the formation until it was perfect.

Campbell was in his place at one corner as he barked, "SAR – THE MARINE DETACHMENT IS ALL PRESENT AND ACCOUNTED FOR!"

Bancroft and Williams looked the ranks over skeptically. They had made it into place within acceptable time, but neither sergeant was going to let them know it.

"Good afternoon, gentlemen," Bancroft roared, "about bloody time ye joined us! Now listen up. According to Corporal Campbell, we have a new Leftenant aboard, and he's . . . " Turning to Campbell, Bancroft asked, "Where exactly is he, Arthur?"

"On th' quarterdeck with Cap'n Hull, Sar!"

"Well, there ye have it. He's on th' quarterdeck with th' Cap'n, and we can safely assume he'll be down momentarily here to see what kind o' sorry examples of manhood he's got fer marines! So, I most humbly beg yer indulgence. Do NOT make a fool of yerselves or ME in the next few minutes, am I understood?"

"SIR, YES SIR!," came the thundering reply.

To Barbary's Far Shore

Turning to Williams, Bancroft said, "Willie, please be kind enough to take a post by th' hatch and let me know when His Nibs is headed this way."

"Certainly, Edmund." Williams strode over to the hatch and was rewarded almost immediately. "On his way," Williams hissed. "Stand to."

Bancroft merely looked at his men with an expression that promised a slow, lingering demise to anyone who thought about making a mistake in the next few minutes.

Hull, joined by Will Allen and O'Bannon, came down the ladder. They were rewarded with a bellowed, "TEN-HUTT!!" from Williams.

Bancroft saluted so sharply that it fairly whistled, and turned to face O'Bannon. "Good afternoon, Sar! The Marine Detachment, United States Brig *Argus,* is at yer disposal!"

Hull, like most naval officers, was of mixed feelings about his marines, but even he had to admit at this moment that they looked, well, military. "Gentlemen," Hull announced, "I have the pleasure of introducing First Leftenant Presley O'Bannon, your new detachment commander. I have no doubt that you shall give him the same loyalty and devotion that you have given me. Leftenant O' Bannon, these are your noncommissioned officers: Sergeant Bancroft . . . "

O'Bannon extended his hand. "A pleasure, Sergeant."

Bancroft, for his part, was momentarily unsure what to do. In his experience, lieutenants were not generally likely to shake your hand and tell you how glad they were to make your acquaintance. But Bancroft, being a reasonable man, was always ready to greet a gentleman in proper fashion, so he returned the handshake; a bit warily, but firmly nevertheless.

" . . . and Sergeant Williams."

"And you, Sergeant," O'Bannon said, shaking his hand as well. "Two sergeants?"

"Without a leftenant here, it's been good to have some extra enlisted rank," Allen said. "On top of that, Sergeant Bancroft is getting ready to retire soon, aren't you?"

Bancroft gave something approximating a smile and a brief nod. "That I am, Mister Allen, on November tenth. Headin' back to Baltimore, I'll be. I've bought a tavern there, near Fort McHenry. I'm figurin' I can't go too far wrong opening up a public house next to a fort."

Under Orders

The three officers laughed, and O'Bannon said, "Well, Sergeant, you'd best reserve a table for this lot some evening. I know they'll be wanting to pay you a visit."

Bancroft's visage went stern again. "Beggin' the Leftenant's pardon, but I'll not be letting this mob into my property. They'll do too much damage and I have two daughters and a wife to protect."

O'Bannon grinned. "It's always a pleasure to meet a man who has his priorities in order. Sergeant, I have to meet with Leftenant Hull right now, but after the evening formation I'd like to sit down with you and Sergeant Williams to discuss the men and anything else that needs to be touched on. If that's acceptable to you, of course?"

Now this was unusual. More than a few officers in Bancroft's experience spent their first, and, for that matter, the rest, of their evenings aboard in a drunken stupor, and most of those who didn't were far from friendly about it. *But a detachment commander who not only wanted to hear about what was going on but asked to meet with you about it? Unheard of; absurd. But,* Bancroft thought, *maybe they were making lieutenants differently now than when he'd come in. Major Nicholas, now there was a marine. Oh well.*

In any event, Bancroft really did have his guard up now. H tried not to show it as he nodded politely and replied. "Sarn't Williams and I are at the Leftenant's disposal; after the evening meal it 'tis, then."

"Looking forward to it. Carry on."

Williams bellowed "TEN-HUTT!" as Bancroft saluted O'Bannon, who returned it and followed Hull to his cabin.

As they watched the two officers, Williams spoke, partly in awe and partly in disbelief. "Now did you ever see the like? 'if it's acceptable to you,' he says."

Bancroft still wasn't entirely sure what to make of Leftenant O'Bannon. Pondering the situation for a moment, Bancroft finally shook his head doubtfully. "Willie, all I can tell ye is that they did not have officers like him in the old Marines. I suppose I'd better come up with some kind of report for him. Get a couple of the lads to bring dinner back and we'll let 'em off easy for the rest of the day." 'Easy' in Sergeant Bancroft's lexicon meant that he would not be personally overseeing the cleaning and weapons details after dinner. He'd be busy trying to make some sense out of Lieutenant O'Bannon.

Hull had a small folder in his hand, neatly labeled O'Bannon, Presley Neville; the good Lieutenant's personnel record. Tossing it on the desk as they entered, Hull turned at the scrape of a chair to see O'Bannon already seated, casually ensconced before his desk. Sitting down in the Captain's quarters before being so bidden was a fairly solid breach of etiquette, but Hull kept his tongue. Sitting down himself, he smiled tightly at O'Bannon. "Please, make yourself comfortable."

O'Bannon merely nodded in acknowledgement, a seemingly oblivious grin on his face.

All right, Hull thought to himself, opening O'Bannon's folder and scanning its contents. After a moment, he looked up at O'Bannon, confusion on his face.

"Something wrong, Captain?"

Hull thought for a moment, then put the reports down. "Leftenant, Mister Carmick was right. Your fitness reports actually say that your best talents are brawling, wenching and playing the fiddle." Leaning back in his chair with his arms crossed, Hull asked, "Please pardon my prying, Leftenant . . . "

O'Bannon smiled like a choirboy. "Please, pry away."

"How in God's name did you ever make it to your present grade?"

Isaac Hull. Source: U.S. Naval Historical Center

O'Bannon thought about that for a heartbeat before replying, "Actually, Captain, that has been a matter of some mystery not only to me but to my superiors as well."

"You've never been decorated for heroism."

"Not at all, Captain, and, with any luck, I'll never have the opportunity to be."

Hull looked back at his record again, then turned back to O'Bannon in complete mystification. "Leftenant, you do not seem to have distinguished yourself in any way whatsoever."

O'Bannon's response was to give a mock bow and say with a self-effacing smile, "Mediocrity, thy name is O'Bannon."

Hull merely looked at O'Bannon for a moment, then spoke, very quietly but very firmly. "Mister O'Bannon, your mediocrity ends as of this moment. As of right now, you will be the finest Marine officer in the Squadron; nay, in the Fleet itself. Am I understood?"

Under Orders

O'Bannon shook his head as if trying to explain a scientific principle to a particularly dim student. "Captain. Please try to understand my position. I made a mistake by joining the Marines. I merely wanted to get away from Fauquier County, Virginia, and I felt that adventure at sea was the way to escape it. I was wrong, sir; more wrong than I have ever been about anything in my life, and that is truly saying something impressive. I have little, nay, strike that, no military talents whatsoever. My men, as a rule, decline to follow me, save to satisfy a morbid curiosity as to what I may do next. My superiors long ago gave up any slim hope of seeing me turn into a competent officer, instead having been reduced to simple prayer to God Almighty that I shall not do too much damage to my ship or my men. For my part, I have tried to accept my lot in life with equanimity and good cheer until such time as I have completed my obligation to the Marines, or I am discharged by some kind soul, whichever comes first."

Hull was now leaning forward on his desk, his jaw resting on his fist as he looked at O'Bannon in sheer amazement. "Leftenant," he finally said, "I will give you this. You shall enable me to die a happy man."

O'Bannon looked at Hull with part confusion and part dread. "And . . . why may that be, Captain?"

"Because I shall meet my maker having indeed, heard everything." Suddenly bursting upwards from his chair, Hull bellowed, "TEN-HUTT!"

O'Bannon may have had few military talents, but coming quickly to attention seemed to be one of them as he snapped upright into an iron brace. Hull walked slowly and deliberately out from behind his desk, looking O'Bannon up and down as if inspecting a green middie on his first day aboard.

"Mister O'Bannon," Hull said menacingly, "I have run across your like before, but never in a position of reasonable authority such as yours. I do not like them, and I certainly do not tolerate them. Am I very clear on that?"

O'Bannon swallowed and cleared his throat. "Sir, yes sir!" His tone had changed from jocular banter to serious and straightforward.

Hull nodded as he continued walking around O'Bannon, leaning over his shoulder. "O'Bannon, from this moment on, you shall be the finest Marine officer on this, or, for that matter, any ocean. You will, not may, not could, but will, take charge of that rabble you have out there and turn them into a trained, capable unit. This ship and I have been assigned a mission that is considered of the utmost importance, and I need a marine detachment par excellence. I will accept nothing less. Understood?"

"Sir, yes sir!"

Hull stood next to O'Bannon, paused for a moment, then hissed in the marine's ear. "If you fail me, O'Bannon, I will take the greatest pleasure in

To Barbary's Far Shore

doing something that I have sworn I would never do to any man, no matter how vicious, wicked or cruel he may be . . . and that is turn you over to Commodore Rodgers for court-martial. That, sir, is no threat; but a promise I make you upon my honor as an officer and a gentleman.

"Now, is there any part of our conversation you do not understand?"

"Sir, no sir!"

The unidentified U.S. ship in this picture is reputed to be the Argus. Source: U.S. Naval Historical Center.

Hull smiled beatifically. "Well then, I for one am glad we could have this little chat to get to know one another. Find Leftenant Allen and have him assign you to some quarters. I am afraid we don't have the kind of accommodations here in our little palace that you may be accustomed to, so you will be sharing a cabin. You are dismissed, Mister O'Bannon."

O'Bannon had the good sense to hold his salute until Hull returned it with a look of utter disdain, then he executed a perfect about-face to leave the cabin. Hull sat down wearily, amazed and angry at what he had just gone through. He had never had to speak to a subordinate officer like that before, and he hoped he never would again.

Under Orders

For his part, Presley O'Bannon strode out of Hull's cabin, his face as dark as a thundercloud. Stomping down the deck, he looked up to see Bancroft and Williams coming back down the ladder to the gun deck.

Snapping to attention, Bancroft started to say, "Beggin' the Leftenant's pardon, but when . . . "

"Quiet!," snapped O'Bannon. "I want every last marine lined up after dinner for an inspection, and God help the both of you if I see one, just one, idiot out of regulations. Understood?"

Bancroft and Williams replied as one. "Sir, yes sir!"

O'Bannon did not deign to acknowledge them, but instead kept tromping down the deck and up a ladder topside. When Bancroft was sure he was out of earshot, he turned to Williams with a self-satisfied smile.

"Well, then," Bancroft said, "that's a wee bit more like it."

To Barbary's Far Shore

SEVEN: PREPARATIONS

As Lieutenant O'Bannon got to know his men, William Eaton was ashore in Syracuse, waiting to see Commodore Barron. Or, rather, trying to get to see Commodore Barron. The young midshipman who had been detailed to look after the Commodore that day was doing his level best, but Eaton was slowly wearing him down.

"Ambassador Eaton," the middie said pleadingly, "please understand my position. The surgeon's orders are that under no circumstances is the Commodore to be disturbed. Are you sure you could not handle this with Commodore Rodgers?"

Eaton smiled benignly, radiating fatherly kindness. *And if kindness doesn't work,* he thought to himself, *I'll just bash him in the skull and go in.* "Young man," Eaton said, "you are to be commended on your devotion to your duty and to the Commodore. He can rest easy, knowing young men such as you serve under his command." With that, the middie stood a little taller. After all, aboard *Constitution*, where he was nominally assigned, compliments were few and far between.

"Rest assured, young man," Eaton continued, "that I would not be so insistent upon seeing the Commodore were it not a matter of the utmost importance. Nay, not mere importance, but crucial to the survival and freedom of three hundred eight of our countrymen!" Eaton said that last with a flourish worthy of the stage, chest thrust out, one finger pointing skyward.

The middie was about to say something, but then he thought for a moment. "The crew of the *Philadelphia*?"

"One and the same, my boy," Eaton replied in a grim, foreboding tone. Leaning in more closely, he asked the middie, "Can you keep a secret?"

"Yes, sir. Of course I can!" The middie nodded eagerly, wanting to prove his worthiness.

167

Preparations

"Good lad," Eaton said, patting him on the shoulder. "Now, what I am about to say is known only to myself and Commodore Barron, so I am taking a grave chance in sharing this with you, but you have shown me that your integrity is such . . . "

"It is, Ambassador, it is!"

" . . . that I feel it is worth the risk." Eaton looked about conspiratorially, and the middie watched intently as Eaton looked about for imagined spies, eavesdroppers, and other unsavory types. Leaning in closely, Eaton said quietly, "Lad, I have been authorized to change my mission. With Commodore Barron ill, I shall be taking charge of the expedition once we reach Alexandria."

The middie's eyes widened. "Ambassador, when were you notified of this? We have received no orders of any kind for the last several days."

Eaton nodded in agreement. "True enough, lad, true enough. They were delivered directly to Commodore Rodgers and myself. He in turn, told me to come over here and get the necessary increase in funds authorized."

The middie looked at Eaton in confusion. "But could not Commodore Rodgers authorize the expenditures?"

Oh dear, Eaton thought, *hadn't realized that*. Thinking quickly, he smiled. "Under normal circumstances, yes. But the orders specified Commodore Barron, and they came from the highest authority."

With that, Eaton tilted his head gravely towards the United States flag that stood in the corner behind Barron's desk. And if the middie hadn't gotten the message yet, Eaton repeated himself with all the dignity and solemnity of a Roman priest. "The . . . highest . . . authority." If the lad took that to mean President Jefferson himself, fine. If he took it to mean the Archbishop of Canterbury, it was equally fine, so long as he let him in to see Barron.

Now, as a rule, junior officers in US service have tended to have a great deal of leeway and authority. After all, they have been highly trained with millions of the taxpayers' dollars, their education has set their parents back an amount that seems equally outrageous, and they are products of the most individualistic society on the planet. But one thing that would have worked as well in 1805 as it can today is the invocation of Higher Authority, with the added implication that failure to acknowledge Said Authority could have Dire Consequences, for the nation as a whole and the young shavetail in particular.

That was enough for the middie. Sure and steadfast in his logic, he understood that if the Highest Authority said that they wanted Commodore Barron to sign the paperwork, then he would do so. Of course, the middie had one last twinge of conscience, which Eaton's appeal to duty quite nicely overrode. Looking back and forth, the middie bade Eaton follow him, and he

To Barbary's Far Shore

knocked softly on the Commodore's door. The heavy oaken door opened on its hinges with a brief squeal of protest, then moved soundlessly to permit a thin beam of sunlight to lance into the darkness beyond.

As quietly as possible, the two walked to Barron's bedside, where a thin stump of candle guttered between light and darkness. In the little bit of light available to them, even Eaton felt a pang of regret. Barron lay stretched out on a thin pallet, a down quilt covering his shivering form. The Commodore's color was that of old parchment that seemed as if it would crackle as Barron's labored breathing filled the room. Quietly, carefully, Eaton stepped past the middie and knelt beside the bed. *Waking this man would be a crime*, Eaton thought, *and to do it on false pretenses would be worse . . . and every second I delay will be that much longer Will Bainbridge and his men would rot in Tripoli.* Closing his eyes and breathing a silent prayer that somewhere, someone would understand, Eaton reached out to gently shake Barron's shoulder.

Commodore Samuel Barron.
Source: U.S. Naval Historical Center

"Commodore?"

Barron's head lolled slightly from side to side, and his eyelids fluttered unevenly open. His eyes seemed not to be attached to anything as they rolled back, then disappeared again as his eyelids slid closed.

Oh my, Eaton thought. He reached out to give Barron a slightly stronger shake. This time, the Commodore moaned; a low, guttural sound that for a moment convinced Eaton that he had provoked the poor man into his death rattle. But, instead of shuffling off this mortal coil, Barron turned his head with an effort that seemed to exhaust him. His eyes opened slightly, and he squinted as he tried to focus on the face next to him. Barron gasped, as much for breath as in surprise. "Why . . .," he wheezed, unable to quite grasp where he was at or what was going on.

"Commodore," Eaton said quietly, "it is Ambassador Eaton. I truly regret having to disturb you . . . "

169

Preparations

Barron's face slid into something that was probably supposed to be a smile. "Eaton ... yes ... it is ... "

"Yes, Commodore, it is I. I wish I did not have to disturb you, but a matter of the utmost urgency has arisen."

Barron closed his eyes and nodded his head. "Urgent ... "

Eaton was staring intently at Barron now, like a mesmerist trying to push his will on a subject. "Yes, Commodore, very urgent. I must ask for your signature for more funds."

With that, Eaton withdrew a folded letter from inside his jacket. He had taken great care to make sure the request for funds was properly worded, clearly stated ... and backdated to just before the Commodore was confined to bed.

Barron squinted once more. "Funds . . .," he said, almost to himself, as if he wasn't quite able to understand the concept.

"Yes, Commodore, the funds. The funds for me to proceed to Alexandria and raise Hamet's army ... do you remember?"

Barron's head rolled for a second, and Eaton was sure he'd lost him this time. Then Barron seemed to revive once more, and his eyes locked onto Eaton. "Yes . . .," Barron gasped. "Hamet's army ... you need funds ... "

Eaton turned to the middie, who was hovering at the doorway and hissed. "A pen, lad, quickly!"

As the middie dashed to get one, Eaton turned back to Barron. "I most certainly do, Commodore. I must have the funds to accomplish my mission."

Barron nodded, swallowing painfully. "Of course," he said, so softly that Eaton almost could not hear him.

The middie trotted back in with a quill pen, which Eaton snatched from him and slipped into Baron's sweating hands. Snapping open the letter, Eaton deftly slipped it under the pen as Barron's hands began to move, more out of instinct than intent. Eaton grasped Barron's wrist as he finished the last letter of his name, lest his signature end in a thin spider web strand of ink trailing off the page. Holding the paper up so the ink would dry, he caught the quill by its end and eased it from Barron's hand, passing it back to the middie. For his part, Barron, seemingly exhausted by the effort, sank back into unconsciousness.

Eaton looked down at Barron and thought to himself, *rest well, sir*. Turning back to the middie, Eaton said, "We should leave the Commodore to his recuperation."

The two stepped gently out of the room and closed the door behind them.

To Barbary's Far Shore

Eaton thanked the middie for his help after swearing him to secrecy yet once more, then bolted out into the pulsing streets surrounding Syracuse harbor. He probably set a record getting to the quay where he had left *Argus'* boat an hour or so earlier. Sure enough, it was still tied up there, with the boat crew flirting with a flock of signorinas and both groups mangling each others' language with cheerful abandon.

Eaton didn't even bother acknowledging the bosun's greeting, instead simply bounding down the stone steps and shouting, "To *Constitution*, and put your backs into it!" The boat crew looked at one another in common bewilderment at the whims of civilians, and pushed off from the quay accompanied by the farewells of the young ladies of Siracusa.

Back aboard *Argus*, Presley O'Bannon was inspecting his men for the first time, starting with Bancroft and Williams. O'Bannon was well and truly annoyed by the events following his arrival, and, as is sadly common in instances like that, it tends to get taken out on the men. Fortunately, any good sergeant has run across instances like this in the past and is prepared. Bancroft and Williams were good sergeants. O'Bannon looked the two of them over as closely as he could without being charged with assault, but found not a thread out of place. It didn't make him any happier, but at least now the sergeants could get a measure of control over things again as they followed him along the ranks.

O'Bannon stopped at Arthur Campbell first. The corporal barely breathed as O'Bannon inspected him. Turning to Williams, O'Bannon raised a questioning eyebrow and the sergeant picked up on it at once.

"Corporal Campbell, sir. An asset to the detachment; exceptional at arranging for provisions and gear."

"Mm. In other words, a thief and scavenger?"

Campbell reddened slightly at that, but Williams stepped in perfectly. "'Thief' and 'scavenger' are terribly harsh words, sir. We prefer to think of Corporal Campbell as a sort of an independent quartermaster."

O'Bannon's expression was dubious at that, but he said nothing and moved on. Two more marines passed under his withering gaze, neither of whom looked like anything to write home about. After inspecting the second one, O'Bannon took man's rifle and spun it around to peer down the barrel, almost clipping Williams in the process. The barrel wasn't dirty, but it could have been a great deal better.

O'Bannon moved on, then stopped, realizing that he was at eye level with one marine's leather collar. O'Bannon looked up, almost to the point where he would have to step back to take the whole man into his sight. The marine was

Preparations

built like a single block of granite, with hair so blonde it was almost white and piercing blue eyes drilling straight ahead.

"Private Bernhard Curtz, Leftenant," Bancroft said discreetly.

O'Bannon nodded at this information. "Where from, Private?"

Curtz' heels snapped together like a gunshot. "Herr Leutnant!" Curtz roared in a thick German accent. "Ze prifate iss from . . . ah . . . um . . ."

Behind O'Bannon, Curtz saw Bancroft silently mouth the word, 'Philadelphia.'

Curtz grinned in triumph and barked "Vhilatelphia, Herr Leutnant!"

O'Bannon merely blinked at this piece of information, smiled tightly, then stepped back while motioning for Bancroft to step closer.

"Sar?" Bancroft asked quietly.

"Sergeant Bancroft. That man is no more from Philadelphia than is King Neptune."

Bancroft looked at Curtz, then back to O'Bannon. "Actually Sar, that had occurred to us. In Private Curtz' situation, he happens to be not from Philadelphia at all . . ."

"Never would have guessed."

" . . . but from Switzerland, where he served with one o' the Grenadier regiments there. He's good, Sar; exceptionally good, though a bit dim occasionally due to th' language barrier."

"Sergeant Bancroft, I don't care if he's William bloody Tell; he's not an American."

Bancroft nodded. "Absolutely true, Sar. And if we hadn't have taken him, we'd be even more short-handed than we are now."

O'Bannon grimaced at the implications of that and continued down the line. A tall, red-haired marine stood locked at attention, in perfect order, except for the fact that his hat was cocked over at a rakish angle. Bancroft gritted his teeth and batted the private alongside the head, then straightened his hat.

O'Bannon raised one eyebrow. Williams answered, "Private Joseph Pritchard, sir."

O'Bannon looked back at Pritchard, who wore a fairly smug grin about the whole thing. O'Bannon wasn't impressed. "Private," he said quietly, "my understanding is that you are taught how to properly wear your uniform. Is that not so?"

To Barbary's Far Shore

"Sir, yes, sir!"

"Then why, pray, was your hat on improperly?"

"Sir, no excuse, sir!"

"Damned right," O'Bannon answered. Three more marines passed under O'Bannon's eye.

"This man's hair is out of regulations."

"Sar."

"This man's uniform is out of regulations."

"Sar."

A pause as O'Bannon reviewed one particularly unkempt marine. "This man does not seem to have ever heard of the regulations."

"Sar."

O'Bannon turned to face the next marine, only to discover that he was looking over the top of the marine's hat. Leaning down, O'Bannon saw a face that could only belong to a child, and from the looks of things, did. Before O'Bannon could even ask, the marine came to attention and fairly squeaked, "Sir, Private Bernard O'Brien, sir!"

O'Bannon looked the miniature marine over, and then shot a glance back at Bancroft and Williams. The two sergeants merely stared straight ahead, doing their best to pretend nothing was happening. O'Bannon leaned back down and asked, "Private, how old are you?"

O'Brien replied, "Sir, eighteen, sir!" in a voice that strongly suggested he was anything but.

O'Bannon could only put his hands on his hips, look O'Brien in the eye, and reply, "You are not."

Before O'Brien could answer again, Williams stepped forward. "Leftenant, Private O'Brien does admittedly appear slightly underage. However, to the best of our ability to determine, his documents are in order."

"No doubt they were filled out by Private Curtz," O'Bannon muttered as he proceeded down the next rank. The first marine in the last rank looked reasonably military, but there was just something a little . . . well . . . scruffy about him.

O'Bannon's glance at Bancroft was answered with, "Private Patrick Duffy, Sar, late of New York."

Preparations

"Mm. See to it Private Duffy gets tightened up a little bit." O'Bannon began to walk away, stopped, then turned back to Duffy, looking at him as if he knew something was wrong but wasn't quite sure what. Looking him over carefully, O'Bannon sniffed once, then again, and stepped back to regard Duffy with near disgust. "Sergeant Bancroft."

"Sar?"

"This man smells like a distillery."

"True enough, Sar. That is Private Duffy's own unique . . . aroma. Were he intoxicated, ye'd be knowing it."

"Good Lord. Private Duffy . . . just how much do you drink?"

Duffy never wavered. "Enough to cheer me up, Sar, but not enough to ignite."

"I see. Please try in the future to reduce the amount, whatever the purpose."

"Aye aye, Sar."

The last two marines were young privates, both in brand new uniforms and perfectly fitted out. O'Bannon inspected both of them closely, then turned to his sergeants. "Now these gentlemen seem to know what 'inspection ready' means. They are . . . ?"

"Privates Stewart and Whitten, Sar. Newly assigned from the United States and not yet corrupted by the ways of the world. I give them about a month at most before they're as bad as the rest of this lot."

O'Bannon did his best to glare at Whitten and Stewart. "Do not let it happen, gentlemen."

"Aye aye, Sir."

"Aye aye, Sir."

With that, the inspection was finished. As Bancroft dismissed the men, O'Bannon stalked off. He turned over his shoulder to say, "Williams, Bancroft, forward." The three men headed for the bow. It was a short walk, and when they were suitably distant from any other ears, O'Bannon turned to face his sergeants. "Gentlemen, I am used to my superiors describing marines as 'rabble,' a 'mob' and an amazingly long and detailed list of other pejoratives, but this is the first time I've ever actually seen one."

Williams cleared his throat. "Well, sir, they are a truly diverse bunch, but."

" . . . but what? The majority of your men look like the pirates we're supposed to be fighting. The remainder are aliens, drunkards, children, and

To Barbary's Far Shore

thieves. And those, I might note, are some of the best looking marines in the formation."

Bancroft folded his arms and lowered his head for a moment, the looked up at O'Bannon. "Leftenant," he said evenly but firmly, "it's possible that you may not quite be understanding our situation out here . . . "

"I understand that you have a great deal of work to do to get these men into shape . . . "

"But the Leftenant does not understand that this detachment is, for the most part, thoroughly wet behind the ears! The majority of 'em are less than two years in the Marines, sar. It takes four years, at th' least, to turn them into something resembling a good marine, and far too many o' them never get the hang o' it. This is a new detachment on a new ship, Sar. It takes time."

"Well, they had better get 'the hang o' it' right quickly. Apparently we have a mission of the utmost importance . . . "

Williams perked up. "Would that be Ambassador Eaton's trip up the Nile, sir?"

O'Bannon looked at Williams as if the sergeant had suddenly grown a second head. "Up the Nile? With this mob?"

Bancroft shook his head with a reassuring smile. "Oh no, sar, not a'tall. Just you and a couple of the midshipmen."

O'Bannon was thoroughly concerned now. "When is that supposed to happen, and why in God's name are we doing it?"

Williams and Bancroft looked at one another, and then Bancroft replied, "My understandin' is that as soon as we get to Alexandria and find someone who knows exactly where Hamet Bey is, we'll be up the river – pardon the expression."

O'Bannon leaned back against the rail in disbelief, then grim curiosity. "And how did you two know about this? Captain Hull hasn't even briefed me on it yet."

Bancroft and Williams grinned. "Tis a small ship, Leftenant," Bancroft answered, "a small ship indeed. Now, with yer permission, Sergeant Williams and I will go below to counsel the lads on their shortcomings." The two sergeants snapped to attention and saluted. All O'Bannon could do was raise his hand in a half-salute, half-wave to dismiss them.

And, speaking of Ambassador Eaton, he had fairly leapt from *Argus'* whaleboat as it bumped up alongside *Constitution*. The frigate's crew knew him well enough, and when he arrived on deck, he snagged the first sailor he saw.

175

Preparations

"You, lad. Take me to the purser's office at once."

The sailor nodded with a smile. "Certainly, Mister Eaton, follow me." The two went below into the warren that made up the belowdecks spaces.

There really hasn't been much change in two hundred years where a ships' layout is concerned. The Navy has, admittedly, developed an extremely precise, detailed system for locating any specific compartment below decks, which is also pretty much incomprehensible to most people. In this case, *Constitution* didn't have a dedicated pursers' office, so that dedicated soul worked out of his own quarters aft. *Constitution*'s purser looked like a clerk, but his responsibility was beyond that of most. Not only did he handle *Constitution*'s accounts, but he also oversaw pay, disbursement, and accounts for the entire Mediterranean Squadron. In addition, he was responsible for something that would stun most modern naval personnel – the Squadron slush fund.

This warrants a brief bit of explanation. As has been pointed out, two hundred years ago a warship captain was a pretty much autonomous authority, a law unto himself. He functioned as a diplomat, military governor, and unofficial ambassador. In addition, he was responsible for the hundreds of little details essential to running a warship: maintenance, upkeep, supply and, last but not least, paying the crew. One thing that has not, however, changed since then is the fact that it takes money to get all those things done. With that in mind, mid-level naval commanders had to carry around a fairly large amount of cash in order to get anything accomplished. In situations where you had large organized naval units, flotillas and larger, running about, the flagships tended to carry very large amounts of cash. In the case of the Mediterranean Squadron, this was aboard *Constitution*, where it is likely that a few hundred thousand dollars were available. A fair amount of this had to be reserved for pay and emergencies, but the rest was available when and as needed; subject to the Commodore's approval, of course. And, as far as William Eaton was concerned, he had the Commodore's approval; just not the one who was conscious.

Knocking politely on the door, Eaton smiled with charm and benevolence as the purser looked up. Not many people came to visit, except on payday, so it was a pleasant surprise. "Mister Ambassador," the purser smiled, "please come in, come in! This is an unexpected pleasure. I had thought you had transferred to *Argus*."

Eaton shook hands with the purser as he replied. "Good day to you, sir. I wish this was a social call; truly I do. But, unfortunately, duty calls. I need to take care of this matter most urgently before I can set off on my mission." Eaton handed the purser the letter signed by Commodore Barron. The purser examined it for a moment before putting on his glasses to read it carefully. Eaton did his best to remain nonchalant as the purser slowly read the letter and looked very carefully at the signature.

To Barbary's Far Shore

After a moment, the purser looked up at Eaton. "Mister Ambassador," he said, "this is a very great deal of money . . . "

"True enough," Eaton replied, his heart skipping a beat. "But given what we have to accomplish . . . "

The purser said nothing as he scanned the letter, squinting slightly as he got to Commodore Barron's signature. Eaton's heart skipped a beat, and he tried to remain nonchalant. Finally, the purser folded the letter neatly and placed it in a drawer of his desk.

"We'll need to keep that for the records," the purser said as he rose to open the massive oaken cabinet that held the squadron's funds. All Eaton could do was nod with a polite smile, thinking to himself that he would have happily given him not only the letter, but the deed to his family's farm and the blessed Constitution itself.

Back aboard *Argus*, Isaac Hull was sitting in his day cabin; an empty plate next to him and a goblet of a rather good port in his hand. Looking out the bay windows in *Argus'* stern, he watched the sun slowly edge below the ragged clouds and down towards the darkening Mediterranean. Its dying rays lit the lush green hillsides of Siracusa with a golden sheen that made every shadow stand out in sharp relief. Around him, the ships of the Mediterranean Squadron were slowly going into their evening routines as everyone prepared for a cool fall evening. The weather looked to be excellent for their departure tomorrow, and, with a bit of a breeze, they would be in Alexandria in a week or so.

In the meantime, Hull thought, leaning back in his chair and putting his boots up onto his desk, *it is a beautiful evening, I am in command of a warship at sea, a small one, true, but a warship nevertheless, and all is right with the world.*

The door to his cabin rattled with two sharp knocks. Rolling his eyes heavenwards, Hull brought his feet back down to the deck and called, "Come!"

The door swung open. Presley O'Bannon stepped through, the very picture of discipline and precision as he snapped to attention and saluted Hull. "Leftenant O'Bannon to see Captain Hull!"

Hull regarded O'Bannon skeptically for a moment, then returned the salute. "Good evening, Leftenant. To what do I owe the pleasure? And please, stand at ease." Hull leaned back again and took another sip of wine.

O'Bannon went to 'at ease,' but there was nothing of ease in his expression. "Begging the Captain's pardon, but I would like to ask about an assignment I have been told I shall be tasked with."

"Mmm. Ambassador Eaton's little excursion?"

Preparations

"One and the same, Captain."

Hull nodded and put the goblet down. "Ask away, Leftenant."

O'Bannon paused for a moment, trying to frame his words. "Captain, it is most . . . unusual for an officer to find out about something like this from his noncommissioned officers."

Hull considered that for a moment then replied, "Can't argue with you on that, O'Bannon. On the other hand, had you been less eager to try and charm your way out of your responsibilities this afternoon, we might have eventually gotten around to it." Hull took another sip of wine as O'Bannon blushed a furious red, but said nothing. "Now," Hull continued, "we can try and fill you in on what you're going to have to do, or you can continue to bumble about until it's time for you to go ashore and then make an ass of yourself. Which do you prefer?"

Presley O'Bannon was no fool, and even he knew when he had to admit error and ask for help. Swallowing hard, he looked Hull in the eye and said, "Captain, I want very much to do this right . . . may I ask your assistance in doing so?"

Hull looked O'Bannon over for a moment with a dubious air, then motioned for him to take a seat. Before O'Bannon could get comfortable, though, Hull looked at him squarely. "Mister O'Bannon, I have a very strict rule. I will never turn down the request of a gentleman or common sailor who comes to me and sincerely asks for help. But, if he lets me down, I will have no compunction whatsoever about throwing him to the sharks. Am I clear on that?"

"As crystal, Captain."

"Good. Pour yourself a glass and I'll fill you in."

"Thank you, Captain." As O'Bannon rose to get his wine, Hull unrolled a map of northern Egypt.

Sitting back down, O'Bannon asked, "Captain, shouldn't Ambassador Eaton be here as well?"

Hull's expression changed to one of bemused surprise. "Good Lord, no, Leftenant. Were we to do that, we might be here until Christmas."

Hull smoothed the map out as O'Bannon leaned over it to get a good look. Hull's finger traced down the Nile from Alexandria as he explained. "You and two of our midshipmen will escort Ambassador Eaton up the Nile until you find Hamet Bey. Once there, the Ambassador will take over and perform whatever diplomatic witchcraft he requires. When he's finished, get him back here in one piece so that we can get on with our real work."

To Barbary's Far Shore

O'Bannon took a sip, then asked, "Do we know where exactly Hamet is, Captain?"

"Absolutely not. We're going to have to do some asking once we get to Alexandria; but we're assured he is indeed in Egypt."

"Ah. My understanding is that the Nile is, to put it gently, a dangerous place these days."

"Your understanding is correct, Leftenant. The cutthroats and vagabonds who rule the Nile past Cairo and Luxor make our friends in Tripoli look like the tamest of nursemaids."

"I see." O'Bannon took another sip, it was closer to a gulp, before asking, "How many marines will I be allowed?"

Hull thought about that for a moment before replying, "Counting yourself, one."

O'Bannon's eyes went wide at that. "Captain Hull, I . . . "

"I know. You will need as many marines as you can take. Unfortunately, so will I. Because, as soon as you're on your way, I have to get back out to sea and guard our merchantmen and that requires marines. Now, Sergeants Williams and Bancroft can handle things until you get back; they shall have to."

A genuine gulp this time. "Who will I get?"

"Two of my midshipmen."

"Oh, Lord." Hull raised an eyebrow at that, and O'Bannon hastened to explain. "Sir, I meant no disrespect, but I should like to point out that the midshipmen probably have even less experience than I do."

"And you would be stating the obvious, Leftenant. However, this is one of those difficult decisions that captains are required to make. I have two missions of equal priority, and only enough manpower to satisfactorily accomplish one. Add to that the fact that a twenty-man detachment of marines would certainly advertise our intent, and the options do not get any better, but they do become much clearer. And, on the subject of your sergeants, as much as I hate to say this, we're going to end up keeping Sergeant Bancroft a bit beyond the end of his term. If I can't have a Lieutenant of Marines on my ship while he's gallivanting up the Nile, rest assured I shall have two sergeants of excellent capability, like Bancroft and Williams."

O'Bannon bit his lip, trying to think of what to say next.

Hull noticed. "Spit it out, O'Bannon. You've been doing well so far tonight."

Preparations

"Sir, delaying a man's release from service..."

Hull shrugged noncommittally. "For starters, Mister O'Bannon, it happens all the time. The winds and waves do not always allow us to be at a certain place at a certain time, so we always encourage those about to leave us to keep their plans flexible. Secondly, Sergeant Bancroft will be paid handsomely for his trouble. And finally, it shouldn't be more than an extra, oh, thirty days or so before we can get him on his way home. Now, I strongly advise you to speak to him and make sure he understands that the needs of the service come first and foremost. He's a good man, and a fine marine, if such a thing is indeed possible, but..."

"Yes, sir."

"Now, on the other hand, the lads are a pretty solid lot, Mann, Danielson, and Peck. Mann is the only one who might give you any trouble. He regards regulations as Holy Writ, and will defend them as devoutly as any Puritan. The other two would go to Hell and back just for the sheer joy of being on the trip. Unfortunately, all three are convinced of the invincibility of youth, and none of them have the experience to let them know just how difficult this might be."

"Have they any formal combat training?"

"You mean with steel and shot?"

"Exactly."

"Not a lick. Don't forget, Leftenant, that they are essentially very junior officers in an academic training status. My job is to teach them the subtleties of command at sea, not the manual of arms. I would guess, however, that they could wield a blade or a musket in extremis."

O'Bannon thought about this for a moment. "I'd like to take some time then to give them some kind of training, so that at the very least they know which end of the weapon to point at the enemy."

Hull nodded in agreement. "An excellent idea, Leftenant. Arrange with for some deck space tomorrow."

"Yes, sir."

O'Bannon's record file was still on Hull's desk, and he picked it up for a moment, weighing it in his hand. "Tell me," Hull asked, "how did you ever end up in the Marines?"

O'Bannon smiled, almost to himself. "Well, Captain, I wish, I truly wish, that I could say it was for adventure and glory and the flag. But that wasn't it at all. The truth be told, I was horrified at the prospect of spending the rest of my life in Fauquier County, Virginia."

To Barbary's Far Shore

"Couldn't have been that bad."

O'Bannon's eyes went out of focus for a moment, and Hull had the impression he was seeing something far and away. "It really wasn't," O'Bannon said quietly. "My father owns a horse farm there."

"Sounds to me as if your family was well off."

"We never lacked, no, sir. Father is a demanding man, but a good one, and my mother is one of God's own angels. I had a good upbringing, sir. The best education Father could afford, and a wonderful place to enjoy it."

"Then why ever did you leave?"

O'Bannon thought hard and long on that. For a moment, Hull thought that he may have touched an unpleasant memory. But then O'Bannon smiled sadly. "The truth of it, sir, is that . . . well, I did have everything, and what little I did not I could have had for the asking. And, may God forgive me for what I say next, sir, but I was bored with it. I thought there had to be something else out there besides horses and grass and empty-headed women. So, one morning, I told Father that I wanted to leave for a while. He was furious, and Mother almost fainted several times. But, in the end, I usually got what I wanted, and this was no exception. Our Congressman was indebted to Father for a few things, so he made the arrangements."

"Which leads you to here." Hull sat back and regarded O'Bannon for a moment before quietly asking, "You truly have no idea what you're in for, do you, Mister O'Bannon?"

For his part, O'Bannon thought about it for a moment, before answering quietly but firmly, "On the contrary Captain, I do indeed. The problem is that I have no idea how to handle it."

Hull had to smile gently at that. "That's not at all unknown among leftenants, Mister O'Bannon. You will never know precisely how to handle any given situation until you've gathered enough experience to look at a problem, evaluate it, and then act on it with the resources at your disposal. In all seriousness, now," Hull continued, draining what was left in his goblet and setting it on the desk with a satisfied smack of his lips, "what kind of experience have you had?"

O'Bannon swallowed, composing his thoughts. "Well, as the Captain has seen, my first assignment was aboard the *Adams* . . . "

"Where you were . . . let us say, less than stellar."

Preparations

O'Bannon nodded in acknowledgement. "I did not show a great deal of martial aptitude, so our captain exchanged me with one of his leftenants on the flagship, and brought me over where, frankly, he could keep an eye on me."

Hull closed his eyes, remembering what O'Bannon's fitness report had read. "'Excellent at administrative functions, less successful at leadership,'" he said to O'Bannon. "Your captain there was a man of considerable tact."

"That he was, sir. Right up to and including the moment he had me reassigned to Headquarters in Washington. If one is to be told of one's failures, one prefers that it be with some decency."

"I know the feeling, Leftenant. Now, once you got to Washington . . . "

"I was an aide to the Commandant. That consisted of primarily being a glorified message boy. On the one occasion they gave me marines to take into the field on maneuvers. I managed to bring everyone home in one piece, but it was a very near-run thing. When the opening came up aboard *President*, I was hustled aboard so quickly I was unable to say farewell to many of my friends."

"Most of whom, apparently, were female."

O'Bannon looked mildly hurt. "The Captain must admit that as of yet, there is still no regulation against it. Nor the playing of the fiddle, I might add."

Hull had to give O'Bannon this; he was right on that score. "True that may be, Mister O'Bannon, but please keep the romantic entanglements to a minimum, and the fiddle playing as well. I am not normally fond of music aboard my vessel, but when I do, I prefer it to be with hornpipe and voice."

Looking back at the clock on his mantelpiece, Hull turned back to O'Bannon. "Leftenant, I must cut this short, as I still have work to attend to. May I assume you will be ready to execute your duties when we arrive in Alexandria?"

O'Bannon stood, straight as an arrow. "Yes, sir." The reply was quiet, yet confident.

Hull folded his hands on the table in front of him and looked straight into O'Bannon's eyes. "Leftenant, let me give you some advice. If you continue to address your duties with the attitude you have brought here this evening, I believe you may grow in your position and become a credit to your service. But, understand this, Mister O'Bannon. Regardless of why or how you got here, you represent your country's interests now. There is no one to fall back on this time, no one to pull you out and send you somewhere else. Fail here, and three hundred good men rot in prison for a very long time.

"Dismissed."

To Barbary's Far Shore

O'Bannon saluted without a word and left Hull's quarters. It was a nice evening, or at least getting that way as *Argus* wound down from her daily routine. Dinner was long past now, and the ship's routine was downshifting only slightly, because it takes just about as many men to run a man o'war at night as it does during the day. But, two centuries ago, before the advent of radar and weapons that can kill from a thousand leagues away in day or night, you still relied on the human eye and those indefinable 'feelings' that those who have been out to sea long enough develop.

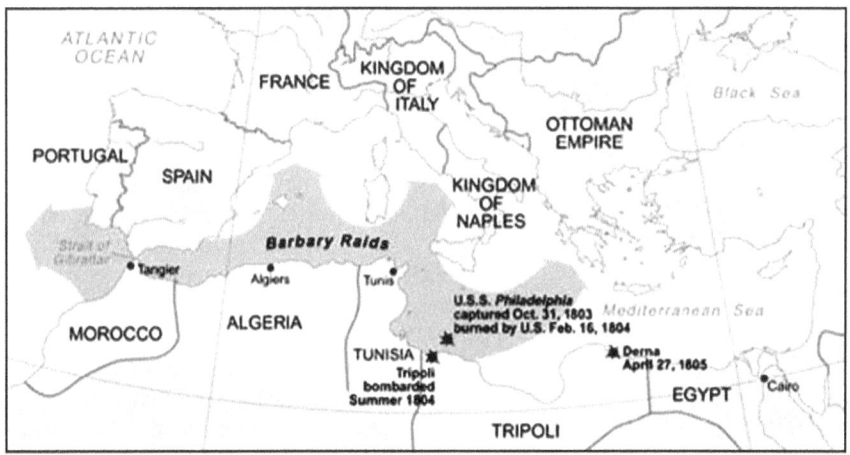

And so the scene is set, the players are in place, and great deeds are soon to be done. Reputations are to be won and stirring legends made. Beyond all that, the U.S. Marine Corps will create its own, inimitable tradition.

O'Bannon stood on *Argus'* deck in the lengthening shadows, trying to figure out what to do next. He absent-mindedly nodded to Will Eaton, who walked briskly past with a nervous smile, trailing a boat crew loaded down with heavy boxes. O'Bannon was about to ask what was going on, but he had other things to worry about, like explaining a few things to Sergeant Bancroft.

We don't know much about the midshipmen aboard United States Brig *Argus*. There were three aboard her, though when and where they showed up is unclear, and, for all we know, there may have been more. We know one, George Washington Mann, was from Annapolis, Maryland, but we do not even know that much about the other two, Eli Danielson, other than the remarkable detail that he was, in fact, William Eaton's stepson, and the improbably named Pascal Paoli Peck. But they were indeed there, and they played vital parts in what was to come next.

183

Preparations

A Midshipman was a young gentleman learning his trade. This miniature expertly captures the youthful earnestness of a typical Middie. Courtesy: Valiant Enterprises Ltd.

The average middie then was young. Often under 18, and sometimes so young that today it would be illegal under more than a few international treaties for them to serve at all. They tended to come from fairly well to do families, or those that at least pretended to the position. In those days, the Navy was considered socially superior to the other services, though today that is not a position one would want to take around a Marine, Army, or Air Force officer. That was in part because it was considered to require far more intelligence and skill to be a successful officer at sea, rather than rutting around in the dirt and mud, waiting to pick off some poor devil at long range before he took you out. It certainly didn't hurt that naval officers also carried with them dashing, swashbuckling reputations, such as those carried down to us by authors such as C.S. Forester and Patrick O'Brien. In legend, myth and sea story, they were dashing, handsome and virile stallions of unutterable gallantry, compassion, and wisdom, whose very names made the enemy shake in fear, inspired allies to cast laurels upon their brows, and fair maidens to quiver in ecstasy. What teenaged boy wouldn't wish to be seen like that?

The reality was, as usual, often less than that. Tom Truxtun was a proud, prickly martinet who often threw temper tantrums if he could not get his way. Richard Morris ran a socially correct Mediterranean Squadron much to the detriment of his actual duties, and we have only examined the printable side of John Rodgers' reputation. Isaac Hull was tall but rotund; most of his portraits are, shall we say, flattering. While Sam Barron and Edward Preble may once have been dashing, they were now stooped and crippled by illness. But, for all that, there was still something about them, noble, chivalrous knights of honor and glory who were somehow more than the sum of their ships and their crews.

Perhaps the best way to describe how these men were viewed is to compare them to the men we consider the Last Great Warriors, the modern fighter pilot. Even in an era of weapons that can kill at four times the speed of sound and aircraft that are increasingly and disturbingly autonomous, we still see them as Tom Wolfe did in his classic <u>The Right Stuff</u>: the champions of single combat, the Nation's last, best hope.

To Barbary's Far Shore

Midshipmen tended, as a rule, to be a bit disappointed by the truth of the life they let themselves in for. Even in the largest warships, their quarters were only marginally better than those of the sailors. Because of their age and inexperience, they were often kept at arms' length by the commissioned officers that they hoped to join some day. They could speak with, and often had, the same authority as the commissioned officers, but most of the time they were in a student status, being taught by, of all people, the ship's chaplain, one of whose duties was that of schoolmaster.

And a hard school it was indeed. The midshipmen had duties of their own when the ship went into combat, and they were exposed to the same risks as the rest of the crew. They were just as much prey to the difficult conditions of disease and injury as anyone else, and the overwhelming majority of them were away from home and hearth for the very first time. And one had to go through, on average, four to six years of this before being awarded a commission. Some went longer, and some, not surprisingly, Stephen Decatur, made it in less. More than a few never finished at all, deciding that the family business or a nice safe college like Yale or Princeton was a much more reasonable bet.

But, in any event, that in a nutshell explains how three teenaged boys found themselves on their way to Alexandria, Egypt. All of them probably had a year or so of service under their belts, and it seems likely that they knew enough about weapons to allay O'Bannon's concerns. After all, when dealing with pirates, it was a good bet that at some point you were going to find yourself face to face with some maniacal Berber who wanted to either ventilate you or cleave you in twain. Having at least a passing knowledge of cold steel and hot lead would have seemed reasonable to even the cockiest teenager with delusions of immortality.

Then there was the matter of Sergeant Bancroft.

Today's marines are familiar, probably far too familiar, with the concept of 'stop loss,' wherein your service tells you that you may not separate or retire on time due to the fact that Uncle Sam needs you just a little while longer. There is almost always a great deal of wailing and gnashing of teeth, followed by a quick review of one's enlistment contract, only to discover that yes, they can indeed do that to you. It apparently did happen from time to time in the old days, and it was that much easier then, especially if you were a thousand miles or so away from home.

In that era, a marine sergeant did not speak back to his superiors, regardless of the provocation. Had it been necessary to tell a long-service noncommissioned officer that he wasn't going home quite yet, the sergeant would have gritted his teeth, saluted, and gone back to work.

But he would not have been happy.

Preparations

The dramatis personae are almost all in place now; Eaton planning, scheming and dealing from his cabin, O'Bannon getting the measure of his marines and they of him, the midshipmen learning that they have managed to stumble into history and Isaac Hull trying to run an orderly ship in the midst of all this. There will be a few more hands joining the parade at Alexandria and beyond. But, until they join us, it is simply the men aboard that lone ship riding at anchor in the blue Mediterranean.

To Barbary's Far Shore

EIGHT: THE POLITICS OF PRESENCE

American soldiers, sailors, airmen, and marines have endured the hardships of captivity since the Revolution. Soldiers of Washington's army and the militia that supported them were jammed into pestilential British prison hulks where they were forced to try and survive under conditions that would be grimly echoed almost two centuries later in the horrors of Japanese POW transports. During the Civil War, neither side could take much pride in the way it treated its prisoners, though the argument can at least be made that to a great extent it was through sheer stupidity and ignorance. But in a dim crevice of our national conscience lies a place called Andersonville, which, in intent if not accomplishment, can lie beside names like Dachau and Auschwitz as examples of what human beings can do to one another and what Americans are capable of in their darkest moments. During the Korean and Vietnam Wars, we discovered that, in addition to the physical miseries that captivity can inflict, there was mental torture of a type and extent never seen before, designed to break the will to resist and refined through decades in the Gulags and tiger cages.

There would be one more unpleasant form of captivity sprung on Americans before the Millennium turned; the great hostage takings, where handfuls of obsessed mullahs, clan chiefs, and deranged religious madmen decided that they could bring down the Great Satan by kidnapping his citizens and holding them for years at a time. And when that failed to have the desired effect, they kidnapped more, proving only that in the end that Santayana was right when he said that a fanatic was someone who redoubled his effort long after he'd forgotten his aim. The failure to bring America to its knees would lead to more anger and frustration and then to greater horrors, but that is another story for another time.

Will Bainbridge and his men, however, were in a unique position. The Dey of Tripoli considered himself at war with the United States of America and was acting accordingly, with one notable exception that we shall examine

187

The Politics of Presence

momentarily. As far as Yusef Bey was concerned, he was at war with the entire United States, Navy and merchant marine alike, and they were all fair game. On the other hand, the United States, in its majesty, glory and wisdom, had not made a reciprocal declaration and therefore had very little to bargain with. In fact, as we have seen, US crews were releasing Tripolitian prisoners.

Yusef's own peculiar outlook on things complicated matters a bit. He did consider himself at war and was acting as such. But, if we were willing to come up with enough money, he was willing to let *Philadelphia's* crew go, whether or not hostilities had been concluded. This was and had been standard operating procedure in that part of the world for centuries. After all, there were only so many people and desert outposts one could conquer, taxes and baksheesh could only go so far and every now and then the navies of the world cracked down on active piracy. That left ransom, an ancient and honored form of royal income that, by the time we encountered it, had been refined to an art form.

It was a pretty simple operation, really. You captured your prisoners and separated them by ability to pay. The poor frequently ended up enslaved, while everyone else was kept in conditions that, while unpleasant, were frequently no worse and occasionally better than those the average subject of the Dey toiled under. This was not a reflection of any compassion on the Dey's part, but rather a simple, hardheaded business decision. Dead captives brought nothing, while live ones were always a bargaining chip. There was no real malice here and it was certainly nothing personal. It was simply business, and on the whole it was reasonably profitable.

There was, however, one drawback that the Dey had not run across before and was now facing every time he so much as thought about the crew of the *Philadelphia*. Put simply, the US wasn't giving in. In the past, everyone played by the rules; everyone was a potential hostage, with the exception of the British, who just did not tolerate that sort of thing and had the will, and firepower, to enforce it. The hostage would send a message home, eventually ransom came back and they would then be put on the first ship out with the Dey's sincere best wishes. But the Americans were sticking to their guns and, what was worse, they were out there blockading the Dey's harbors. True, it wasn't the tightest of blockades, but it was starting to pinch the belts of the average Tripolitian and, worse yet, it was starting to pinch Yusef's purse. And it cost money to keep those damned infidels, three hundred of them, fed and housed in a near derelict building near the harbor.

Captain William Bainbridge, however, would have probably taken great exception to any claim by the Dey that he was caring for his unfortunate guests. Their quarters, we are told, were in a vermin-infested warehouse close to or actually on Tripoli Harbor. They would have certainly heard the commotion when Stephen Decatur came to visit and they would have known when *Intrepid*

To Barbary's Far Shore

made her last, doomsday ride. Little question about that, in fact. The Dey was livid when he found that his prize had been snatched out from under his nose, and made sure Bainbridge and his men knew it. *Philadelphia's* hulk was still smoldering when the Dey's guards dragged them from their garret and beat them unmercifully, before spitting on them and then declaring that from now on they would do more work for less food. From wherever he looks upon mankind, the Pharaoh of the Exodus would have smiled.

On the subject of their rations, they were, by the book, two twelve-ounce loaves of bread a day, although 'twelve ounces' was probably a very flexible concept, and, if whatever they were given bore any resemblance to loaves, it would have been purely coincidental. That little bit of nourishment was intended to keep them alive while they worked from dawn to dusk, the luckier ones as stevedores on the Tripoli waterfront, unloading contraband from the blockade-runners that had made it in. The rest got to work as street cleaners. Both groups were shoeless by now and, usually, close to naked.

Still, they held together as best they could, and for the most part they made it from day to day. Bainbridge and his officers instituted something they called the 'University of the Penitentiary,' where they shared their higher education with the rest of the crew. It was a very smart move on Bainbridge's part. It gave the men something to rally around every day and drew them into a tight, cohesive unit, something we have since discovered is vitally important to survival under those kind of conditions.

Not everyone made it, though. A handful simply could not survive the conditions, poor food and forced labor. They died without ever seeing home again. Five crewmen 'turned Turk;' that is, they threw their lot in with the Dey and became overseers with marginally better food. According to Robert Wallace in Smithsonian, the worst of these turncoats was *Philadelphia's* quartermaster. In the best tradition of such stories, he got his; the traitor was cornered by one of *Philadelphia's* marine sergeants and beaten within an inch of his life. The sergeant, in turn, took almost five hundred with the bastinado, but there is a certain satisfaction in such acts that can alleviate a great deal of pain. However, at the end of the day, and there would be nineteen months worth of days, the officers and crew of the *Philadelphia* were on their own in dealing with whatever fresh torments the Dey and his minions could come up with.

They did have one bright spot, and it was a bright one indeed, His Excellency Nicholas Nissen, Consul of Denmark to the Dey of Tripoli. Consul Nissen, as has been previously noted, dedicated himself to doing whatever he could to alleviate the suffering of the Americans and, in doing so, frequently put his own hide at risk. The Deys, Bashaws, and etc. were not at all averse to holding diplomats hostage or outright prisoner, and, had they known that Nissen was also passing messages from the Mediterranean Squadron to and from the

The Politics of Presence

prisoners, he may very well have joined those poor souls. But, as the Danes would prove during the Nazi occupation, they are a remarkably tough and inventive people when dealing with tyranny and Nissen was a fine example. He did his best to visit them as often as he could, knowing that just the sight of another Westerner who cared would be immensely comforting to the prisoners. He often spent his own money to get them just a little bit more food, a blanket or two, perhaps a Bible, anything that would help them hold out just one more day.

Nissen would stride in with all the authority that Their Majesties back in far away Denmark could give him, and with all the dignity God could give a man. Even the Dey's guards, hardened souls who reveled in the cruelty they could inflict upon prisoners, trod carefully in his presence, treating him with all the deference they gave their own commanders. Nissen, no fool, took advantage of it at every turn, pulling the Dey's beard every chance he got.

One can only imagine what he would think as he approached that grim stockade to visit his charges. Almost certainly, the smell, sweat, sickness, sewage, and yes, fear, was what you noticed first; though it could be argued that it was only marginally worse than anywhere else within that benighted kingdom. The odors combined to create a physical presence, a warning more powerful than any sign that something worse than usual lurked in that ramshackle old building, one that back in Copenhagen would have been torn down long ago. Two guards stood in a relaxed manner beside a door made of weather-beaten planks, looking as if it was an afterthought rather than part of the building.

Nissen put his shoulders back and stood up as straight and tall as he could. He gripped the book under his arm a little more tightly and made to walk straight through the door. One of the guards, a new one, from the looks of him, blocked the door with his arm and growled.

"Aish, what?"

Nissen looked him dead in the eye. "I am here to see the prisoners. I have the permission of the Dey."

The guard gave a snort of laughter. "Well, go away, kufr. The Dey has not told me you may visit."

The other guard smiled gently. "Friend, I advise you to let him in. He indeed speaks with the Dey's authority."

That, of course, would get the attention of any Berber with an instinct for self-preservation. The first guard looked Nissen up and down for a moment, and then decided that discretion would indeed be the better part of valor. Of course, there was nothing that said he couldn't make things difficult for the infidel, so he reached out to grab the book from under Nissen's arm. Nissen felt a brief

To Barbary's Far Shore

twinge of fear as the guard turned it in his hand, looking at it carefully, but slowly took a deep breath and stayed calm.

"What is this book?" the first guard asked suspiciously.

Nissen smiled with a serenity he didn't feel. "That, sir, is the Holy Bible, the true and inerrant word of God."

With that, the first guard gave a start and practically threw the book back to Nissen. He caught it with a smooth motion and tucked it back under his arm. Like any good Muslim, the guard wanted no truck with blasphemy, something Nissen had been counting on in the clutch. Looking at Nissen with a barely suppressed snarl, he opened the door with a shove.

What Nissen would have seen on each visit probably made him wonder whether or not mankind had ever truly progressed beyond barbarism. The building that housed most of the Americans, still nearly three hundred of them, was only twenty-five by fifty feet. Pace off that distance, and then imagine yourself immured there with two hundred and ninety-something friends. It always took a few seconds to adjust his eyes to the darkness. The few tallow candles that sputtered and dripped from sconces on the walls did little to help.

"Mister Nissen," came a croak from nearby. "It's Mister Nissen!"

The cry went up, moving through the room like a wind and growing into a cheer. One form, taller than the rest, rose and stepped nimbly towards Nissen. Will Bainbridge had been built like a barrel, and would be again, but now he was lean and wiry. The remains of a set of uniform trousers were his only concession to modesty. His hair was badly cut, Nissen could only guess how, and a few days beard darkened his features, but, behind the grime and sweat, one could still see the proud United States Navy captain.

"Mister Nissen," Bainbridge rumbled, taking the consul's hand in a grip that was still firm. "An unexpected surprise, sir. I had not expected to see you for a few more days."

Nissen smiled and bowed slightly. He was about to answer when the door swung open and the first guard stepped in, alerted by the noise. The chatter and noise stopped as suddenly, as if on command. The men closest to the door skittered away to avoid whatever wrath was to be directed at them.

Nissen turned to the guard and snapped, "What is it?"

The guard said nothing. He just crossed his arms and stood watching the consul and the captain, certain that if he stood and watched them long enough, he would catch them at something.

The Politics of Presence

Nissen shook his head and turned back to Bainbridge, holding the Bible up to the captain. "Captain Bainbridge," Nissen said with a sly smile, "will you join me in prayer?"

The guard, seeing the book once more, took a small step backwards; just enough that he could still see what was going on, but far enough away that the infidel's words would not hinder his journey to Paradise. That was far enough for Nissen. For his part, Bainbridge blinked for a moment, not entirely sure what was going on, but certainly willing to find out.

His eyes locked on Nissen, Bainbridge took the Bible from his hand and gently kissed it, saying, "Of course, Mister Nissen. I have always found it to be a great comfort in times of trouble."

Nissen merely nodded, saying quietly, "As have I, Captain. You may find the Thirty-Fifth Psalm to be of particular solace."

Bainbridge slowly opened the Bible to Psalms, with its hymns to the glory and grace of God and promises of comfort and protection. Quietly reciting the Doxology, Bainbridge turned one last page to behold the Thirty-fifth Psalm and its seventeenth verse:

Lord, how long wilt thou look on? Rescue my soul from their destructions, my darling from the lions.

And beneath it, in small but precise script, a letter. Bainbridge looked up once more to see Nissen smiling. "As I said," the consul said, "particular solace."

Captain Bainbridge;

Your ordeal shall soon come to an end. As we speak, planning is underway for an expedition to overthrow Yusef Bey and secure your release.

Bainbridge grinned from ear to ear, suppressing the urge to leap like a schoolboy. Regaining his composure, he read on:

Be of good cheer and exhort your crew to be likewise. More information to follow.

Bainbridge continued to smile, until he got to the signature. His head snapped upward, a look of such surprise upon his face. For a moment, Nissen thought that Bainbridge's heart was giving out.

"Captain," Nissen said with quiet concern, "are you ill?"

Bainbridge closed the Bible, handing it back to Nissen. "Mister Nissen," Bainbridge said quietly, "Are you aware of who that . . . ah . . . psalm was written by?"

To Barbary's Far Shore

Nissen shot a quick glance over his shoulder. The guard was trying to stare down several marines, who glared back defiantly. Turning back to Bainbridge, Nissen replied. "I have not had the honor of meeting the psalmist, but I know who he is."

"But you do not know of his reputation?"

"No."

Bainbridge shook his head almost despairingly. "Mister Nissen, I have feared for the lives of my crew every day we have remained here. But nothing, no beating, no illness, no cruelty, has terrified me to the extent that William Eaton's signature has."

Nissen's face showed puzzlement as he tucked the Bible back under his arm. "Captain, I am not sure I understand your concern."

"Mister Nissen, I have spent most of my life at sea, where I sincerely believe one may find the most arrogant, overbearing blowhards on the face of the planet, save for William Eaton. If he has convinced the powers that be that he has a reasonable plan with even a remote chance of success, I fear we are all doomed, sir."

Nissen placed a hand on Bainbridge's shoulder. "Captain, I understand that what you face here may be daunting. But, surely, you must feel joy that at least an effort is to be made?"

Bainbridge shook his head sadly. "I feel nothing but dread, Mister Nissen. So much so, that I do not intend to even inform my men of this."

Nissen shook his head. "Captain, what kind of man is this? Surely anyone who can lead such an effort must be a man of courage and bravery."

Bainbridge replied skeptically. "Courage and bravery are one thing, Mister Nissen; and, from what I know of Mister Eaton, he lacks neither. It is the intelligence and wisdom that must go along with it that I fear for."

Bainbridge paused to look around the room, where his men tried to rest after another day of misery. Turning back to Nissen, he spoke. "The truth be told, Mister Nissen, our lives are in the hands of a lunatic. God help us all."

Nissen, unsure what to make of Bainbridge's comments, took the Bible back from the captain and gently closed it. "Thus endeth today's lesson," the diplomat said softly.

Argus' voyage to Alexandria had been quiet in terms of combat, but things had not exactly been quiet aboard the ship. William Eaton, of course, was happily planning his conquest of Tripoli, even though he hadn't even found Hamet Bey yet. From the records we have available, Isaac Hull was becoming

The Politics of Presence

progressively less fond of Consul Eaton, especially Eaton's propensity to speak and act as if he was in supreme command of everything. However, Hull kept his temper for the time being.

O'Bannon and his men were slowly, carefully becoming a tight, coherent unit, training and drilling every day. It was probably just as well, because it was keeping everyone busy. That went especially for the midshipmen, and for good reason. Take any average seventeen-year-old boy you know, tell him you're going to send him on a secret mission and he will almost invariably assume that he already knows enough to survive and prevail. A veteran marine sergeant will quite rightly tell them otherwise, but they won't believe it until they see it. That is how Edmund Bancroft and William Williams found themselves with two eager pupils and one reluctant one.

It was difficult at best for men who had already twenty years in the service of their nation to accept these teenage boys as their betters, but that was the time-honored tradition, and it could not be ignored. Therefore, the two sergeants had to remember at all times that they were training and addressing their military and social superiors, no matter how badly they would have liked to have simply cuffed them alongside the head as an incentive to learning.

As it turned out, the lads were actually familiar enough with the cutlasses that Bancroft decided no further training was necessary. Contrary to popular belief and Hollywood myth, although the cutlass, the saber, the rapier and the epee were common weapons of the time, the overwhelming majority of the men who used them were not highly trained swordsmen, capable of cutting the buttons from your blouse with a flick of the wrist. Some, of course, were better than others, but they were the exceptions rather than the rule. There were a few basic moves that everyone was taught; to thrust, or parry. But, for the most part, cold steel was used in a hack-and-slash style that was far more appropriate for a butcher shop than a duel.

That, of course, left firearms.

The personal weapons of the time were very basic; the musket and the flintlock pistol, both fairly inaccurate, muzzle-loading single-shot weapons. Most Americans had experience with firearms, as most Americans in 1804 lived on farms or on the frontier, where a firearm was a required tool of survival. However, as with the sword, the level of skill with a given weapon was not what legend would have it.

The standard muzzle loading long arm of the time was the famed Kentucky rifle, the rifle that literally won this nation's independence. It was simple, sturdy, and almost impossible to damage or break. The US Model 1803 Harper's Ferry, the first US rifle to be produced at a government arsenal, was ever so slowly replacing it. However, for all the Kentucky rifle's fame, it was not terribly accurate at even medium range for the time. Its rate of fire was only about two rounds per minute and that only in the hands of highly trained and drilled professionals. That is the reason for those tightly packed lines one sees in paintings and films. In order to have any effect at all on an enemy, the troops had to fire *en masse* at fairly close range. Not firing until you saw the whites of their eyes wasn't bloodthirsty rage, but rather sound tactical doctrine. One final drawback for the rifle was that, in the hands of the midshipmen, it was usually longer by some distance than those who were going to carry it, an almost insurmountable problem in the field. Almost by default, the midshipmen were going to have to carry pistols.

The flintlock pistol was even more limited in its capabilities than the Kentucky rifle. Although it packed an impressive wallop, a typical ball size was roughly ½," or .50 caliber, it only did so at close range, perhaps about thirty to fifty feet at best. Since it too was a muzzleloader, the rate of fire was as abysmal as that of its larger counterpart. For the pistol, that was an even bigger disadvantage. The infantryman fired in ranks, one reloading, another firing, and so on; so, at least you had some covering fire while you were desperately trying to get another shot ready. But since the pistol was intended for very close in and personal work, the odds were not good that you'd get another round off if you missed or if there was more than one opponent.

On top of those very real drawbacks, ammunition and powder of the time was not always trustworthy. Ammunition, of course, were lead balls, provided by Government arsenals for the military, occasionally store-bought by the civilian, and frequently cast by hand on the frontier. Military shot tended to be of good quality, made from lead that was reasonably free from impurities and usually even fairly round. But civilian and homemade ammo varied wildly in composition and quality. It was not unknown for it to come in highly irregular shapes that would fragment as soon as it left the muzzle.

Powder was also a question mark. Mister du Pont's factory in Delaware was known even then for probably the finest and best quality black powder in the world; but, as it was so, it was comparatively expensive. In those days, as today, the lowest bidder always got the contract, so not a great deal of du Pont's products made it to the typical infantryman. What did get to them was often coarse, poorly milled, filled with impurities and highly sensitive to damp or wet conditions. The result was that when you actually tried to fire a weapon while using the stuff, you could get a round that would go twice as far as you'd expected, go more or less where you wanted it to, do nothing at all, or just roll out of the muzzle and land on the ground with an exhausted plop.

All of the above, plus the image cold steel brought with it, insured that for close-in combat, the firearm would be a more of a backup weapon rather than a primary one for some years to come. But the Marines have always believed that

The Politics of Presence

if you are going to carry a weapon, by God you'll know how to use it and use it right, which is what led Sergeants Bancroft and Williams to be guiding the midshipmen through a bit of practice one day.

The midshipmen, Eli Danielson, Pascal Peck, and George Mann, would have been gathered forward with their sidearms, black powder, ball shot and firing caps when Bancroft and Williams arrived. Coming to attention, the two sergeants saluted and were saluted in return.

"G'd mornin' to ye, gentlemen," Bancroft said in the most pleasant voice he could muster. "For those of ye who've not met me already, I am Sergeant Bancroft, and this is my associate, Sergeant Williams." Williams, for his part, snapped his heels and gave a short, sharp little bow.

"We're gathered here this mornin' to instruct ye in the use and handling of the pistol. Now, I see that . . . " Bancroft paused, looking at George Mann.

The middie was leaning against the rail, looking as bored with the proceedings as he could possibly be. Peck and Danielson tried to look somewhere else. Mann did not appear to have any weapon on his person other than his sword, firmly buckled around his waist. Had one of Bancroft's marines shown up without their weapons, he would have probably kicked them around the deck a few times. It was not possible to do that with the midshipmen. Bancroft then elected to try a reasonable approach.

"Mister Mann, sar?"

Mann looked up at Bancroft with an expression that went past mere contempt and bordered on hatred. "What," the Middie replied in a venomous tone.

Bancroft took a deep breath, forcing himself to remain calm. "Sar, I believe we are here to train you in the use of the rifle and pistol."

"And?"

"And you do not appear to have either of those items, whereas Misters Danielson and Peck do."

Mann gave a snort of laughter in reply. "For your information, Sergeant, the weapon of a gentleman is the sword, though I expect you would not be aware of that."

The other two midshipmen' eyes widened in shock. Although they knew and understood themselves to be Bancroft's betters, they would have thought long and hard before insulting a man with his reputation as a fighter. For his part, the sergeant's jaw merely tightened, while he took a deep breath and stayed calm.

To Barbary's Far Shore

"Be that as it may, Mister Mann," Bancroft replied evenly, "Cap'n Hull has directed that I give ye some knowledge of how to use them, and give it to ye, I shall." Turning to Williams, he said, "Sergeant Williams, would ye be kind enough to run down to the armory and procure another set of weapons for Mister Mann?"

"Of course, Sergeant." Williams trotted aft towards the armory.

Bancroft returned to his teachings. Turning to Danielson, he asked, "May I use yer rifle for just a moment, sar?"

Danielson smiled and nodded eagerly. "Certainly, Sergeant." Danielson handed over the Brown Bess, very nearly as long as he was tall.

Bancroft grasped it in knowing, experienced hands as he turned back to the group. "Gentlemen," he said, "this is a rifle. There are many like it, but these are yours. They are yer very life in combat. To survive, ye must know how to keep them working, how to load them, and how to fire them. Once you have those mastered, ye will be able to take on any adversary at any time. Now, what is the first thing ye think ye must do when receiving yer weapon? Mister Peck?"

Peck, a gangly, black-haired teenager, thought for a moment, then answered. "Load it?"

Bancroft shook his head. "Sar, that is not correct. The first thing ye must do is make sure it is clean and in good repair. A dirty musket is a useless musket, gentlemen, and a useless musket will get ye killed six ways to next Sunday. Let's begin with yer musket, Mister Peck."

Handing the musket back to Danielson, Bancroft took Peck's musket and held it up to inspect the bore. He was rewarded with a small cloud of dust, dirt and unidentifiable specks rolling out of the muzzle and into his face. Bancroft gave a quick cough and lowered the musket, handing it back to Peck.

Bancroft tried to remain calm. "Sadly, sar, that musket is not clean. Ye really should put a bit more work into it."

For his part, Peck stood straight, smiling as he answered. "I shall, Sergeant."

Bancroft gave a tight little smile. "Glad to hear it, sar. Mister Danielson; yours again, please?"

Danielson handed the musket back, and Bancroft again raised it for inspection. It was certainly clean; that much Bancroft was able to tell before the iron ramrod slipped out of its brass tube beneath the barrel and smacked Bancroft squarely between the eyes. He staggered backwards into Williams,

The Politics of Presence

who had just returned with another musket and pistol. Danielson recoiled in horror. Then he leapt forward to pick up the ramrod now loose on the deck.

"Dear Lord, Sergeant, I'm sorry," the middie exclaimed as he took the musket back and secured the ramrod.

Bancroft's first instinct was to take the musket and swat the little idiot with it, but his training overrode it. Placing his hand between crossed eyes and bringing it away unbloodied, he smiled, more grimace than smile. "Think nothing of it, Mister Danielson, but ye must be more aware o' it in future."

Danielson nodded nervously. "I shall, Sergeant, I shall, and I am so very sorry . . . "

"Mister Danielson?"

"Yes, Sergeant?"

"Please stop apologizin.' "

"Yes, Sergeant."

Once Bancroft had his bearings back, he blinked a couple times to orient himself and started again. "Now . . . where were we . . . "

Williams spoke in a stage whisper. "Clean musket, Sergeant Bancroft."

"Right. A clean musket, gentlemen, is how you insure accuracy an' dependability. Therefore, I respectfully request that ye clean yer weapons daily while we are en route to Alexandria. That way, ye'll be in the habit. Now, I'm assumin' ye know how to load one?"

Danielson and Peck nodded eagerly, but Mann continued to look everywhere else on the ship. This, of course, was Bancroft's cue, and he took it.

"Mister Mann," he asked gently, "would it be possible that, how shall I say this, ye aren't familiar with firearms?"

Mann reddened slightly and still refused to look at Bancroft, but mumbled something that sounded like "Yes."

Bancroft smiled like kindness itself. "Now sar, there's no needin' to be bothered by it. All o'us have a first time for everything, no matter what it is; and I'm sure that even ye'd admit that this is somethin' ye'll be needin' to know someday, am I right?"

Mann regarded Bancroft skeptically for a moment. "Well . . . I suppose that might be the case . . . "

"Well then," Bancroft replied equitably, "there's no time like the present to learn. Sergeant Williams?"

To Barbary's Far Shore

Williams handed Bancroft the flintlock pistol he'd drawn from the armory. The pistol was loaded and fired in the same manner as the musket, but was far easier to handle in the process, making it ideal for someone who'd never done it before to learn. Handing Mann the pistol, Bancroft took his own musket. "Now, sar, just follow my lead and we'll have you shooting in no time."

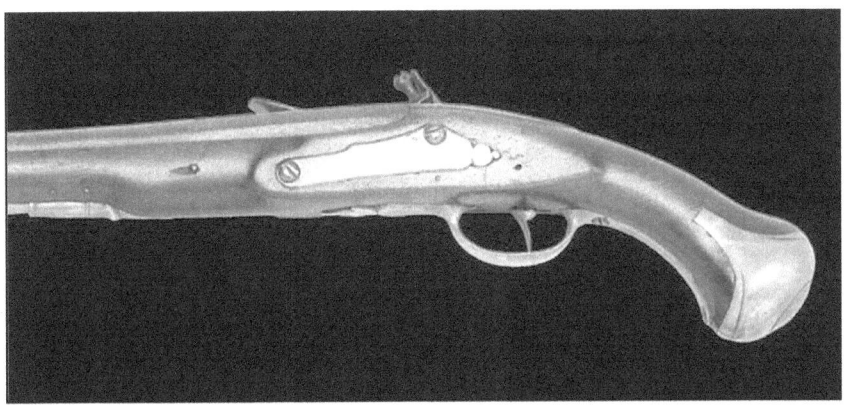

A black powder pistol is a deceptively difficult thing to shoot. Many a pistolman accustomed to modern weapons has managed a negligent discharge when firing black powder for the first time. Source: U.S. Navy History and Heritage Command

Mann took the pistol, but still regarded it suspiciously. "I've never handled a firearm before, Sergeant. Are you sure?"

"Trust me, sar. Ye'll do fine. Now, take the powder horn here and give her just a wee dram."

Bancroft filled his with the smooth motions of a practiced expert. Mann took the cap off his powder horn and delicately poured in a thin trickle of powder.

"There y'be sar, should be quite sufficient. Now, take yer wadding, that'd be the wee thin strips of cloth there, and push it a bit into the barrel."

Mann watched as Bancroft went through the same motions, taking the thin cotton cloth and easing it into the barrel. Bancroft smiled. "Now, Mister Mann, ye can be a wee bit more aggressive, the barrel won't be breaking. Next thing we do is insert the ball itself."

Bancroft held up the dull gray metal ball as if it was a rare gem, and then dropped it into the barrel. Mann dropped his in as well, this time with a flourish. Bancroft smiled encouragingly.

The Politics of Presence

"There now. One last thing. . . " Like a maestro preparing to lead an orchestra, Bancroft slipped the ramrod from its protective sleeve, and motioned for Mann to do the same. Mann gave a tug and the pistol's ramrod eased from its place beneath the barrel.

Bancroft nodded in approval. He almost twirled the ramrod in his fingers as he spun it into the bore and tamped the ball in with three swift motions. Mann was watching intently, and he followed Bancroft's moves flawlessly. Replacing the ramrod, Bancroft grasped the musket by the stock, just behind the flintlock and held it up for display. For his part, George Mann held the pistol in his hand, barrel pointing straight up, and looked at it with a mixture of apprehension and pride.

"There ye are, sar," Bancroft said. "Loaded and ready. Now, if I may ask you to remain still for just a moment, I would greatly appreciate it if th' other gentlemen would load theirs as I have just demonstrated."

Peck and Danielson turned to their task with gusto and alacrity, if not a great deal of precision. Strips of wadding and small dots of powder drifted down onto the deck.

Some distance aft, Joshua Blake stood on the quarterdeck just behind the helmsman. He had the deck at that moment, and it would have been difficult to find anyone happier aboard *Argus*. It was a lovely day. The ship and her crew were working flawlessly, and with a little luck they might happen across a runner before the sun went down.

Scanning the deck, Blake's eyes were drawn by movement forward. Squinting just a bit, he saw the two marine sergeants and three of his midshipmen gathered together, and it seemed they were loading pistols.

Turning to a marine at rigid port arms a few feet away, Blake said, "You, marine?"

Pat Duffy came to attention with a clatter of gear. "Sar!"

Blake winced at the cry. "Do you have any idea what's going on forward?"

Duffy peered down the deck for a moment. "Sar, that's Sergeants Bancroft and Williams. They're teachin' some of the midshipmen how to load and shoot pistols."

Blake absorbed this for a moment, looked forward again, then back to Duffy. "Weapons training?"

"Aye, sar."

"For the midshipmen?"

"Aye, sar."

200

To Barbary's Far Shore

"Hmm." Blake paused for a moment. "Go fetch Surgeon Cable then; we'll be needing him momentarily."

Duffy was off at a dead run before the echoes of Blake's words died away.

Other officers on board *Argus* had working spaces of their own, but the Lieutenant of Marines had to do his administrative work in his own quarters, cramping the already tiny space even further. O'Bannon was scratching away on a report when he heard boots pounding down the passageway and the door rattled in its frame.

"Come."

Duffy poked his head in, the aroma of something vaguely alcoholic preceding him. "G'mornin,' sar. Mister Blake's compliments for Surgeon Cable."

O'Bannon put down his pen and looked at Duffy with mild disbelief. "Private Duffy, do you see Surgeon Cable here?"

Duffy smiled. "No, sar!"

O'Bannon could only blink at the equanimity with which Duffy replied. "Duffy, let me rephrase the question. During duty hours, why would Surgeon Cable be here, instead of the sick bay?"

Duffy came to attention and answered, "No reason at all, sar; excepting that I don't know where the sick bay is, so I thought I'd try here first and work my way over."

O'Bannon's jaw dropped slightly at that one. "You don't know where the sick bay is?"

"No, sar. Anytime I've not felt properly, I just take a bit o' the hair o' the dog, if you catch my drift."

O'Bannon sniffed the air and grimaced. "Catching your drift is the first thing I did, Private. You will find Surgeon Cable in the sick bay. Just keep going forward until you cannot any longer."

"Thank you, Sar!" Duffy saluted and began to about face.

"Wait a moment . . . "

Duffy turned. "Sar?"

"Why exactly did Mister Blake need the surgeon?"

"Well, sar, I'm expecting that any moment now, Sergeant Bancroft is about to be shot by one of the midshipmen."

201

The Politics of Presence

Before O'Bannon could reply, the unmistakable sound of a pistol shot came from over their heads. O'Bannon jumped up from his chair, looking up then back to Duffy, who for his part smiled.

"You'll be excusin' me, sar; all the way forward, y'say?"

O'Bannon nodded in stunned surprise as Duffy nodded politely and headed forward. For his part, O'Bannon's reflexes kicked in and he leapt out of his cabin, racing up the ladder topside. All he had to do was follow the sounds of Sergeant Williams' enraged bellows. There was Bancroft sitting on the deck and white as a ghost. His shako lay on the deck beside him, with a ragged black hole on either side. George Mann was literally hopping up and down, screaming something unintelligible. Danielson and Peck were trying to restrain him, and Josh Blake was roaring at Mann. O'Bannon also saw a pistol, spinning slowly on the deck like some demented toy.

Be damned, O'Bannon thought, *one of the midshipmen really did shoot Bancroft.*

"Williams, what in God's name happened?"

Williams' normally gentle demeanor was replaced with red-faced, furious anger as he saluted O'Bannon. "Begging the Leftenant's pardon, but Mister Mann just tried ventilating Sergeant Bancroft!"

Mann was screaming. "I told you I have never handled a firearm, you jackass! When your betters tell you something, listen!"

Mann drew back a leg as if to kick Bancroft. Josh Blake wouldn't have any of that on his deck; he swung a lightning-fast backhand at Mann that knocked him cleanly out of the grasp of the other two midshipmen and sent him spinning down to the deck.

"There'll be none of that, Mister Mann! If there's any swinging to be done here, I'll do it."

"Or I."

Everyone snapped to at the sound of Isaac Hull's voice, as he suddenly appeared behind them. Even Bancroft suddenly remembered where he was and scrambled to his feet. Behind Hull was Richard Cable, *Argus'* ships' surgeon. His black bag rattled as he passed Hull and moved immediately to Bancroft's side. The sergeant was pale and still swaying just a little as Cable looked him over with a calm, professional eye. While Cable tended to his patient, Hull turned to the little group.

"What happened?"

To Barbary's Far Shore

O'Bannon looked at Williams and nodded. Williams took a deep breath. "Beggin' the Captain's pardon, Sergeant Bancroft was showing Mister Mann the proper procedure for loading and firing his pistol. Mister Mann . . . "

Mann shot Williams a deadly look, but said nothing.

". . . ah . . . seemed to lose track of where he was at . . . and was examining the weapon when he fired it."

Hull for his part simply fixed the middie with a glare that would have penetrated solid oak and bellowed. "WELL?"

Give him this, Mann wasn't prepared to give in just yet and snapped. "I told that ass that I never . . . "

Hull took one step towards Mann, towering over him like a ship of the line over a rowboat. "I didn't ask for your excuse, you brat! IS IT TRUE?"

Mann went as white as Bancroft had been and simply squeaked "Yessir."

For a moment, there was dead silence on the deck except for the waves. Hull surveyed the entire scene with a thoroughly displeased expression before turning to Blake. "Confine him to his quarters. Let him study the manual of arms someplace where he's not likely to kill anyone valuable."

Mann inhaled as if to say something, but thought better of it as Hull looked at him again, this time with murder in his eyes. The other midshipmen, knowing that part of the battle was knowing when to leave, quickly hustled Mann away from the captain and headed for the hatch as quickly as they could.

Hull stepped over to Cable, who was handing Bancroft's shako back to him. Looking at Bancroft, Hull tried to stay serious but couldn't completely suppress a grin.

"Sergeant, I have to confess it would have done nothing for your reputation had Mister Mann actually killed you."

Bancroft's voice was still a little shaky as he replied. "Th' Capn' has a point, but I'm thinkin' that it wouldn't have made much difference to me."

Hull grinned, asking Cable. "Any damage?"

Cable closed his bag. "None to the Sergeant that I can tell, Captain. However, I don't believe that shako is long for the world."

At that, O'Bannon stepped over to peer at the shako. The ball from Mann's pistol, roughly sixty-nine caliber in size, had punched neatly through the shako about two fingers' width above the visor, then sailed into the Mediterranean sky to drop into the water. Fortunately, it did no more harm than scaring a few fish. It was pretty clear that Bancroft had been remarkably lucky; had Mann been a

The Politics of Presence

bit taller or holding the pistol at a slightly lower angle, it would have gone cleanly through Bancroft's forehead and retired him most abruptly.

Shaking his head in amazement, O'Bannon took a good hard look at his senior sergeant. Bancroft was still pale, but looked to be getting his bearings back. The sergeant was at rigid attention though, proof if any was needed that training and reflex could overcome the most unnerving event.

"You all right, Sergeant?"

Bancroft smiled with just a hint of nervousness. "Bless the Leftenant fer askin', but I'm fine, sar; really. I am, however, thinkin' that teacher will not be one o' my choices for future work."

O'Bannon grinned. "Wasn't your fault, Bancroft. I get the feeling that Mister Mann isn't the best pupil in the world."

A Marine Shako. Mann's negligent discharge must have literally parted Bancroft's hair. Source: U.S. Navy History and Heritage Command

At that, Hull spoke to O'Bannon. "Looks to me as if everyone hasn't any lasting harm. Blake, I'll be back in my quarters; O'Bannon, get this sorted out."

Hull spun on one heel and strode off before anyone could even reply. O'Bannon turned to Blake. "I'm going to let Williams take the duty for the rest of the day. I think Bancroft could stand with a few hours off."

Blake looked at them skeptically for a moment, and then nodded. "Sounds right enough." The three marines watched as Blake headed aft again, and then O'Bannon turned to Bancroft. "All right. First, you are relieved of duty for the remainder of the day."

"Sar . . ."

" 'Sar' nothing. Go below and stay there. If I see you above decks again today, I'll see to it that Mann gets another crack at you. Before you go, however, I want to know what happened up here."

Bancroft's face hardened. "They sent me a bloody idiot for a pupil is what happened."

To Barbary's Far Shore

"Thank you for stating the obvious, but I was hoping for a more technical explanation. Williams?"

Williams cleared his throat. "Sergeant Bancroft was doing everything by the book, sir. Mister Mann was apparently not familiar with the loading and handling of firearms and was extremely nervous about the whole thing. I'm guessing he got a bit too nervous . . . and that was that."

"Are those pistols on a hair trigger?"

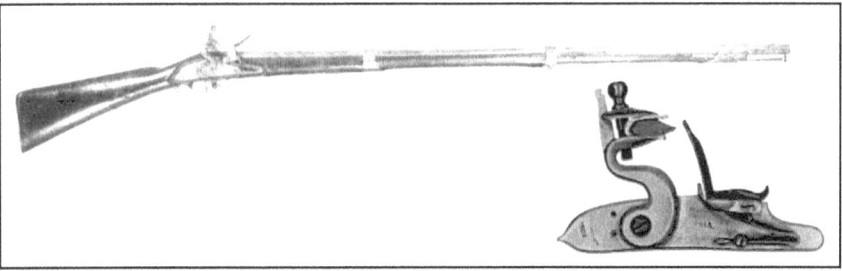

Model 1802 "Virginia" Musket. Source: U.S. Navy History and Heritage Command

Williams shook his head grimly. "Not intentionally, no, sir. They're quite worn, and the mechanisms are rather touchy. We can't get replacements, as the companies on the frigates get 'em first."

That wasn't a surprise to O'Bannon; it was a tradition then and still is now. The units down at the far end of the supply chain have the toughest time getting anything. On the other hand, it didn't make his life any easier; all the more so because he and his men were going to have to rely on those weapons. *Well, this one was easy enough to fix*, O'Bannon thought.

"Right then," O'Bannon said. "Get them to the armorer and have them tightened up at once."

Bancroft discreetly cleared his throat. "Beggin' the Leftenant's pardon, but *Argus* doesn't have an armorer. Nearest armorer is w' Mister Carmick aboard *Constitution*."

O'Bannon took that equitably enough, nodding in acceptance. "Do you have any armory experience?"

Williams nodded. "A wee bit, Leftenant."

"Good. See to it yourself, then. I'll be inspecting them in the morning." O'Bannon started to walk away then stopped and turned to the sergeants one more time. "If you need any assistance, Bancroft can help you. He's not doing

The Politics of Presence

anything for the rest of the day." O'Bannon headed back down the ladder below decks as the two sergeants watched.

Williams turned to Bancroft. "Edmund, when I said a wee bit, I meant. . "

"Not bloody much at all. As I'm recallin,' yer total experience is limited to standing by th' armory door and making sure they were still intact after the lads were done playin' with them."

Williams winced, but nodded in agreement. "That's not far off."

"Fair enough, then," Bancroft replied. "Let's get to it then. And if we see Mister Mann again, I'll pass on the bloody pistols and just beat 'th hell out of him."

To Barbary's Far Shore

NINE: GOOD KING HAMET AND THE HOUSE OF RENDEZVOUS

Alexandria was a typical Mediterranean port when *Argus* pulled in on November 26th, 1804, but it had not always been typical. It had been a tiny fishing village called Rhakotis but three hundred and thirty one years before the birth of Christ. Alexander the Great saw the little peninsula that extended into the Mediterranean and the two circular bays that rested on either side of it and decided that it would be a perfect place for a regional capital.

Alexander was so enchanted with the project that he himself laid out the outline for the new city's walls. Alexander never saw the completed city; he died ruling the known world in 323 BC. But Alexandria would grow, a beacon of learning with the most legendary collection of knowledge in the ancient world, the Library of Alexandria. It was a literal beacon in the form of the Lighthouse of Pharos, one of the Seven Wonders, almost fifty stories high, and crowned with a statue of Poseidon. Alexandria was the world's second city after Rome, and, in some ways, might have been greater. It became the capital of Egypt under Ptolemaic rule, and Julius Caesar himself came to visit it in 48 BC. There, the queen of Egypt, a young, ambitious woman named Cleopatra, gave Caesar the gift of a carpet.

With her in it.

She may have intended to rule at Caesar's side, or, perhaps, as a co-equal over northern Africa. She would do neither. When his own senators murdered Caesar in March of 44 BC, Cleopatra chose the wrong side in the fourteen-year civil war that followed. She was captured by Caesar's successor Octavian, who is believed to have had a most spectacular, and unpleasant, end planned for her. She cheated him out of that bit of triumph by applying an asp to her breast, and that was the end of Egypt as an even theoretically independent nation for the

207

Good King Hamet and the House of Rendezvous

next nineteen centuries. With that, Alexandria started a slow decline into obscurity. The magnificent Library would burn at the hands of conquerors; the Lighthouse would slowly deteriorate and collapse. Alexandria's new rulers, the Muslim Arabs, built the fortress of Qait-Bay atop its foundations.

The Arabs cared nothing for Alexandria. Their outlook was that of the horseman and they faced inwards towards the great golden seas of sand that had been their cradle. They built mosques, great hulking structures like the Al-Mursi that overlooks the eastern harbor, but that was about all. The town continued to shrink. As the world moved on, Alexandria began to revert to its origins and became once again a sleepy fishing town.

Being inconspicuous, however, is no guarantee of not being a strategic target. Alexandria still had one of the finest harbors on the North African coast, and, as such, provided superb access to the Egyptian interior for a man who considered himself Alexander's rightful heir: Napoleon Bonaparte. His Majesty brought the French Army ashore at Alexandria on July 1st, 1798 with negligible resistance from the Ottoman garrison, which did its best to get away without actually trying to defend the place. What Bonaparte found when he got there was eight thousand new subjects living in mostly ruined buildings. The walls, fortifications and palaces poked up from sand dunes that were encroaching into the center of the city itself. Napoleon's rule lasted only a year before Horatio Lord Nelson defeated the French at Abou-Qir, later Anglicized to Aboukir. The Ottomans established a tenuous rule over Egypt, and the city was under the authority of Kourshek Pasha, an easy-going and remarkably amiable sort who more-or-less represented Ottoman power there.

The Ottomans may have been starting their final decline, but they weren't fools and they knew a good thing when they saw it. Alexandria was a good thing with a lot of potential, and the Ottomans immediately started clearing the harbor of a millennium or so's worth of rubbish and garbage and rebuilding the docks and waterfront, along with tidying up the town a little as well. By the time Isaac Hull brought his ship into the Great Harbor, Alexandria was starting to come back. The problem was that between Napoleon's departure and Kourshek's arrival, civil authority had pretty much vanished. The result was what we would later see in Afghanistan and Iraq after the collapse of civil (such as it was) authority. The local gangs, strongmen, and assorted godfathers had taken control, each in their own part of town, making sure the business was taken care of and nobody got too big for their britches. The overall effect was that Alexandria was a busy, bustling center of international commerce, and an almost lawless city with no effective government. There were several consulates there, British, French and Spanish; all of whom engaged in low-grade spying on one another while trying to oversee their respective nations' interests.

To Barbary's Far Shore

Hull would have seen the warehouses and shops along the curved wall of the inner harbor, with carts and donkeys and stevedores unloading the ships tied up along the seawall. Fishing boats were headed outbound in ones and twos, rough feluccas that could have been brand new or a hundred years old with brown nets trailing alongside. Two obelisks known as Cleopatra's Needles marked the site of the Caesareum, a temple constructed by Augustus after he had disposed of Cleopatra. They were partially buried; one actually lay on its side, but the other still pointed skyward in the center of town. One will end up in London and the other in of all places Central Park, where they stand to this day.

Hull was able to enjoy the view; Blake was navigating the ship in while the captain kept an eye on everything. Of course, Hull, and everyone else on the ship, was getting the unique smell of Alexandria. Livestock, human sewage, and the usual odors of the seashore all assaulted the noses of Hull and his men as they approached. Depending on what time of day it was, Josh Blake's commands as the *Argus* came alongside the seawall might have been a counterpoint to the keening cry of the muzzeins at Al-Mursi: "There is no God but Allah, and Mohammed is his Prophet . . . "

Hull stood silent on the quarterdeck as Blake took the conn and carefully guided *Argus* alongside the seawall. The line handlers stood along her starboard side. O'Bannon had his marines topside and turned out as well, with Bancroft and Williams patrolling the deck. O'Bannon, hands clasped behind him, walked across the deck and got Bancroft's attention.

"Sar?"

O'Bannon looked across the narrowing strip of water between *Argus* and the seawall. "Tell me about Alexandria."

"Charmin' place, sar. I can think of better places to go ashore."

O'Bannon considered this for a moment. "Any ideas if we'll be able to find someone who can help us locate Hamet?"

Bancroft thought about this for a moment. "Sar, I've no doubt we'll be findin' all kinds of people who will help us find good king Hamet. Whether or not they can is the question. After th' Frogs pulled out, a great many gentlemen of questionable character took up residence here, and they're more than happy to tell you anything ye want to hear as long as th' money's right."

O'Bannon shook his head dubiously. "And Mister Eaton has both a willingness to listen and the money to spend."

"That he does, sar. And, with respect, it won't take long for th' word to get around that he's got th' money. At that point, sar, there will be a great many unsavory types willin' to relieve him of it by whatever means necessary."

Good King Hamet and the House of Rendezvous

O'Bannon frowned at that, keeping one eye on the dockhands ashore who were now dragging the *Argus* over the last few yards of murky water. "Sergeant Bancroft, you have a very good point. When Eaton goes into town, he is to have at least two marines with him at all times. It's not perfect, but it'll keep the more unpleasant types away."

"Aye aye, sar. I'll put Curtz and Pritchard on it."

One of O'Bannon's eyebrows tried to lift clean off his forehead. "I can see Curtz; the man's built like a damned mountain. But Pritchard . . . ?"

Bancroft nodded. "Private Pritchard is not th' most military of th' men aboard, sar. But he does have a certain swagger to him, and that's somethin' th' good people of Alexandria respect. If they think yer' insane enough to take 'em on, they tend to leave you alone."

There was a bump as *Argus* settled against the seawall and a flurry of activity ashore as the ropes were pulled in and made fast around ragged wood bollards. Blake regarded the operation with mild disdain for the grimy men ashore, and then turned to Hull. "Captain, we are secure to the dock. Your orders?"

Hull surveyed the whole scene for a moment and then looked to O'Bannon. "Leftenant O'Bannon, please be kind enough to insure our friends ashore do not steal my ship. Mister Blake, set the watch and notify the quartermaster to start replenishing. I am going to . . . "

"At last!!" Everyone, ashore and aboard, looked up to see William Eaton standing on the quarterdeck, looking at Alexandria like it was a prize to be plundered. "Captain Hull," Eaton announced, "I will need to go ashore at once to locate a guide to going downriver, and then we'll get a list of supplies to the quartermaster as well. With a little luck, we'll be on our way in a day or so."

Hull's smile was a gentle and affectionate one. "No."

Eaton stopped as if he had walked straight into a bulkhead. "I beg your pardon, Captain?"

"I said, 'no,' Consul Eaton. It is a word used to express dissent, denial or refusal, as in response to a question, request or *et cetera*. It is a word I strongly suggest you accept hearing on a regular basis."

Eaton started to speak up, but Hull shook his head and cut him off before he could make a sound. "Consul Eaton, we are in a part of the world where the normal rules do not apply. Authority ashore in Alexandria extends from the Pasha's throne to the outside of his palace wall, and that is as far as it goes. And, lurking between those palace walls and this ship, is one of the grimmest collections of thieves, cutthroats, brigands and other criminals I have ever had

To Barbary's Far Shore

the misfortune to run across; that includes, by the by, the crew of my own ship. I am responsible for your safety, and you shall go nowhere until I can arrange for said safety ashore."

With that, Hull motioned for Eaton to run along, but Eaton was having none of it. "Captain, I must protest! Our time is limited . . . "

"As is the number of men I have to support you with, and, in any event, the needs of my ship and crew must come first. Now, I strongly suggest that you sit down with Leftenant O'Bannon and make your plans; but do not implement them until I tell you it is safe to do so."

With that, Hull spun on his heel and strode aft, leaving Eaton momentarily nonplussed; at least, until he spotted O'Bannon, and took off after him. Short of jumping overboard, O'Bannon had nowhere to go, and prepared to weather Eaton's onslaught. *On the whole*, O'Bannon thought, *I'd rather be in Washington* . . .

The goal was quite simple: find Hamet Karamanli, get him to approve, support and join the project, then return to Alexandria to organize the assault force. The first problem was, as Eaton had pointed out, nobody knew exactly where Hamet was. So the first effort that Eaton had to make was to find someone reasonably trustworthy who knew where to find the presumptive Dey of Tripoli.

Easier said than done, unfortunately. The embassies and consulates there weren't really interested in helping Eaton. The truth be told, they'd rather see him fail. A change in leadership in Tripoli could only lead to instability, not to mention the end of whatever cozy arrangements those nations had set up there over time. So there really was only one possible way to go: the locals.

Alexandria, like any Mediterranean port, would have had literally dozens of taverns, inns, shops, brothels and assorted traders who would have incredible amounts of information available. Whether or not that information was true was something else entirely; but that would be up to Eaton to decide.

It would have been a few hours and quite possibly the next day before Hull finally let anyone ashore, and Eaton would have gone down the gangplank followed by his escorts. They would have had very simple and direct orders: do not, under any conceivable circumstances, allow any harm to come to Consul Eaton, lest ye find yourself facing what will remain of a short, miserable career with the United States Marines. They were not to partake of the many and varied pleasures of the Alexandria waterfront, nor were they to observe the cultural and social attractions. They were to keep Consul Eaton warm, safe and dry . . . or else.

Good King Hamet and the House of Rendezvous

As it turned out, the marines weren't needed. Sometime late that afternoon or early that evening, the British Consul at Alexandria showed up with what Professor Fowler refers to as 'a number of Turks.' King George's gentlemen had, quite clearly, done their homework. It was apparently at this point that the word was passed to Eaton that Good King Hamet was in a tiny village near Cairo, just about a hundred and twenty miles away. With the many streams of the Nile Delta easily navigable by small boat or barge, it would be an easy matter to get there in only a day or so.

Although it is not explicitly stated, there may have been another Briton with His Majesty's Consul as he visited *Argus*. His name has come down to us as Richard Winston Farquhar, Esquire and Gentleman.

The truthfulness of his name is questionable; the accuracy of his rank highly doubtful. From what we know of Mr. Farquhar, he probably was indeed a former officer and gentleman in the service of His Majesty King George III, and an experienced one at that. Farquhar was a relatively young man, and, at that time, one achieved a full colonelcy one of two ways. First, one worked his way up through the ranks, which could take decades under the best of conditions. The other way was to have your family buy one or obtain one through what was delicately referred to as 'connexion.'

Either way, Farquhar didn't seem to come under either of those headings. He was probably a leftenant or junior captain who helped drive Napoleon out of Egypt and simply enjoyed the thrill of combat more than the boredom of garrison life. There is also the darker possibility that, like many soldiers of fortune, he had been invited to leave military service for one unpleasant reason or another, or been flat-out cashiered for that dread and awful phrase, "conduct unbecoming an officer." In those days, that covered a great many things from cheating at cards, to failure to pay debts, to 'ruining' a young lady. In any event, Farquhar had promoted himself high enough up the ladder that he could pass as a reasonably competent officer, and he also had the one mandatory prerequisite of any gentleman: a manservant. Aletti the Manservant, to be precise, described as a massive and silent Turk who was constantly at Farquhar's side.

We do know that, no matter what happened, Captain Hull was quietly grateful that Eaton would be out of his hair for at least a few days. Eaton's behavior was, by all accounts, beginning to not only grate on Hull but was also beginning to seem a bit unusual, even for William Eaton. The records are maddeningly vague on this, but Eaton was starting to refer to himself as 'General' Eaton. We know Eaton hadn't been above the occasional inflation of his military record, but going from captain, a quite respectable rank in those days, to general had to have been noticed, commented on, and made a matter of some concern. Eaton was probably assuming that with the expedition he had in

To Barbary's Far Shore

mind, whoever led it was going to be a general, and he was going to lead it, therefore . . .

A few days after *Argus* had tied up at Alexandria, Isaac Hull was up on deck, taking the air, such as it was, and enjoying a few minutes out of his cabin and away from the paperwork. It had, at the very least, been quiet since the British Consul's visit. Eaton had been ashore at the Consulate or in his quarters the whole time, plotting his next move in the great game. *Well, let him,* Hull thought to himself. In the meantime, Hull had real work to do. Crews had to be fed and cared for, ships had to be manned, and then, once all of that was accomplished, he still had to go out to sea and deal with the weather, the pirates, and all the other things that could possibly kill him, his men, and his ship. Far more serious and important matters than chasing after some comic-opera prince.

The sounds of deckhands screaming at one another in a handful of different languages got Hull's attention. But, at first, he brushed it off. Sounds like that were the very music of a port like Alexandria, and, if you stopped for every sudden eruption of bellows and roars, you'd never get anything done. But Hull began to realize that this particular cacophony wasn't fading out. instead, it was growing and, for that matter, getting closer.

Hull spun on one heel and froze in amazement at the sight before him. Inching closer to the *Argus* was a battered, weatherbeaten barge; it wasn't so much moving through the turbid waters of Alexandria Harbor as wallowing, and it wasn't even wallowing well. On its deck were a half-dozen, or so, Egyptian deckhands, each trying to keep it under control while screaming and gesturing at one another. On its bow were two poles about a man's height. One had a Union Jack posted to it, and the other a brand-new American flag, the Stars and Stripes trimmed with a yellow fringe running along its edges. And standing between them, one foot on a gunwale and arms folded across his chest like a conquering titan, was William Eaton, grinning from ear to ear and looking almost deliriously happy.

Hull stood quietly and watched the apparition stagger past. There was a small deckhouse, with crates and sacks piled high around it; on each side of the deckhouse were two shiny new twelve-pounders, a total of four polished and mounted on freshly painted US Navy carriages. The Egyptians were now running to and fro between the cannon with the poles they were using to help steer the barge, arguing over exactly which way to push so as to keep it moving in something vaguely resembling a straight line. Eaton, however, seemed blissfully ignorant of the drama behind him and continued to gaze nobly straight ahead.

Within a minute or so, the barge had somehow managed to pass *Argus* without colliding with her and was being shakily guided to a bumpy mooring just ahead of the warship. As Hull watched the barge, he heard footsteps; Josh

Good King Hamet and the House of Rendezvous

Blake came up beside him. The two men stood silently for a moment, unable to take their eyes off the barge, before Blake spoke.

"Captain."

Hull nodded in acknowledgement. "Leftenant Blake."

A pause. "Sir, have you ever . . . "

"No, Leftenant, never in all my travels."

"Ah. Why on earth would he have the Union Jack flying?"

"Utterly beyond me, Leftenant."

"I see, sir."

Pause. "Leftenant, two things."

"Of course, sir."

"First, make sure the watch knows about . . . that . . . and is especially vigilant."

"Of course, sir. And the other?"

Blake detected just the ghost of a tic at the corner of his captain's eye as Hull spoke. "Please be good enough to count our cannon." With that, Hull strode slowly away, shaking his head in amazement.

Isaac Hull was a big man with a voice to match. Right now, it was rattling the bulkheads in his cabin as he tried to impress his concerns upon William Eaton, while Presley O'Bannon and Joshua Blake stood as discreetly as possible in a corner. "Mister Eaton, I have . . . "

Eaton's expression was peace itself. "Consul."

Hull stopped abruptly in mid-rant; his eyes wide, his voice going to a stunned whisper. "I beg your pardon?" Hull asked in surprise.

"'Consul,' Captain Hull; 'Consul.' My title is 'Consul.'" Eaton smiled beatifically, as if bestowing grace and understanding upon a dim student. "And, quite frankly, I'm not at all sure why you're so upset. All I did was avail myself of a few spare military assets that were surplus to requirements . . . "

"'Surplus to requirements?' For God's sake, Mister, pardon me, Consul, Eaton, those cannon were spare because, well, because they were spares! They were aboard in the event we needed extra weapons or replacements! They were most certainly not there to be used as bribes!"

Eaton grinned and gave a dismissive wave. "Captain, what more proper use could they serve than to help convince Hamet of our bonafides? And, in any

To Barbary's Far Shore

event, your skills in combat are most impressive; you'll have no need of anything like extra cannon. Besides, we do need some armament. Our friend, the British Consul, advises that the Nile above Cairo is . . . shall we say, less than secure, and with our cargo we will need some kind of protection."

Hull raised his finger to make a point, then suddenly stopped and looked quizzically at Eaton. "What cargo . . . ?"

Eaton blinked, nonplussed. "Why, our gifts for Hamet Bey, of course; some firearms, swords . . . "

"Out of my armory, no doubt!"

"Of course not, Captain. Let me see, now; firearms, swords, an exquisite Bible . . . "

It was O'Bannon's turn to be amazed. "First, where did you manage to find a Bible in Alexandria, and second, what possible use could a Mohammedan have for a Bible?"

"Well, Mister O'Bannon, one can find anything if one looks hard enough, and Corporal Campbell was so very, very helpful, and to see the source of our strength and confidence, of course; but, where was I? Oh yes . . . firearms, swords, the Bible, the silver . . . "

Three other voices echoed in return. "Silver?"

Eaton looked as if he'd been physically challenged. "Well, one will need to pay one's crew . . . "

"I'd like an explanation of how, when and under whose authority you commandeered one of my non-commissioned officers," snarled O'Bannon.

"Reasonable," replied Blake.

" . . . some minor compensation may be needed along the way . . . "

"Prudent; but, about Campbell," commented O'Bannon.

"And, of course, we will need to get Hamet Bey started in his endeavors."

That was the most reasonable thing Eaton had said so far, but Hull looked more skeptical than ever. His jaw rested in his hand as he simply looked at Eaton the way he would a recalcitrant midshipman. After a pause that seemed to go on forever, Hull quietly asked, "Consul Eaton . . . how much money were you authorized for this expedition?"

Eaton smiled with a calm he didn't feel. "Two thousand dollars, Captain."

Hull's fingers tapped on his desktop like the slow beat of a drum leading mourners to a funeral. "Consul Eaton," Hull finally said, "are you telling me that

Good King Hamet and the House of Rendezvous

you have hired that barge and its highly trained and experienced crew, provisioned and equipped said barge and crew and still have more than sufficient funds remaining to accomplish your mission . . . for two thousand dollars?"

O'Bannon did some very quick estimates in his head and realized that, even with his limited experience, Eaton's accomplishment was most impressive, and highly unlikely.

For his part, though, Eaton remained utterly calm and self-assured as he literally purred. "My dear Captain Hull, you forget that I am a veteran of some years in this part of the world; I do know how to haggle. When you add to that the most gracious assistance of my colleagues at the British Consulate, it is surprisingly easy to accomplish."

Hull remained silent and glaring at the consul in response to this information; even in the cool November morning, Eaton felt a thin, cold trickle of sweat begin to roll down between his shoulder blades.

After what seemed like an eternity, Hull finally spoke in a low, menacing tone. "Consul Eaton . . . I do not believe you, sir. The only things preventing me from confining you to your quarters until I may conduct a full investigation are the facts that first, this mission, as lunatic as it is, is of the highest priority to my superiors. Secondly, my direct superior is a man who is not only the most unpleasant man in the United States Navy but who is also as deranged as you are."

"Captain," Eaton burst out in a wounded tone as he leapt up out of his seat, "I must protest in the strongest possible terms . . . "

"Sit down and shut up, Consul Eaton." Hull's voice was a sharp, quiet blade in the room, and Eaton decided that it would be wise to obey the Captain. "I cannot prove that you have misappropriated funds, Consul, and I do not have the time nor ability to inquire further. Let me simply say that if I again have reason to suspect that you have indeed defrauded the United States, and tried to deceive me, I will happily send you back to Washington in chains and take my chances with Acting Commodore Rodgers. Am I most clear on this?"

Eaton swallowed deeply. "Captain, you shall have no reason to suspect anything of the sort."

Hull's quick smile was a predatory one. "I am most gratified to hear it, Consul. Please feel free to go about your business."

Eaton stood and straightened his coat, bowing slightly. "Thank you, Captain." Turning to Blake and O'Bannon, Eaton merely inclined his head with a quiet "gentlemen," and left the cabin.

To Barbary's Far Shore

The two officers decided to remain prudently silent until Hull spoke, and they didn't need to wait long. "Two thousand dollars, my ass," Hull muttered quietly. "O'Bannon, how much do you think he really had?"

O'Bannon shook his head. "Captain, I don't know for sure. I remember, while we were still anchored in Siracusa, he brought aboard several small chests that I assumed contained money. I didn't inquire any further."

"Mmm. Let that be a lesson. Ask about everything, Leftenant. When is His Nibs planning on leaving?"

"First thing in the morning, sir. I would be guessing at an arrival sometime the next day or two."

Hull leaned back in his chair. "O'Bannon, I truly have no idea what Eaton is up to. At best, I believe he has badly misrepresented what he has been directed to do. At worst . . . "

Hull threw his hands in the air and shook his head. "Your role is to protect Eaton and serve as an ambassador to Hamet Bey. I am confident that you will be able to accomplish that much; but, whatever you do, do not provide aid and comfort to Eaton's fantasies. With any luck, Hamet will not be there, Eaton will give up, and we may find some practical way of freeing the crew of the *Philadelphia.*"

Blake looked at Hull with mild confusion. "Captain, if I didn't know better, I'd say you were hoping Eaton would fail."

"At the rate we're going, I am beginning to think that the only thing worse than his failure might indeed be his success. Dismissed, gentlemen."

And so the voyage up the Nile began. Eaton finished transferring his belongings over to the barge, and O'Bannon led the official detail aboard, himself, another one of *Argus*' junior officers, Lieutenant Joshua Blake, Midshipmen Peck and Danielson, Farquhar, Aletti, an Egyptian named Ali whom Eaton had hired as his personal servant, a Turk named Seid Selim and a half-dozen other assorted servants, batmen, and lackeys. The barge would have had its two flags flying in the breeze that was coming in off the Mediterranean as they slowly wandered out into the shallows, then headed east to pick up one of the many tributaries of the Nile that led back towards Cairo.

The voyage would have been mostly quiet, with the exception of Eaton continuing on about his plans and machinations and the sound of the rhythmic chanting of the Egyptian crew as they kept the barge heading in a straight line. The town they were headed for, a nondescript little village called Minieh, was a hundred and fifty miles south of Cairo. That meant that the expedition would have gotten a superb tour of the many wonders of ancient Egypt.

Good King Hamet and the House of Rendezvous

They wouldn't have known of many of them, mind you. The great wave of exploration hadn't quite started yet. The Rosetta Stone, the key that would unlock the otherworldly hieroglyphs of the Pharaohs, had only been discovered a few years before and would not be deciphered for nearly two more decades. But, the things they would have seen, they would have been marvelous. They would have passed Gizeh, where the Great Pyramid had stood noble for four thousand years, and where the Sphinx, at that time buried almost up to his chin and with a chipped face from where Napoleon's troops had used him for target practice, posed his silent, eternal question. One would like to think that William Eaton probably thought he could answer it, but he wouldn't have had the time to stop and try his luck. As they slid past Al-Kahira, for as such was Cairo known then, they would have seen a bustling Islamic city, with traders and farmers, fishermen and feluccas, all of them overflowing with goods that came up and down the Nile. There was a brief stop in Al-Kahira, where Eaton picked up another recruit for his mission, one of the most . . . unique . . . men to ever serve with the United States military.

His name, at least the one he gave Eaton, was Eugene Leitensdorfer, but he went by at least one other name, and certainly many, many more. A classic soldier of fortune, Leitensdorfer said he had served in, and deserted from, three armies, been a Capuchin monk, and finally a Muslim. For whatever truth may have faintly struggled beneath all of that, Leitensdorfer did show a positive genius for languages, especially the dialects of the North African coast. And, since he seemed to have an outstanding grasp of the sort of people they would be dealing with, Leitensdorfer probably seemed to be a good match for their adventure. A quick handshake, and almost certainly a few coins changing hands, and the deal was struck.

Finally, they would have passed the necropolis at Memphis; once the capital of Egypt, now a cemetery that stretched two miles deep along the west bank of the Nile and went south for another sixty; with nearly a hundred pyramids and uncounted millions of graves. Even in Biblical times, Memphis' reputation was an awesome one. The prophet Hosea, decrying the wickedness and sinfulness of the Israelites, thundered that Egypt would gather them up once more, and Memphis would bury them. The prophet knew from whence he spoke; the six hundred thousand who journeyed from Rameses to Succoth would easily have been swallowed up at Memphis.

Downstream, they approached a remote little village called Fiaume. It would have been at about that point, assuming the crew of the barge was paying attention to their surroundings, that they would have noticed watchful eyes upon them from the reeds and mud-brick buildings of the shore. Reasonable enough. Eaton and his expedition were outsiders deep inside a part of the world where outsiders are, at best, considered with extreme suspicion; at worst, as threats that needed to be eliminated with extreme prejudice. The many traditions in that part

To Barbary's Far Shore

of the world of hospitality to travelers were mostly honored in the breach, if at all, even when other Egyptians or fellow Muslims came through the area.

Those watchful eyes would have belonged to a race of warriors known as the Mamelukes, whose history long predated the deys, beys, and *et cetera* whose actions precipitated Eaton's visit. They had originated as a tribe in the Caucasus, known for their skills in the saddle and ferocity in combat. Their specialty was cavalry warfare, at which they had a terrifying reputation, and their hallmark was a long, curved saber that was remarkably well adapted for its purpose. Over time, their reputation made them highly valued both for military use and in the internecine political warfare of the Ottoman Empire. Over time, they eventually became a warrior caste much like the famed Janissaries; officially slaves, but eventually assuming true power, both military and political. But unlike the Janissaries, they remained loyal, more or less, to the Empire.

By the beginning of the 1800s, they were the military presence of the Empire in North Africa, and they were in political, social, and military control of Egypt. In fact, Cairo itself was for the most part a Mameluke citadel, built by and for them. A great many of Cairo's greatest mosques were built under Mameluke supervision and are noted to this day for their beauty and grace. By and large, the Mamelukes ran a fairly loose and tolerant operation for the Ottomans, making sure that sufficient tribute made it back to Constantinople and keeping the locals well behaved and compliant.

All that changed, however, when Napoleon arrived. At first, the French cavalry, marginally equipped, poorly trained, inexperienced and relying more on élan than tactics, was cut to ribbons on a regular basis by the Mamelukes, who pretty much had things their own way. This, of course, greatly displeased His Majesty, who decided that this course of events simply would not do.

On July 21st, 1798, the French army was deployed across the desert roughly seven miles north of the Pyramids at Gizeh, ready to meet the Mamelukes. Napoleon was in personal command of the army that morning, and say this for him, he was in rare form that day. With the Great Pyramid catching the glare of the rising sun, Napoleon turned to his men and roared "Soldiers, from the height of these pyramids, forty centuries look down upon you!!" That was enough for the poilus, who bellowed their approval and devotion back with a thunder unlike anything the Pharaohs in their long slumber had ever heard before.

For their part, the Mamelukes seem to have thought that this would be just another day at the office. They could not have been more mistaken. They charged the French formations, five huge squares. The Mamelukes were all flowing robes, glittering swords and blood-curdling ululations, but this time there was no French cavalry to be seen. The infantrymen grimly stood their ground. As the Mamelukes closed in at full gallop, they must have suddenly

Good King Hamet and the House of Rendezvous

realized that this battle was going to end far differently than all the others; but they did not retreat, they did not turn back. They kept charging right up to the moment when the French ranks and French artillery opened fire.

That first volley must have been murderous; the ones that followed were worse. The Mamelukes were funneled between the squares and picked off as they tried to get through and regroup. Regroup they did; instead of leaving the field, they charged again. This time, a handful of survivors left nearly six thousand dead on the field and fled for Cairo. The French lost perhaps a hundred dead and wounded.

In just a little over an hour, the power of the Mamelukes in Egypt was broken beyond hope. Three days later, Napoleon entered Cairo, and the survivors scattered. A few Mameluke clans went to ground south of Cairo and waited out Napoleon's rule, which ended just over a year later. That is where they were when Hamet Bey showed up on the run from his dear brother. Now, there was, however, one part of their military history that survived their defeat on the plateau of Gizeh; their sword.

The Mameluke sword is an awesomely beautiful weapon; long, slender, and slightly curved, like the scimitars it is related to. It is actually much closer to the Turkish killij, a dramatically curved saber, that in turn came down from the primitive blades that the great Khan and his hordes swung as they invaded what would later be the Ottoman Empire. These swords, light but strong, were passed down from father to son, or occasionally taken as a prize, and they were the swords (the Mamelukes were familiar with firearms, but preferred cold steel) that defended Hamet Bey. They were also the ancestors, after a fashion, of the current Marine sword.

The sword that Marine NCOs carry today is known officially by the prosaic moniker of sword, M1859, and is the oldest continuously carried weapon in US military service; the officer's sword is known simply as the Mameluke. Both have the unofficial nickname of the BFK, or Big Fucking Knife. The two of them have been a part of the Marine legend since that long-ago adventure, and how it came to be so is a remarkable story in and of itself. Without question, the first Mamelukes found their way into Marine service on the Derne expedition, but how they came into standard service after that is a bit unclear. In a fascinating article some years ago in the USMC historical journal Fortitudine, Brigadier General Edwin Simmons tried to trace the lineage of this noble piece. That will be looked at later; but, for now, we will simply say that it was here, along with so much of the Marine epic, that the story of the sword began.

In any event, back to Hamet's protectors. The Mamelukes at this point were much like the ronin, Japanese samurai without masters. When Hamet Bey arrived, he had only his closest advisors and retainers, and not much in the way of personal security. Hamet had plenty of money, at least sufficient to take care

To Barbary's Far Shore

of the battered Mamelukes, and they were at liberty, so it was a natural fit. And it was amid the quiet fierceness of the Mamelukes, the buried lost glory of the Pharaohs, and the timeless murmur of the Nile that Eaton, O'Bannon and party met Hamet Bashaw Bey Karamanli, the rightful Dey of Tripoli.

Though one would not know it from more recent combat, the Middle East has produced more than a few great military commanders. King Nebuchadnezzar, the man who built Babylon, was a legendary king and general. He took Egypt and Syria and subdued Palestine, leading Israel into a captivity that would be mourned by the chosen people as they sat beside the Tigris and the Euphrates. He built monuments to his campaigns with his name inscribed on every brick, and turned Babylon into a world power, with all its attendant risks, and became a man who held utter, complete, and absolute power over its inhabitants. And, as with more recent rulers of that ancient land, he would be brought low; not by another enraged superpower or by his own people, but by the hand of God, who turned him into a monster whose "hairs were grown like eagles' feathers, and his nails like birds' claws."

There was Salah ad-Din Yusuf Ibn Ayyub, a brilliant and noble Kurd who has come down to us as Saladin. His armies ruled Syria, conquered Egypt and defeated the Crusader armies so regularly that the Third Crusade was to a great extent paid for by a 'Saladin Tithe' levied on the faithful of England. For all his skill and leadership in combat, he was every bit the compassionate and merciful adversary and ruler that popular history makes him out to be. So much so that, when he died quite peacefully, by the way, in 1193 at his Damascus palace, it was found that he had given most of his personal fortune to charity. This could not, of course, have pleased his heirs.

Sadly, as the Islamic empires deteriorated, so did their level of military competence. The desert tribes slid slowly back into raiding, and badly organized raiding at that. Even the largest surviving Islamic army, that of the Ottoman Empire, had slowly fallen into the dereliction of a corrupt, unused garrison force. Military skill tended to take a back seat to political reliability and family connections, none of which had ever been shown to win wars or fight campaigns in any organized fashion that would be recognizable to a modern Western observer. Or a Western observer in 1805, for that matter. What passed for Arab, African or Berber military expertise was considered no more than mere banditry among the civilized militaries of the world. All things considered, that was what Yusef Bey was doing back in Tripoli, albeit on a larger scale.

Sadly, Hamet Bey was not one of the commanders whose skill was needed to retake Tripoli.

Mister Kemal might well have been one of the first to come bounding down the rickety dock to greet the delegation. Eaton would have strode forward to take command of things from the start. Once the pleasantries had been

Good King Hamet and the House of Rendezvous

dispensed with, and there would have been many of them, the delegation was led to meet the man they had traveled so far to meet.

We should, perhaps, take a small detour to provide a bit of background about the man who the United States had decided to support in its first foreign entanglement. Hamet, born Ahmed, but Yankee accents kept mangling this to Hamet, Bey Karamanli was born in the late 18th century as the middle son of Ali Karamanli, second of the Karamanli dynasty. Hamet's grandfather Ahmed had founded the line in 1711 after convincing the Ottomans that he was in fact the best person to run the place. As it turned out, he probably was; he ran Tripoli fairly well for that era and added a large amount of territory. His son Ali took over around 1754, which meant that Ahmed had a pretty good run and ended up dying quite quietly in bed.

Ali, sadly, was not cut from quite the same cloth as his father, but he made a good effort. He had three sons, Hassan, Hamet, and Yusuf, of whom we have already become acquainted. Ali had every intention of turning the throne over to Hassan, but Yusuf apparently believed that he deserved to be Pasha instead of his two older brothers. The fact that those two older brothers were still very much alive and well was but a small obstacle. Around 1780, Hassan was killed by Yusuf, who in turn made his desire to be Bey about as clear as he possibly could. However, for the next decade or so, Hamet and his court managed to hold Yusef off. Ali died thirteen years later, and Hamet was being measured for his coronation clothes when the Ottomans quite unexpectedly showed up and announced they were taking direct control of Tripoli once more. Hamet was, understandably, disappointed; Yusuf was furious, as he had intended to take power himself as soon as a few last minute arrangements could be made. The Ottomans did not really infringe upon the Karamanli family intrigues too much; instead, they just insured that things stayed quiet and the tribute flowed smoothly. It would have been well had the Ottomans paid just a bit more attention to what was going on behind the walls of the Karamanli palace.

One fine morning in 1795, Yusuf showed up with a small but well-trained and disciplined force borrowed from the Bey of Tunis and overthrew the Ottoman viceroy. Wisely, he let him live; to do otherwise would have brought the wrath of the Sultan directly down upon him. Yusuf sent him back to Constantinople with a message that the money would continue to flow, as always.

The second thing Yusuf did was to go after brother Hamet, the rightful heir. Hamet got out of Tripoli just a few steps ahead of Yusef's men, and completely without his family, who remained behind as hostages under Yusef's watchful eyes. Hamet retreated, oddly enough, to Tunis, which had provided the manpower to prevent him from taking power; far enough from Yusuf to appear less threatening, close enough to keep an eye on things. And apparently, Hamet

To Barbary's Far Shore

and the Tunisians worked on the understanding that this was nothing personal, just business. He was still residing there when Eaton first met him in 1801.

According to Richard Zacks in the superb *The Pirate Coast*, Eaton had come across intelligence that Yusuf was planning to eliminate Hamet once and for all by luring him back to Tripoli with the promise of partial power, then, quite reasonably, killing him at the earliest opportunity. Eaton warned Hamet, who was suitably grateful. So much so, Eaton came up with a very early version of the plan he would carry out four years later; let the US Navy escort you back and its firepower will help you regain your throne. In return, you shall promise to never take another American ship hostage.

On the face of it, it was a good idea, but it had some flaws. First, Eaton seemed to be quite sure that Hamet would never betray him. Hamet was, and would forever remain, personally grateful to Eaton for saving his life, but a great many assumptions were being made here. Second, the Navy was less than eager to assist. No one wanted to authorize the Mediterranean Squadron to start a war, most of all Commodore Richard Morris, who, as we have seen, liked remaining in port and avoiding conflict just fine, thank you very much. And, as it turns out, Morris was less than enchanted with Eaton, in any event, and said so in official reports. Be that as it may, Eaton had his mania now and intended to follow it. It would simply take a while.

In 1802, Hamet decided to try and take back his throne, or, at the very least, annoy his brother intensely. He went to the city of Derne, Tripoli; then, as today, a town that for some reason has a tendency to be displeased with whoever sits in authority in Tripoli. Yusuf either didn't know of or decided to overlook his brother's presence, believing him to be no threat. To his credit, Hamet proved him wrong. In the fall of 1803 when he gathered a small force together and marched on Tripoli. Early gains turned into reverses once Yusuf got rolling, and, in early 1804, Hamet was forced to make a run for Egypt, where Eaton found him hiding out some distance below Cairo.

At least, Eaton thought he had found him. Hamet, it seems, was hiding in fear of his life and didn't wish to speak to anyone; not even people who wanted to make him a king. Surrounded by the Mamelukes, who were in no mood to be pleasant to any uninvited guests, Eaton entered into a weird game of tag. The goal was to eventually make contact with Hamet Bey. But, in any event, Eaton could talk with the people there he could see, and make his pitch in the hope Hamet would come out to see him. The trouble was, even this would take time. In Muslim cultures, nothing political is straightforward, and it takes a great deal of time to get to the point of a visit. Eventually, however, Eaton got to the point, backed up by O'Bannon's military expertise, such as it was, and the presence of the midshipmen, whose sole task was to look vaguely military. Eaton wanted to form an army, ideally the smart, disciplined, and capable Mamelukes, to march

Good King Hamet and the House of Rendezvous

on Tripoli and place Hamet back into power. Eaton, of course, would command this proud force. Hamet would then promise to never, ever, ever molest American shipping again, once he was back in his rightful place.

As with Eaton's original plan from a few years before, it had some genuinely good points to it; not least of which was the fact that, if Tripoli could be taken out of the piracy equation, it would make things a great deal tougher for the Tunisians, the other big pirate power on the Barbary coast. The Tunisians would be effectively isolated, with the British to their west at Gibraltar and the Americans, at least nominally, to their east at Tripoli. American basing rights there were an unspoken condition of support. Now, short of an invasion of Tunisia, piracy wouldn't completely stop, but the Dey would have to think long and hard about just how hard he wanted to push matters.

An American friend in Tripoli would be a wonderful thing, especially if he owed us his safety and his throne. It was assumed that Hamet was at least slightly more decent than his brother Yusef, so our government would have had at least some hope that Hamet would have kept his word and stayed reasonably friendly to us. No matter what, an American-Tripolitian alliance would have had a great many benefits for all concerned, even if it were more to one side's benefit than the other.

On the other hand, and from all accounts, no one seems to have taken this point into consideration, unless we were committed to putting Yusuf's head on a pike outside of the Dey's palace in Tripoli, something that was not explicitly discussed, he would have been more than able to run for refuge in Tunis and bide his time until he was ready to recapture the throne he had stolen fair and square in the first place. All that meant that the United States, in order to keep the benefits it was working so hard to get, would have had to commit to an alliance with Hamet to insure he stayed on that throne once we put him there. This almost certainly would not have happened. Far too many people in Washington had been present to hear that city's namesake speak against 'foreign entanglements.' Even if they had been so inclined to make this sole exception, an unlikely event for a great many reasons, Thomas Jefferson would have sooner quit the President's House in disgust than ever sign such a treaty. Remember always that Jefferson looked inward to what we could be, not outward to what we had only so recently thrown off.

Hamet's people listened to Eaton and decided to pass the request back up the chain of command to Hamet, wherever he was. In the meantime, Eaton decided to look for Hamet on his own. He sent Leitensdorfer and at least one other courier out in the general direction of where they thought Hamet might be, with a plea to meet at the little port of Rashid, Anglicized to Rosetta, the finding place of the famous Stone, about sixty miles northeast of Alexandria.

To Barbary's Far Shore

This actually wouldn't have been as hard as one would think. Hamet was still going to live in as much kingly splendor as possible, and that sort of lifestyle would stand out like a resurrected Pharaoh in the Nile Valley. With that, Eaton and his crew moved back up the Nile to Alexandria and plan their next move.

The good news, once they got back to Alexandria and returned Captain Hull's cannon, was that the emissaries had indeed found Hamet Bey, and had brought letters. The bad news was that Hamet was insisting on a meeting at a point roughly two hundred miles away . . . back in the direction they had just come. There must have been a great deal of gnashing of teeth and rolling of eyes among the Americans, but there was nothing for it but to go back, this time over land. Horses were hired, packs filled and maps purchased. Once more Eaton rode into Egypt. He still had company; Joshua Blake, who was almost certainly given orders to keep a close watch upon the Consul, even if he could not keep a tight rein upon him, and George Mann, who appears to have been along simply for the hell of it.

On the chilly morning of January 22, 1805, Eaton, Blake, and George Mann headed south once again. They made it just about forty-one miles before running into a garrison of Ottoman infantry at the miserable little town of Damanhour, who promptly surrounded the infidel invaders and took them into custody. Eaton, brandishing his letters of passage from the Viceroy and applying all his formidable powers of persuasion, and probably a few more pieces of gold, not only convinced the local commander to let them continue, but to send out a local tribal chief who promised to bring Hamet back in about a week.

Given how much could have gone wrong with this particular piece of improvisation, but also being more than familiar with Eaton's luck so far, it should not surprise the gentle reader that, sure enough, on Sunday, February 3rd, the tribal chief came riding proudly back into Damanhour beside Hamet Bey Karamanli. There was a cheerful, but brief, reunion. On Monday the 4th, Eaton, Hamet, and the others headed back for Alexandria once more. There the plan was to put Hamet aboard *Argus* and sail for Derne, while the force they had raised marched there overland to meet him. There would be a wild, triumphal welcome, Hamet and his men would march on Tripoli, overthrow Yusuf and all would be well, *inshallah*.

At this point, international intrigue rears its lovely head.

The reader will remember that, prior to the unpleasantness upon the Barbary Coast, the United States was engaged in a conflict with France that has come down to us as the Quasi-War; a mostly naval and diplomatic war where the US Navy and the infant Marines gave the first inklings of what they were capable of, and not incidentally, handed the French their heads.

Good King Hamet and the House of Rendezvous

It should not be any great shock to therefore discover that the French were less than happy with the United States and would take whatever actions they could short of war to confound, irritate and just generally make life difficult for the Americans wherever they could. The French consul at Alexandria was apparently one of those who held a grudge. For, as soon as Eaton arrived back in town, he began dropping hints to the local Ottoman commanders and the governor that the Americans were not what they seemed to be. In fact, they were British intelligence agents sent to undermine Ottoman authority and power.

Sadly, reports of conspiracies and plots have always found fertile soil in that part of the world, mostly because they offer those who believe in them a chance to blame someone else for their misfortunes. The French consul was probably banking on this. The fact that Richard Farquhar, a true and genuine son of Albion but, nonetheless, quite likely one of the King's Gentlemen, was along for the ride could not have helped but aid his case.

The Ottomans were neither foolish nor stupid. They were most certainly glad that someone was going to go into Tripoli and relieve them of this turbulent Bey without the need for Ottoman blood and treasure to do so. Too, the Americans had been honest; and, truth be told, quite generous up to this point. On the other hand, one couldn't simply ignore the statements of a man like the French consul, who also had the ears of people who had the ears of people with great power and authority. So, after much consideration, Something Had To Be Done.

The Ottoman leadership at Alexandria decided not to allow Hamet aboard *Argus*. What exactly this was thought to accomplish is unclear, but it was not for nothing the Ottomans and the term 'Byzantine' became eternally linked. But someone in Alexandria must have thought that keeping Hamet away from that little ship would sufficiently cover their own backsides if an emissary from Constantinople showed up asking difficult questions, and they were sure enough of it to simply leave the matter at that.

In the event, it turned out to be a good thing for the expedition. The soldiers they were assembling, including quite a few of Hamet's own men, having traveled back up from their bolt holes in the Nile Valley, turned out to have some qualms about following anyone besides Hamet down the road from Alexandria. This is not terribly surprising; during Operation Desert Storm, some of our Arab allies were less than eager to be under the command of anyone who wasn't a politically correct Muslim, whether or not he actually knew what he was doing in combat. But here this was somewhat understandable; they were fighting for Hamet, not the Americans, and without Hamet there the little army may well have fallen apart before it even got started. So, making a virtue from necessity, it was decided that Hamet would march across the desert at the head

To Barbary's Far Shore

of his army. Honor and punctilio were served; the Ottomans were satisfied, and the French were balked.

Of course, this now changed the basic nature of the operation, which required Eaton to once more go to Higher Authority and request assistance in light of the new circumstances. With that in mind, Hull got underway with *Argus* on Valentine's Day, headed for Malta with dispatches and demands from Eaton for more men, ships, and money than were available to the entire naval service, much less his own needs. One can only imagine the response those demands would receive, but Eaton was serene in his belief that he would get every last bit of it.

Hull was instructed to meet Eaton at Bomba Bay, along the coast road, with all the supplies and reinforcements he could bring. Undoubtedly Eaton had visions of the entire Mediterranean Squadron appearing over the horizon and disgorging weapons, food, munitions, and more men than he could count as they rallied and marched to deliver divine justice to Yusuf Bey Karamanli. Hull, for his part, was undoubtedly was glad to see William Eaton fade into the distance.

A few days later, on February 23rd, Eaton and Hamet sat down in Alexandria to lay out the nuts and bolts of the joint American-Tripolitan effort to overthrow Yusuf and place Hamet on the throne. One can easily imagine that Eaton did a great deal of the talking and Hamet simply made sure that all of it ended in the words "your return to the throne." The agreement that the two men signed can be summarized thusly:

- The United States Government would do everything in its power to put Hamet back on his throne. This was probably far more than Eaton had any right to claim, but never mind.

- Eaton would receive the title of "General in Chief" and be in overall military command of the mission, but Hamet would be its titular leader.

- Hamet, for his part, would promise to repay every dime the United States had or would spend in this effort, by paying the United States a share of the tribute paid by other nations to Tripoli.

The mind boggles.

The efforts to build an army were still underway, and it will be appreciated that this was no easy task. The hard core of Eaton's force would be the few trained troops that Hamet still had with him and under his direct command, approximately one hundred cavalry whose skill was assumed to be up to the task at hand. Hamet also had about another one hundred men as a combination of personal bodyguard and servants. These were his most loyal followers and unquestionably loyal to him unto death. They would follow whatever orders

Good King Hamet and the House of Rendezvous

their bey gave them, though how well they would respond to William Eaton was still up in the air.

There would also be the marines, Presley O'Bannon, Edmund Bancroft, Joseph Pritchard, Bernhard Curtz, Patrick Duffy, Bernard O'Brien, John Whitten, and Ned Stewart, and the midshipmen, in the form of George Mann, Eli Danielson, and Pascal Peck. The staff would be General-in-Chief Eaton, with Richard Farquhar and Eugene Leitensdorfer appointed as full colonels, and a certain Doctor Mendrici as the entire medical team. Well and good, but approximately two hundred and twenty-five men do not an army make. Eaton had, fortunately, foreseen this. While he was off trying to find Hamet in the first place, he had left behind instructions to raise sufficient forces to make the whole effort worthwhile.

At this time the concept of the mercenary was, if not exactly new, not one that most commanders would embrace unless it was a last resort. The first large use of mercenaries was during the religious wars that tore Europe apart in the 1500s. The concept of the standing army had not yet been created and wouldn't be, really, for a century or so yet. Princes who wished to go to war had to find the manpower somewhere. The Swiss, strong, smart, disciplined, and very, very good in combat, were always favorites; so too were Irish, and as were the Germans. That tradition went from the 1500s to the 1770s, when the Landgrave of Hesse-Kassel offered to provide infantry units to the British for use fighting in the Colonies. As it was for some reason always easier for the British to pay for mercenaries than to find enough of their own citizens to fight, all the more so because a surprising number of British citizens and members of the ruling classes were in sympathy with the Americans, but never mind, the British Army accepted with alacrity. They were good soldiers, if a bit dim and notably brutish, but their only real effect on the war was to push a great many conditional Loyalists off the fence and onto the side of the Rebels.

The last resort aspect should be evident. When a military unit's allegiance is money instead of king and country, and often without even the unifying tie of a common language, there is always the fear lurking in the back of one's mind that they are almost as much a liability as a blessing; a potential disaster in the making if everything goes suddenly wrong. For William Eaton, the use of a hired army was at best his second resort.

He seems to have believed that a great many more men had followed Hamet into exile than actually had, though it is not hard to understand. Hamet had actually made some fairly impressive gains against Yusuf while operating from Derne, and it would have been reasonable, though prudent to check it out thoroughly, to assume he had kept his army together. He had not; most of them likely went to ground between Derne and Alexandria, and were making do as simple civilians.

To Barbary's Far Shore

In any event, Eaton had to find soldiers and find them quickly. He always thought big. One author says he was confident that he could raise an army of twenty to thirty thousand men if he'd had the money; it should be pointed out that this was probably twice the number of men on active duty with the Army, Navy, and Marines in 1805. But, no matter how high voltage his sales pitches and pleas for assistance were, he knew he was limited in what he could actually pull together. So, around Christmas of 1804, Eaton directed the opening in Alexandria of a recruiting center where men who were interested in taking Mister Jefferson's Dollar could meet, hear the terms, and sign their name. Places of this sort were called, for reasons unknown to us at this distance in history, houses of rendezvous. Likely that was a metaphor for recruiting station at some time in the past, and the term was still used.

Posters would have gone up in Alexandria touting the fortunes to be made and the excitement to be had simply for signing one's name or whatever mark one made. The word would have gone out by caravan and ship to Rosetta, Malta, Cyprus and other Mediterranean ports that men were needed. The men interested would have slowly made their way to Alexandria, to see what they could find. Military recruiters today, this author having had remarkably unpleasant experience in that particular form of misery, are very closely restricted in what they can tell potential recruits; but, in those long-gone days, it was a different story. Truth was an elastic concept, if it existed at all, and just about any lie, untruth, deceit, falsity, prevarication and whopper was fair game if it brought warm bodies in the door.

From the side of those signing up for the grand adventure, it was just as bad, if not worse. There were, quite simply, damned good reasons these kinds of people weren't otherwise occupied in good, honest trade. Now, for instance, the Hessians were, as a rule, reasonably honest men who were hired out by der Landgrave and were decent, hardworking men who happened to be in a hard profession. But, for the most part, someone who walked into a house of rendezvous and signed on to fight for a cause he was not only uninformed of, but also, most likely, unable to even understand, usually had a good reason for not doing something else a great deal calmer and safer. In many cases, these men were former soldiers; paid off or frequently shown the door for misconduct, and in those days, that would have been misconduct indeed. They had developed a taste for the soldiering life and simply couldn't adjust to the tedium of the civilian world. Quite a few were, to put it bluntly, on the run for one reason or another, ruined women, abandoned families, confidence men, brushes with one or more local magistrates and/or sheriffs and more than the occasional robbery, assault or flat out violation of the Sixth Commandment.

Now, eventually, someone would have the good sense to figure out a way to put those seeking to be lost to good use. The French, practical souls they are, came up with the Foreign Legion twenty-six years later, where the rule was that

Good King Hamet and the House of Rendezvous

no matter what you were or had done, all was forgiven and forgotten. Of course, this would be at the cost of total devotion to the Legion and a better than average chance of being killed in the process, but never mind. But before the Legion, there were only the houses of rendezvous.

Between Christmas, 1804 and March 7th, 1805, approximately four hundred men signed up to take Hamet Bey to Derne and then on to Tripoli. Gardner Allen describes them marvelously and succinctly as follows:

" . . . twenty-five cannoniers of various nationalities, with three officers; thirty-eight Greeks with two officers, . . . a party of Arab cavalry under Shiek al-Tahib and another chief, altogether about four hundred men and a caravan of one hundred and seven camels and some asses . . . "

Everything was set to get underway early on the morning of Friday, March 8th. There was one last matter to take care of first, however.

On the morning of March 2nd, the supplies they had were being loaded onto a small vessel to get them to the small village of Marabout, likely today's al-Shaikh Mabrouk, about twenty miles southwest of Alexandria. Hamet and the mercenaries were gathered there to keep the whole thing as far away from the eyes of the authorities as possible. It wasn't that the Ottoman leadership didn't know about the planned invasion; as has been mentioned previously, they knew and more-or-less approved, but probably preferred to cast a blind eye upon the whole enterprise. But, just as the supplies, all rather conveniently in one place, were secure aboard the ship, Turkish troops showed up and seized it, along with everything aboard. At the same time, another detachment showed up at Marabout to arrest Hamet, who showed his mettle by wanting to make a run for it and abandoning his troops. O'Bannon and his men, at Marabout for exactly this kind of emergency as well as security, collared Hamet and, only through the most strenuous argument, managed to keep him in camp.

What happened and why is one of those details about this mission we will never be sure about. It could have been the French causing mischief again. It could have been Yusuf trying to stop his dear brother. Or it could have been a local official who felt that he had been denied his proper share of the baksheesh being so liberally spread around. This was Eaton's belief; in the event, he offered said official more money and the problems magically disappeared. This being the Middle East, however, it does not completely rule out any of the other two possibilities. But, in any event, the matter was more or less settled.

The new problem this left in Eaton's lap was that the money was now starting to run out. One had to be generous with bribes; that was the way of things in North Africa. Eaton had allowed for that, but he had cut it close; he had just enough money now to get the expedition underway. Hopefully, Hull and the

To Barbary's Far Shore

rest of the Mediterranean Squadron would show up with men, munitions and money in time to avoid a complete collapse.

Hopefully.

It is now the late evening of Thursday, February 7th, 1805. The sun is setting, a gorgeous red/pink band down along the horizon as the sun gently disappears. Presley O'Bannon and Edmund Bancroft are taking one last stroll through the encampments and checking on their own men, while Farquhar, Aletti and Leitensdorfer do the same for their little force. The equipment has been checked, the supplies packed and everything checked and rechecked a dozen times, if only to keep busy. There is a murmur of men and animals, but not the cacophony one would expect. Rather, it is a firm, solid tone that says these men mean business, and they are ready for it. There is some laughter around the campfires, some singing, but these are the exceptions.

O'Bannon stopped and looked around for a moment, before turning to Bancroft. "You know, one could almost enjoy this sort of thing. The sense of adventure; the feeling that one is part of something larger than one's self."

Bancroft never took his eyes off the camels and their herders in the makeshift pen that had been put up some yards away. "Beggin' the Leftenant's pardon, but I'm going to guess that he's never actually been shot at."

O'Bannon was glad for the dusk, because it meant Bancroft couldn't see him blush. "No, I haven't, Sergeant, but . . . "

"But nothin', sar." Bancroft replied in a tone that bore some mild reproach. "For the record, sar, a boardin' party is miserable enough, w' a great many people ye don't know trying to kill ye on general principle. A landin' party is just as bad, w' the exception of the fact that the land doesn't quite move as much. But this . . . "

Bancroft moved one hand in a sweeping gesture. "Dear God in Heaven above, sar . . . I've served as a marine since 1775, and I've never seen anythin' quite like this. We're goin' to be travelin,' assumin' General Eaton's calls are correct, of . . . F'r pity's sake, what am I sayin'? Of course, he's correct. He's always correct; just ask him. Travelin' about five hundred miles, on foot, with . . . this . . . "

And we're led by . . . "

"Halloooo!!!!!"

With that, General William Eaton came galloping up the road, his robes flowing like clouds behind him as he came into the camp. He headed for the well-lit and guarded tent that marked his field headquarters, next to an even more ornate tent where Hamet Bey pondered the fate of thrones behind a

Good King Hamet and the House of Rendezvous

phalanx of very heavily armed men; after all, one could never be too sure. A cheer went up from the gathered men, but interestingly enough, none from the small marine camp where Obie O'Brien had drawn first guard duty.

Standing beside Joe Pritchard's small tent, Obie asked quizzically, "tell me, Joe; whaddya think of the General?"

Pritchard thought for a moment, then simply nodded. "He's quite mad, you know." He pulled his head in and tried to get some sleep, despite the slow, lilting tune the lieuenant was playing on his fiddle. *Oh well*, Pritchard thought. *From the sounds they are making, the donkeys like it . .*

To Barbary's Far Shore

TEN: THE MARCH

Friday, February Eighth, Eighteen Hundred And Five

They had a surfeit of drummers in the group; or, at least, a great many people who claimed to know how to play the drums. The one who was awake this early was sufficiently talented to play something that resembled "To Quarters." The marines knew that one intimately, and, before the staccato drum roll had echoed into the distance, they were already up, getting a quick breakfast, and brewing coffee. Now, for what it was worth, breakfast would have been problematic. Today's Marine Corps says that every man will have three hot meals a day, and if that is not possible, they will get meals, ready to eat, also known as the MRE; a meal intended to give a marine all his daily nutritional requirements in a single meal, and at the same time give him or her something that tastes close to what they would get at home.

O'Bannon and his marines, sadly, did not have anything quite so welcoming. The available food for breakfast that morning, and, as it turned out, every other meal on every other day, was either rice or what was optimistically called 'biscuit.' We know it as hardtack, a simple cracker made of flour, water, and if one was lucky, salt or other seasoning. It was rarely ever baked to a standardized recipe; even if it was, its quality could, to be charitable, vary pretty wildly. It was likely to be moldy; not a serious problem, one simply took the corner of one's uniform blouse and wiped it off. It was frequently infested with weevils and could be as hard as a rock. There were a few barrels of beans and some grain, but, on the whole, this was what the expedition had to eat.

Richard Farquhar, who was in charge of provisioning, did make sure Hamet Bey had plenty of decent food and drink; fresh meats and plenty of spiritus fermenti, which, on reflection, should have been off limits to a good

The March

Muslim like Hamet, but never mind. For the record, Farquhar was also doing a pretty substantial amount of skimming off the provisioning accounts, for which he would be relieved of those responsibilities in the very near future. But, for now, he was making sure that His Majesty was more than comfortable.

William Eaton, in all his dreams of glory, had never imagined anything like this, even for a moment. Beside him rode Richard Farquhar, Colonel of Volunteers and his servant Aletti, massive and mountainous on two coal black chargers. Next to them, Colonel Eugene Leitensdorfer grinned like a fiend as his horse wheeled about in the dust. Behind him, Lieutenant of Marines Presley Neville O'Bannon, in the best uniform he could muster, sat on his own horse before the midshipmen and, behind them, the marines. And finally, behind the marines, Eaton's army; the force through which he would administer righteous justice upon Yusef Bey Karamanli, restore Hamet Bey, who was somewhere in the middle of that army, surrounded by a hundred or so bodyguards, to his throne, free three hundred odd of his countrymen from a captivity of medieval torture.

Eaton searched his thoughts for something noble to say, but decided that there would be plenty of time to come up with an appropriate thought later. With that, he reared his horse, its front hooves beating the air, and waved his right arm forward. There was a roar of elation as the mass of men slowly, ponderously began to move down the coast road, southbound from Marabout.

"Well," Sergeant Bancroft said, mostly to himself, "we really are going on with this, then."

Campbell spoke not too loudly, lest O'Bannon hear. "That we are, Sarn't . . . and that means you owe me five dollars. A bet's a bet."

"Shove it up yer arse, Arthur," Bancroft said with a grinning snarl. "A noncommissioned officer isn't allowed to gamble."

The dust was already rising in a slow, steady tide as they tramped forward. Obie O'Brien looked around for a moment in wonder. "Like nothin' I've ever seen, I tell you," he said with a fascinated smile. "Must of looked like this on the Exodus, it did."

John Whitten gave a snort of laughter. "The Israelites had the good sense to be headin' the other way."

Bernhard Curtz nodded lugubriously. "Und dey hadt vimmen."

The planning for the invasion of Tripoli, for regardless of the scale, that was certainly their intent, was to make about thirty-four miles a day.

The first day they made fifteen miles from Marabout, roughly to the little town of el-Hamam. Keep in mind, the average walking speed of a man is around three and a half miles an hour; so, clearly, they seem to have been behind the curve from the start. There are a great many possible reasons, not least of which was that this was a group of most unmilitary individuals trying to move in a military fashion, and failing miserably. Had there been the time or inclination to even try and practice moving as a unit, and units of far larger size moved faster on a regular basis, they may have done better, but this was about the best they could do.

"Leftenant."

O'Bannon knew that Bancroft waking him up while it was still dark couldn't be good. He quickly rolled over to see the sergeant looking out towards the camel herds, where some kind of argument could be heard from there.

"I'm awake, Sergeant. What's the matter?"

Bancroft shook his head. "The Leftenant might want to go over to the camel pens, and right quick; the damn' drivers are threatening to leave."

"WHAT? We've only gone one day."

"I know that, the Leftenant knows that, and so does everyone else, but they're trying to leave, and General His Bloody Majesty Eaton is trying to convince them to stay." Bancroft cocked one eyebrow as he said that, and O'Bannon knew his duty. Scrambling from his bedroll, he grabbed his shako and jacket and headed for the camel pen.

This was the situation. Just after the fajr call to prayer at around 4:15, the camel drivers sat down with their charges and stayed sat down, refusing to move. Alert members of the force notified the leadership. Leitensdorfer, master of languages, got there first, with Eaton hot upon his heels. There was much shouting and gesticulating, with the camel drivers appearing not only aggrieved but insulted.

"Leitensdorfer, what in God's name are they saying?" Eaton demanded.

Leitensdorfer, still trying to sort out individual complaints from the mob of Egyptians who faced him, motioned for quiet. He was nominally successful as he turned to Eaton. "Herr General," the Colonel explained, "zey say zey hadt no idea zat zey would be going zizz far . . . "

"'This far??'" Eaton asked, dumbfounded. "We have only traveled fifteen miles!"

"I am not sure ze distance iss relevant to our friendts, but I shall ask." Leitensdorfer and the Arabs traded incoherent shouts for a moment, then the Colonel turned to Eaton. "Zey say it does not matter; zey do not belief you intendt to pay zem."

The March

Eaton's expression was now one of hurt anger. "Of course I'm going to pay, them, I'm a man of my word, a Christian . . . "

"Vich may be ze actual problem here, Herr General. Zey zay you are an infidel not deserving of zere trust, und zey vill not moof another step unless they are paid in full now."

"They cannot do that!" Eaton roared, then he turned to the camel drivers and bellowed. "RETURN TO YOUR ANIMALS AND FOLLOW ME, OR I SHALL HAVE YOU SHOT!!"

Leitensdorfer gave a short, sharp bow. "I applaudt the Herr General's choice of disciplinary messures! I shouldt, however, point out zat it iss extremely unlikely zat any of zem understood a zingle vordt you saidt."

As it turns out, someone may have understood the word, and proceeded to make things even more difficult: Sheikh al-Tahib, leader of the Arab cavalry. It appears that the Shiekh and his cavalry were, in fact, not far removed from a fairly competent band of bandits on horseback, but one takes employment where one can get it. Along with that, the Shiekh, the word is actually 'Shaykh', which means 'elder,' generally signifying the head of a family or clan, or someone deserving of exceptional respect, does not seem to have liked infidels one little bit, and the infidel Eaton even less than most.

From the beginning of the expedition, he seems to have been belittling Eaton at every turn, and especially so to the camel drivers. Al-Tahib was now instigating the camel drivers, it's not clear whether or not Eaton knew this, and they were digging in their heels; no money, no movement. As the camels were carrying all their supplies, from beans to bullets, this became a matter of some urgency. A messenger was eventually sent to Hamet Bey's tent with an urgent request that His Majesty please come out to assist with the negotiations. Sadly, His Majesty was unable to be of any aid, but he wished General Eaton well in his efforts. If anything gave Eaton second thoughts about the man he was backing, this should have. It certainly did nothing to raise the general opinion of Hamet through the rest of the army.

The standoff lasted until about noon on March 10th, almost thirty-six hours later. Eaton was negotiating with the camel drivers, who in turn were parroting whatever al-Tahib was feeding them, and Hamet Bey was of no help at all. So Eaton hit on the idea of giving up the whole thing; or, at least, pretending to. Forming up the marines and a few others he believed reliable, Eaton let it be known he was heading back to Alexandria.

This placed the camel drivers in an awkward position. Going on offered them the possibility of at least some extra money; whereas going back to Alexandria on their own guaranteed them nothing. Apparently, al-Tahib never entertained the possibility that Eaton might just tell them all to go to Hell. Eaton

To Barbary's Far Shore

and his loyalists actually ended up going a short distance back before some hushed, hurried conversations among the camel drivers resulted in a decision to press on. But they would be keeping an eye on Eaton. After many shaken hands and bows, the army ended up making another twelve miles westward before camping for the night.

Twenty miles went by on March 11th, and then early in the morning of the 12th, a single rider came out of the west with news that electrified everyone. Derne was rising! Hamet's many friends there had decided to support his efforts, or, more properly, Eaton's efforts, to overthrow Yusef. This was stunning, but also welcome. It proved, if true, that there was a base of support and manpower in Tripoli that would be beyond priceless to their plans. Everyone welcomed this news, most of all the Arab cavalry, who proceeded to open fire into the air in the odd tradition of armed celebration in that part of the world.

This, in turn, convinced the already suspicious camel drivers that they were under attack. Since, by their way of thinking, such an attack could only be a fiendish infidel plot to kill them and avoid any necessity for paying them, they drew down on anyone they could identify as a Christian. That in turn led to an exceptionally tense armed standoff that lasted several hours before chiefs on both sides of the religious divide were able to settle things down.

In the Middle East, religion is always the determining factor, no matter what else there may be. Eaton had just gotten a very strong lesson in that. He might have been able to deal with it on an individual basis, but his men, the ones who would have to sleep, march, and fight beside each other, would have a very different outlook from that moment on.

It began to rain on the 15th. Most people tend to be surprised when hearing about rain in the desert; but, keep in mind that 'desert' only means less than sixteen inches of rainfall a year. If it all comes at once, or even a part of that, life can come to a complete halt until Nature returns to normal. The ground saturated quickly, and little depressions became big puddles that one had to try and work around, but everyone gave it a good try and marched on. By March 18th, they had made it about one hundred and fifty miles from Alexandria; or, to be more precise, about twelve or thirteen miles past the present day site of Zawya Sidi Mousa, Egypt.

"Leftenant . . . they're at it again."

This time, O'Bannon had the good sense to keep everything within arm's reach and his boots on as he and Bancroft took off running.

This is what had happened: the camel drivers awoke that morning, saddled their charges, and began to head back to Alexandria. Colonel Leitensdorfer's linguistic talents came in handy once more, but this time he could bring no good

The March

news; in fact, only bad. The camel drivers said that His Majesty Hamet Bey had only hired them to go this far and no further, and that was that.

This time, it does not appear that Shiek al-Tahib was responsible for the situation, though he did little enough to help resolve it. Hamet had indeed paid the camel drivers to go just one hundred and fifty miles, and their native sense of direction and location would have done credit to a GPS unit. Hamet, due, of course, to pressing matters of state, was unable to step out and try to convince the drivers to proceed further, but it is certain that he expressed complete and total confidence in General Eaton's ability to resolve the matter.

This was blackmail in its most naked form. One is reminded of the old joke where a criminal says to his victim "Blackmail is such a harsh word; I prefer 'extortion,'" and it worked. Eaton literally passed the hat and came up with enough money to keep them for another forty-eight hours; or, at least, until they got to, more or less, present day Mersa Matruh, and still in Egypt, by-the-by.

The promises of fealty, the protestations of loyalty, and the protection bought by Eaton's last dollars lasted until very early on the mornings of the 19th and 20th, wherein the camel drivers, once more under the sway of Shiek al-Tahib, decided that they had had quite enough and simply started returning up the coast road to Alexandria.

As if this were not bad enough, it got worse. All the Arab chiefs, al-Tahib at their head, now demanded that they halt in place until someone could be sent to Bomba and back to see if the US Navy had ever shown up. Once again, however, it was made clear that al-Tahib, for all his fighting and riding skills, had never quite dealt with anyone like William Eaton. For when the camel drivers went east, Eaton put his most reliable men on guard duty around all of the supplies those drivers had left upon the ground. Eaton then proceeded to cut off the food of everyone who was challenging him. That definitely included the chiefs, al-Tahib and, for good measure, Hamet Bey, who now emerged as a champion of negotiation and tolerance.

Now the desert Bedouin, of which al-Tahib surely sprang, are a hardy people who need little food and water to live their lives. But they most assuredly need some, and any effort to take it from Eaton and his guards would result in a gunfight that no one would win. More time was lost in face saving negotiation and parley, three long days, to be exact, with the end result that the chiefs and about half of the camel drivers decided to press on.

Two days after this, on or about March 22nd, the army came upon a sight that must have seemed like something from the Arabian Nights, a huge Bedouin encampment of the Eu ed Ali tribe, a group of people who intensely disliked Yusef Bey and had nothing but respect and admiration for his deposed brother; a respect that went considerably deeper than the mere financial concerns that

To Barbary's Far Shore

many in the army had joined for. And, most importantly, they showed it by enlisting with Hamet Bey for the duration; nearly one hundred more cavalry, another hundred and fifty infantry, and, most blessedly, more than sufficient camels to make good the desertions of the previous days. Eaton had the good sense at this point to bivouac here for a few days, let the army rest and resupply, and then move on.

The few days became the better part of a week, until yet another messenger came riding out of the desert. There was much hope that this was a dispatch from Bomba, where the United States Navy had arrived in force with whole ships bursting at the seams with supplies, fierce soldiers champing at the bit to get ashore, and chests overflowing with gold.

The reader who has stayed with us this far will have already come to the conclusion that it was no such luck. Instead, the message was that Yusuf Bey Karamanli was annoyed and he had finally bestirred himself to action.

Of course, Yusuf knew that his brother was coming back with an army. It was the most open of open secrets in Alexandria and Rosetta, and, if Yusuf was not exactly getting updates in real time, he was certainly being kept well informed. The problem, surprisingly, was what exactly to do about it.

Yusuf was no fool; he knew this was a threat. But he also had to take into account a great many other factors. First, the whole plan might just fall apart in the gritty sand of the coast road. If it didn't, he was going to have to do something, but that had its own risks.

First, armies were expensive; a lesson that Will Eaton took considerable time to learn. Although Yusuf Bey was, praise Allah, a wealthy man, it was his money and not to be distributed to a bunch of illiterate savages, even in so worthy a cause.

Second, and perhaps more important in this context, armies can change their minds. Hamet may have never actually ruled a day in his life, but that was beside the point. He was in the field with a large group of men with guns, and that made him a viable option to Yusuf's continued rule. All it would take would be for just one disaffected commander to decide that he was turning coats, and that would be it. At least a third of Tripoli would be in Hamet's hands, and the rest would be up for grabs. But since, at the end of it all, Yusef wasn't willing to bet that Hamet would be tolerant of him a second time, there was nothing else for it other than to open the treasure houses and send his men east.

Now, Eaton, O'Bannon, Farquhar, Leitensdorfer and etc. were not apprehensive, but rather ready to do what needed to be done. What needed to be done was to get to Derne with all possible speed. The sooner they held the town, the harder it would be for Yusef's men to dislodge them. On the other hand, if they got there and Yusef's forces had reinforced the town, that would be it. Even

The March

with their enlarged force, they wouldn't be able to take the place. The plan was still quite doable and, if they got moving now, there was no reason they could not pull it off.

There were, however, some monumental flies in the ointment at this point. First, a large number of Arabs, led by the courageous Shiek al-Tahib, made a run for it that night, believing that they were in imminent peril. Many of the newly hired camel drivers went with him. Hamet at first demanded that Eaton go after him with money and bring him back; but, for once, Eaton stood his ground and refused to do so.

Hamet was in a near state of collapse by this point. Even when al-Tahib showed back up less than a day later, it did nothing to buck him up. By that day, however, the 28th, Hamet was a nervous wreck and announced that he was striking his tents and heading back to Alexandria, refusing to have anything else to do with the plan and wanting nothing more than to go back to a life of modest safety. Eaton, surprisingly, agreed at first and allowed Hamet to begin on his way. Eaton had the good sense to keep all of Hamet's baggage and supplies, which almost certainly contained a great deal of money and other valuable property.

His Majesty reconsidered his position and returned just a few hours later.

Things now began to descend into chaos. The Eu ed Alis, now looking at the possibility of actual combat, began to reconsider their commitment to Hamet Bey. They decided that, with Yusef's army inbound, it would be perhaps better if they rejoined their families; their sense of urgency in this matter was reinforced by the counsel of Sheik al-Tahib. They left the night of the 28th. 'An officer was sent after them,' likely the multilingual Leitensdorfer, and they returned before dawn.

The fun however, was not yet over. Al-Tahib and another Arab leader managed to get into a brawl, which ended in the second sheik and his people taking their scimitars and going home. For once, Hamet stepped up and personally went to talk to the wandering Arab and convince him to come back, which he succeeded in doing.

Four days later.

In the meantime, al-Tahib had come to Eaton's tent and asked, no, rather demanded, more food and supplies for his men. Eaton, now almost berserk with anger at the recalcitrant sheik, not only said no but promised that if he did not mend his ways immediately he would have the sheik arrested, tried for mutiny, and beheaded in the classic manner.

This was serious stuff at this point. Eaton was no longer in any mood to put up with al-Tahib's subversions, and he meant it. There is no doubt whatsoever

To Barbary's Far Shore

that, if al-Tahib had tried one more stunt, Eaton was ready to send his reliable Christians after him. On this matter, they were of the same mind as Eaton.

On the other hand, the Arabs would have joined forces against the infidel, no matter how deserving of chastise al-Tahib was, and the expedition would have ended in a fratricidal brawl out there in the desert. The Arabs might well have temporarily retreated, but after that all bets would have been off. Trying to get his loyalists back to Alexandria would have been a nightmare of harassing raids and ambushes all the way back; assuming they got all the way back, which would have been unlikely. Al-Tahib saved history this unpleasant ending by saddling up with two other sheiks and making a run for it. However, for reasons unclear, possibly including the fact that they had not yet been paid in full, and the plunder of Derne awaited them, they came back later that evening and swore to never again challenge Eaton's authority and to follow him unto death. Will Eaton would have been at least politely skeptical, but he shook hands and everyone went home.

Hamet Bey returned on or about April 1st and a monumentally furious General William Eaton was waiting for him. Eaton's patience with Hamet Bey and his putative allies had finally, completely and utterly run out, and he told him so in no uncertain terms. With that in mind, Eaton called a meeting of all the clan leaders and officers, Muslim and Christian alike, in Hamet's tent after dinner on April 2nd. Doing something he should have done before they ever took a single step from Alexandria, Eaton made an impassioned speech to the gathered leaders and convinced them to swear an oath of loyalty to Hamet Bey Karamanli.

For his part, Hamet stood tall and courageous this once and demanded that the leaders swear that oath. They did. With a great shouting of oaths and weapons fired into the air, all returned to their encampments. Eaton gave orders for everyone to prepare for departure in the morning. Although men worked like Trojans throughout the night to prepare, the General slept better than he had in weeks.

They were now three weeks into the expedition, with damned little to show for it.

The new loyalty and strength lasted exactly ten miles on April 3rd, which was when the majority of the Arabs refused to take a single step further until they had been provided with fresh dates. The nearest source of said dates was a ten-day round trip. It must have taken every bit of self control Eaton had to keep from executing the Arab chiefs on the spot, but he agreed to it, only if everyone else would press on to Bomba Bay. The date convoy, clearly so vital in the Arabs' eyes to the success of the mission, would meet them there. The Arabs reflected upon this and agreed, heading off towards the oases where the dates could be found.

The March

Eaton marched on. By now supplies were starting to be a serious problem once more. The reader will remember that they were barely provisioned in the first place, and the new Bedouin recruits brought little if anything with them, having almost nothing in the first place. The water started to become critical on the 4th, but by April 8th they had found a good and plentiful source that should keep them for some days longer yet. However, by the evening of the 8th, General William Eaton had a great deal more to worry about than mere water.

The dry goods were almost gone. There was sufficient water, but only enough food for another week, and that would have been stretching it. There was no sense in keeping this secret; everyone knew it perfectly well, including His Majesty Hamet Bey Karamanli. That night, he refused to go any further, quoting the well-known supply shortage. Hamet called Eaton to his quarters and told him that he, and his newly loyal followers, were stopping right there and would go into bivouac until a courier could be sent to Bomba to see if the Navy had finally shown up. Eaton responded by going to his fallback option, cutting off what rations were left to Hamet and his chiefs. But this time, the ruse failed, and did so spectacularly. Hamet rallied his men and got back on the road for Alexandria, apparently planning to forage until they got back to Eu ed Ali territory and then strike out for Egypt. Eaton knew that, without the Arabs, he and his men were far too deep into enemy territory to survive for long, if at all. He resorted to a most drastic measure. His loyalists set up a skirmish line across the road with weapons loaded and aimed squarely at the Arabs. The Arabs, of course, formed up their own line. The two sides stared at one another for an hour or so, before the Arabs blinked first and began to slowly straggle back westward.

What happened next could only be explained by General Eaton himself, and, since he has long since gone to sleep in the house of his fathers, we will never know for sure. We can guess from his previous actions that he was quite probably thinking that, with the Arabs backing down, he could rub their faces in it just a little. For, at this moment, Eaton decided to have his troops, lined up for combat, practice the Manual of Arms. This display of martial skill convinced the Arabs that they were about to be shot in the back. They not only returned to their skirmish line, but began making preparations to charge the Christian line, with Hamet himself at their head.

It is possible that more than a few men, on both sides, commented on the fact that the first time Hamet had mounted up and drawn his sword was in an impending attack on his own allies, but never mind. This time it was very serious. Hamet got his men formed up and began to lead them in a charge across the space between them and Eaton's loyalists. Eaton's people stood firm. The Arabs kept coming, at one point ordering their men to open fire on Eaton's officers. Whatever else those men may have been like, they had no intention of just standing there and taking it. They dropped to one knee and lined up on

To Barbary's Far Shore

Hamet and his sheiks. That appeared to have gotten everyone refocused on why they were there in the first place; after a few more tense moments, some of Hamet's officers rode out between the two lines at considerable danger to themselves, and got the sides separated and calmed down. Within a few hours, everyone was back in their encampments and Hamet and Eaton had mended their quarrel. It cannot, however, be doubted that, from that moment on, everyone looked at each other with a great deal of suspicion.

By the tenth of April, the army was on its way again and on the verge of starvation. The only food left was the rice and the water was being cut back once more. Although Hamet had behaved himself since the near shootout a few days before, he was growing increasingly morose; he began to suspect that the Americans were simply using him to draw his brother into an agreement to free Will Bainbridge and his men, and would then throw him aside. Gardner Allen dryly comments that this was 'a suspicion not wholly unreasonable.'

It continued to get worse; the artillerymen planned to mutiny for full rations later that day.

Eaton by now had the good sense to keep informers throughout the army. He saw this one coming, but now there was damned little that he could do about it. Food and water were at a critical stage; even if he had wanted to give the cannoniers a full day's ration, he couldn't. That night, after lights out, Eaton called O'Bannon into his tent and told him just how bad the situation was. O'Bannon brought in his marines and made sure they knew the full extent of the problem. If it came down to it, there was a good chance that it would come down to Eaton, the marines, the two colonels, and perhaps a handful of other Christians; and, if it was that bad, perhaps not even then. The mutiny, Eaton knew, would come the next morning. There was nothing to do but go to sleep and be ready.

In the event, unnecessarily. Just after dinner that evening, more rice, less water, a rider came thumping up the road from the west with news that had to have sent Eaton into transports of ecstasy: the Navy was at Bomba! With clarification, that became 'there were ships at Bomba.' Eaton does not seem to have inquired too closely into the report, but instead spread word throughout the camp. The artillerymen were almost immediately mollified, and everyone else decided that they were close enough now to make one last effort. On the morning of April 11th, Eaton led them out again, now with their spirits high and eyes keen. Bomba was close now, exactly how close, well, never mind, but everyone, most of all Will Eaton, felt it close enough to touch.

The last rice was issued on April 12th. Camels and sheep were slaughtered, and the troops were foraging for whatever greenery they could find. Everyone staggered ahead, believing that over the next rise, around the next curve, would

The March

be the blue plain of the Bay of Bomba and the white sails of the Navy. It was getting warmer now too, and the water was disappearing even faster.

On the early morning of April 15th, there would have been a stretch of three or four miles blocking their view of the bay until they got to the site of present day Ayn al Ulaymah, Libya. Eaton would have mounted one of the low hills in the area with his telescope, looked out to the blue Mediterranean and seen . . . nothing.

No ships, no camp, no supply depot, no anything. Eaton had advanced them as far as he could, hoping beyond hope that, any second now, sails would ease over the horizon and their deliverance would be assured.

Nothing.

They marched up to the ocean itself now, and the word had moved like lightning throughout the army that the Americans had not shown up.

They had been abandoned.

Eaton pressed on to Bomba itself, now the site of a Libyan military base and airfield, as well as the best anchorage in the bay. There was nothing to be seen as the night of April 15th closed in. The troops were roaring in fury. All of the conflicts, arguments and dissensions of the last thirty-seven days came back with a vengeance, and it was all aimed directly at the head of William Eaton. It got so bad that Eaton, the marines, the colonels and as many Christians as could fit ended up pulling back onto a low mountaintop and digging in behind huge bonfires. There were three possibilities now. One: a ferocious mutiny that ended in a glorious and brutal death for the Americans and the Christians. Two: a complete abandonment of the mission by Hamet and the Arabs.

Or three: all of the above.

The night of April 15-16, 1805 passed long and unpleasantly. Every single man on that mountaintop was unable to sleep; they crouched behind their weapons, listening for the prolonged, ululating roar that signaled the final and irrevocable termination of their mission. Eaton sat looking down past the glare of the fires, watched the pulsing movement in the camps below and wondered just how he was going to extract his men from this.

The sun came up around 6:40AM. It revealed the Arabs milling around in complete disarray, with random pillars of smoke rising at several points. Word had made it up the mountain that the Arabs were planning on scattering later that morning. Eaton began to feel that, at least, he stood a good chance of surviving that day. He still had to figure out how to get back to safety with the marines and —

"SAIL HO!!! SAIL HO!!!"

To Barbary's Far Shore

The call echoed down from the mountaintop. One of the men pointed out to sea, at a small white square that had without warning poked up over the horizon. A dull, hollow boom from far out to sea followed.

Without regard for his own safety, Eaton ran madly down to the encampment and convinced the army to fire just one, just ONE, signal shot out to sea. Anger and fury suddenly gave way to renewed hope. The muskets let go over Bomba's blue-brown water; a semi-circular ripple quickly expanding, then vanishing into the ocean. There was a pause, a long one. Eaton's heart stopped and he held his breath, knowing that the next moment or two held his fate. He wondered how he would face it if these men turned on him. Would he have a few glorious moments fighting to the end or would he just suddenly drop with a ball or blade in his back?

BOOM!

Or would he like an equestrian statue in Washington, or simply one of him holding a sword, gazing nobly into the distance?

It was *Argus*, arriving with timing that would be dismissed as too far fetched in a movie or television show. Isaac Hull sent a boat for him as soon as they were anchored. Eaton went aboard around noon to find out just how well fate had favored him.

The answer, it turned out, was quite nicely. Commodore Barron had sent a letter dated just a few days after they left Alexandria, and in it Eaton got almost everything he had wanted. There were tons of fresh supplies straining *Argus'* seams. Beef, mutton, grain, rice, biscuit and wine, enough for them to march all the way back to bloody Norfolk if they chose. There were thousands of dollars of gold; enough to keep every last damned Arab happy for the next year, much less the next few weeks. There were no extra men. Barron had to explain to Eaton one last time that asking for more men than existed in American military service would be less than useful, request denied.

It mattered not to Eaton; his joy was unconfined as he distributed money and food to everyone there. They'd also found a good-sized cistern at Bomba, so now they had water as well; enough to even risk trying to clean up a little bit. By this point, Yusuf Bey could probably smell them coming. Things got even better when *Hornet* pulled in on the 17th with even more provisions. Eaton then made the decision to remain at Bomba for a few more days while they got things sorted out and everyone had a chance to settle down and relax a bit.

The army stayed at Bomba until the 23rd, when Eaton finally gave them the word to get back on the road. Derne was now just shy of fifty miles away, two day's good march at worst. The Navy now had their right flank. They were more flush with supplies and morale than they had been since they left Alexandria; more so, in fact. There was now a sense that they could pull this off; that all it

The March

was going to take was one last concerted push, and Derne would be theirs. And then what? Push on to Tripoli? Topple Yusef himself from his throne, liberate the crew of the *Philadelphia*, possibly even rid the entire Barbary Coast of the curse of the pirates. Anything was possible that day.

Right up to the moment on the 24th that a courier, dusty and tired, came bounding up the Derne Road. His news was enough to stop the army in its tracks once more, and with good reason. Yusuf's men were closing in on Derne, perhaps only twenty or thirty miles to the west of the city. At their current rate of march, there was every likelihood that they would beat Eaton to the city. The spirit and good feeling of the last few days evaporated even more quickly than it had appeared, and most of all in those two pillars of the expedition, Sheik al-Tahib and Hamet Bey.

Al-Tahib screamed mountains of imprecations down upon Eaton, blaming him for the delays that let Yusuf get to the city first, conveniently forgetting that a great many of the delays could be laid at the feet of the sheik himself. And, of course, Hamet retreated to his tent in abject terror and sadness, now facing the possibility that he would actually have to face down Yusef's forces in combat and not march into an undefended city. Long discussions wound through the night of April 24-25, assisted by deliveries of funds and promises of more. By the morning of the 25th, all concerned were once more sufficiently motivated to press on.

Monday, March 25th 1805.

Eaton is at the head of his little army, moving down the coast road towards Derne, or Darna, or Darnah, whatever one wished to call it. The place has as many spellings as its future ruler, Moammar al-Gaddafi, had for his. The temperature will zoom from cold, easily down into the thirties, to the seventies in no time at all. The heavy wool uniforms, relatively new and clean since the rendezvous with *Argus*, of the Americans would have become uncomfortable just as quickly, and, of course, the dust they were kicking up would have stuck to everything. There is a reasonable breeze coming off the ocean, visible to their right about a thousand yards away. Everyone is far more silent than they have been for days; the only consistent sound the rattle and creak of gear and the braying and calls of the animals.

The coast road is still there, by the by. It is paralleled a few hundred yards to the south by a decent blacktop road; mostly decent, anyways. It peters out just before it would cross the coast road at the wadi east of town, probably a victim of the nightmarish civil war there. There is a dip of about ten feet through the wadi; men and animals stagger, slip and slide a little as they head down into it. Some will wonder idly about the river that once ran through it, probably in a time before there were even men here, but most will not even think about it. It is one more obstacle in a landscape filled with them; just one more thing to make

To Barbary's Far Shore

life difficult, *inshallah*. Eaton, however, is having the time of his life. He charges his horse up the west side of the wadi and stops at the top, then shouts for his officers to follow him. O'Bannon, Farquhar and Leitensdorfer spur their horses up the slope and pause at the top beside Eaton.

The prize lies before them.

They would be at the very eastern corner of modern Derne; a rise that gently rears up about two hundred and fifty feet above the rumbling blue ocean, about two and a half miles away from the 1805 city, whose eastern border is the wadi that today runs neatly through its center. The look of the city hasn't truly changed except for its size. Irregular clusters of dirty beige or concrete-block gray buildings seem to be moored in the sand and scrub; their only truly common feature is the dust.

It is open space between Eaton's army and the city they wish to take, and that truly worries the three professional warriors present. Eaton sees only a glorious charge with banners flying and shouts of courage and bravery. O'Bannon and the others see a killing ground that they will have to cross before they can even get within range of their target. The plain before the city has the Mediterranean to their right. Hills rise two hundred and fifty feet above their heads to the left, an almost unbroken wall that will funnel them right towards whoever might be preparing a welcome.

It was roughly dinnertime by this point. Even William Eaton was loath to launch an assault on an empty stomach, so the little army sat tight for the evening while its leadership decided what to do next. As the men ate and the officers deliberated, the army received visitors. More than a few local sheiks came riding out to meet the invaders and pay their respects to Hamet Bey, who they pledged undying loyalty to. Eaton had heard that one before, but never mind. After tea and polite conversation, everyone got down to business.

First, the good news. Hamet Bey had been a favorite of Derne, and the earlier report that it would back him seemed to be accurate, although the uprising reported back on March 12th does not appear to have materialized. Rather, His Majesty could count on the firm moral support of two thirds of the city. Better than nothing, one supposes, until the assembled worthies explained about that other third. The coastal district of the city, where the finest homes, businesses, mosques and, not incidentally, where the fort would be, was firmly in the hands of Yusuf's loyalists. The fort itself was not exactly what the Western-trained soldiers were expecting. To Leitensdorfer, who claimed to be a military engineer in an earlier career, it made perfect sense.

There was no fort as we understand it. No timbered palisade or monumental brick enclosure, but a series of low earthen works just northwest of the harbor, complimented by various and sundry buildings that had essentially

The March

been turned into part of the fort's wall. Each one in turn became an individual strongpoint. That would have even given Will Eaton pause; the idea of house-to-house fighting would have filled any reasonable man with dread. The fort's Sunday punch, its artillery, was comprised of eight nine-pound cannon covering the harbor proper. Now, in theory, at least, these weapons could be moved and brought to bear on the advancing force, so they needed to take a good hard look at their capabilities.

The nine-pounder was a fairly familiar weapon. The light cannon was usually found as a 'bow chaser' on warships; one of the cannon mounted in the extreme bow of the ship to fire on vessels being pursued. Most cannon of that era, regardless of size, had an extreme range of about two and a half miles, or three miles if properly mounted in a professionally built shore battery; hence, the concept of the 'three mile limit.'

Realistically, however, there were pretty severe limits to their abilities. Those abilities came down to training, the condition of the gun itself and the quality/condition of the propellant. Training, even to the extent of what the militia back home got, was probably nonexistent; the Derne crews were probably all operating on distant memory rather than recent practice. The condition of the cannon could not have been good. Sea air and lack of use would have equaled corrosion and corrosion in a cannon is an invitation to a disaster. Finally, the quality of the powder and shot would have been questionable. Trained, professional gun crews take care of such things, but it is not likely the Derne gun crews dealt with such things; the guns would work when they needed them to, *inshallah*. It is not at all likely they were stored in proper magazines, just whatever reasonably protected area close to the cannon they could find. Now, quite truthfully, all of these applied to Eaton and his men just as much as they did to the Derne fort. Just as Eaton's artillerymen had to have faith in their abilities and their weapons, Yusef's men behind the low pile of sand and scrub had the same confidence. And not even Will Eaton would have dismissed those eight cannon.

There was one other large-caliber weapon that could be brought to bear, and it was in an unusual place. Eaton spotted something he called a "ten-inch howitzer" that was mounted atop the governor's palace. Eaton, though not an artilleryman by trade, knew his weapons and is not likely to have misidentified it. However, there are a couple things here that do not sound quite right. A howitzer is defined as having a comparatively short barrel and small propellant charge, making it ideal for mobile forces, and that makes it an ideal weapon for the sort of fighting that one could reasonably expect if one was charged with the defense of the eastern gateway to Tripoli. But lugging a wheeled howitzer up to the top of a home sounds slightly out of kilter. The carriage would have had to been strongly chocked, as recoil forces would have been unpredictable, and there would have been little room on that roof for the carriage to roll back. On

To Barbary's Far Shore

the other hand, taking the carriage off and giving it a fixed mounting would have taken away a lot of the flexibility gained by getting it to such a wonderful vantage point, probably the best in town.

This author would then suggest that Eaton knew exactly what he was looking at, only that he didn't go into quite enough detail. It was indeed a howitzer, and may very well have been a minor curiosity known as an Abus gun. Developed by the Ottomans, it was a small, marginally portable weapon that could deliver a remarkable punch, a roughly five-pound projectile fired to quite some distance. It was mounted on a tripod of sorts that allowed for mobility and firepower. It came in several different calibers up to nine inches, so, allowing for the distance between Eaton and the gun and his telescope, a guess of ten inches is reasonable. Regardless of its precise size or identity, it was a large gun under the direct command of the area's leader, and, as such, probably had better trained gunners and better quality powder.

And then there was the matter of the fort's garrison.

Eaton's visitors gravely assured him that, to the best of their knowledge, there were approximately eight hundred of Yusuf's men in and around the fort. The attentive reader will remember that Eaton had, at best, roughly six hundred men. Military doctrine, then as now, called for at least a two-to-one superiority when on the offensive. As Eaton did not have sixteen hundred men or eighteen cannon, in fact, he had no cannon at all, though more than sufficient men to operate them, he had to have known that this would, at best, be tricky.

On the other hand, he had the thirty-two cannon aboard *Argus* and *Hornet*, though they had their own limitations. Still, in terms of artillery, at best it would be close. At least it would be, if the ships were here. The two vessels had to deal with the vagaries of wind and water, and that meant that on a bad day the army could cover more distance on foot than they could at sea. Since leaving Bomba, there had been a succession of bad days. The army had far outdistanced them, and, as of the evening of the 25th, the ships were nowhere to be seen.

Eaton mused over the maps and the notes they'd written upon them. For a moment, he was Caesar in his tent, musing over the fate of empires. "We could do it . . . and just charge . . . they'll never expect it . . . "

Leitensdorfer leaned in close, almost nose to nose with the Herr General. "Zir, I enchoy a gut fight as much as ze next man . . . but it is vell past time I zaid 'nein' to you. I am going to point out to you zat ve are outnumbert und any attack unter our present zircumstanzes vould be foolhardy at best, unt zuicidal at vorst. I vill not, I repeat, not, make vun schtep towards that fort unless ve haff a plan."

Eaton reacted as if slapped, and began to protest. "Colonel Leitensdorfer, you are about to find yourself at the wrong end of a noose . . . "

The March

"General Eaton," O'Bannon said quietly but firmly, "I am most definitely with the Colonel. Simply putting our heads down and charging would end up with you and most of us dead. I have sincerely followed and served you for more than a month now, but enough is enough. We have one more night; let us make a plan."

Eaton looked around with murder in his eyes, then whipped his head towards Farquhar. Before he could say anything, Farquhar simply shook his head. "I am on their side, General. I do not mind dying in combat, but dying in combat to no good purpose is something I will not do. Let us sit down and make a plan and stick to it. With everything else that has gone wrong, this is one chance I will not take."

Eaton was about to respond, and most unpleasantly, before Aletti, standing behind his master, just slowly shook his head. The assembled sheiks sat quietly, taking the whole show in and wondering just how it was going to end. In their culture, this would almost certainly finish with someone drawing a sword or pistol and the whole argument ending in a most spectacular fashion. They were, however, to be disappointed. Eaton took a few deep breaths, and closed his eyes for a moment, before clenching his teeth as he spoke. "Perhaps I should step back for a moment and consider my options . . . "

"A vise moof, Herr General," Leitensdorfer said. He turned and said something in Arabic to the sheiks. They responded with nods, comments and murmurs of assent. Leitensdorfer waited until they finished. "Zey say zat you are wise to vait until our schips appear, eefen iff it means Yusuf gets here first. Zey also suchest zat ve giffe ze vort a chance to zurrender. If zey feel zey can do zo mit out. how do you zay . . . loosing face, zey may do zo und completely affoid a battle."

Eaton contemplated this for a few heartbeats. His face brightened. "Colonel, they may indeed have a point. We shall work on our plan tonight, but, in the morning, we shall first give them a chance to surrender with honor. Fair enough?"

Leitensdorfer translated again, and this time the sheiks clapped their hands and voiced their approval.

Glasses were raised and protestations of loyalty and best wishes made. As the sheiks made their way back to Derne in the dark O'Bannon and Bancroft watched the small caravan move out in the waning last quarter moon.

"Beggin' the Leftenant's pardon," Bancroft asked, "but what did ye take away from that particular meetin'?"

O'Bannon reflected on that for a moment. "Well . . . First, they seem to want to avoid a fight by asking the fort for its surrender."

To Barbary's Far Shore

Images of the 'opposition' in the Barbary Wars are rare. This very poor quality image is entitled "The Bey of Tripoli" but it is not certain whether this is Good King Hamet or the treacherous Bey Yussuf. Source: National Archives.

Bancroft nodded sagely. "True enough, Leftenant; true enough. I'd also be pointin' out that they don't seem to be stayin.'"

It took O'Bannon a moment to realize exactly what Bancroft meant, and then he gave a start. "They're just going off to organize things . . . rise up from inside the town."

Bancroft shook his head. "No, sar. They're going to keep their heads down until they see who's winnin'. If it's us, they'll come out once we've won, an' tell us how they were on our side all the bloody time. If it's the other side . . . well, they'll do things to us that would leave ye sleepless for a month were I to tell ye

251

The March

of 'em, and then they'll tell Yusuf Bloody Bey that they were just waitin' for the right moment to rise up an' defeat the infidel."

O'Bannon said nothing for a moment, then tried to, moving his mouth but finding nothing would emerge.

Bancroft straightened his jacket. "Sar, 'tis goin' to be a truly long day tomorrow. I'd advise we get to bed and stay that way as long as we can."

Nodding, O'Bannon finally found his voice. "Of course, Edmund, of course. I'll be along shortly . . . good night."

Bancroft snapped to attention and gave O'Bannon a sharp, precise salute. "G' nite, sar." O'Bannon stiffened and returned the salute. Bancroft lowered his hand and spun on one heel to stride off into the night. As he disappeared into the shapeless mass that was Eaton's army, O'Bannon heard Bancroft mutter. " . . . all this bloody effort for nothin' more than a wide spot in the damned road . . . "

Bancroft went to sleep quickly, the way he always did before a fight, if he knew one was coming. The only thought in his mind as he drifted off was that, for the first night since Bomba, he couldn't hear O'Bannon's fiddle. *Mostly a blessing,* he thought.

To Barbary's Far Shore

ELEVEN: A WIDE SPOT IN THE ROAD

Friday, March Twenty-Sixth, Eighteen Hundred and Five. Two Miles East of Derne, Tripoli

R eveille, such as it was, was promptly at 6AM, though the Muslims had been up for some time following their first prayers. There were some mutterings about going to war on behalf of the infidels on *jumu'ah*, the Muslim Sabbath, but they were quickly overcome by the remembrance of what could be won this day. Surely Allah would forgive them. Huge bonfires were set, where they got the kindling is anyone's guess, to let the Navy, if they felt like appearing, know that they were here. That would probably be a good thing if the fight got underway today.

The marines of course, had been up as soon as their Muslim allies stirred, just to be on the safe side, and the midshipmen were up with them. Coffee and breakfast were fast, quiet occasions today. No one spoke much at all as they got their gear on and checked to make sure that everyone as ready as they could be. Of course, one is never ready for combat. There are plans and options and constant checklists. You will go over them once, twice, a dozen times and you can be convinced that you will be fine; nothing can possibly happen to you, you are invulnerable. At the same time you'll be just as utterly convinced that today, you will meet your Maker, and there is not a damned thing you can do about it. In the meantime, you have to appear as secure as possible in the knowledge that you will surely prevail because your cause is just.

The marines, with the midshipmen beside them, were lined up at Bancroft's command. O'Bannon was off at the command tent with Eaton, Hamet, and the other madmen in charge of this floating asylum. The routine was written in stone then; with few changes, it's come down to the Marines of the 21st century pretty much intact. Bancroft took one pass behind them, checking their packs, the forty pound canvas and brass-buckle bags they would live and fight from for however long it would take to finish the job over the next day or so. Then he took another

A Wide Spot In the Road

pass down the front of the line, a quick look at each man, straightening Pritchard's shako, checking each weapon for cleanliness. It only took a few minutes.

Bancroft stood in front of his men and went to parade rest. "All right, ye mob, listen up! With a little luck, we'll be over an' done with before bloody lunch, assumin' ye' can follow orders! Now this is th' drill. We stay around th' General an' the Leftenant, an' keep his arse safe at all costs. 'Tis a simple request I'm makin', an' I'm trustin' ye NOT to let it go wrong! If ye do, if Eaton or the Leftenant get so much as ONE SINGLE SCRATCH, ye'd better hope the bloody wogs get ye, because I will have no mercy upon ye when I get ye. AM I UNDERSTOOD?"

"SERGEANT, YES SERGEANT!"

There was a clatter of hooves as O'Bannon rode up and exchanged salutes with Bancroft. "G'mornin' sar, an' a lovely mornin' i'tis!"

O'Bannon could only grin as he replied. "Sergeant, I was under the impression noncommissioned officers were not allowed to lie."

Bancroft could only smile. "Beggin' the Leftenant's pardon, but I may simply be mistaken about the quality of the mornin'! Do we have any word on when we're movin' out?"

O'Bannon shook his head. "Stand by for a bit, Sergeant. General Eaton is about to try his diplomatic skills once more."

Bancroft rolled his eyes and shook his head. "As I remember, sar, the General waren't all that good at diplomacy, which is why we're here in the first damned place."

All O'Bannon could do was shrug and smile weakly in acknowledgement. "Stand the men down for a bit. we'll have plenty of warning when we're going in."

Bancroft saluted smartly as O'Bannon rode off, then turned to his men and barked. "You heard the man. Stand by 'til I tell you to move out. DISMISSED!" As the camp had already been broken, there really was nowhere to go, so the Americans simply stood around and looked at each other before sitting down and going to sleep. After all, if the gods were going to give you spare time, you might as well get some sleep. No sense in dying tired.

Tradition is that Eaton himself marched to the gates of the Derne fort to demand its surrender. One has to admit that there is a certain charm to that vision, not to mention a certain insanity that is hard to attribute to even the General and Special Naval Consul. But, in fact, he sent a delegation with a white

To Barbary's Far Shore

flag to its gates instead, with a note that he wrote himself, no doubt after several drafts. Richard Zacks reproduces the note in full:

> Sir,
>
> I want no territory. Advancing with me is the legitimate Sovereign of your country - Give me a passage through your city; and for the supplies we need you shall have fair compensation - Let no differences of religion induce us to shed the blood of harmless men who think little and know nothing.
>
> If you are a man of liberal mind, you will not balance on the propositions I offer. Hamet Bashaw pledges that you shall be established in your government. I shall see you tomorrow in the way of your choice.
>
> WILLIAM EATON

Now, written communications between the marines and their enemy are not unknown. It is said, and sadly, may indeed be no more than legend, that, during the Korean War, when the war had settled down to a grim standoff, a Marine unit relieved a US Army unit on the front line. That evening, they were attacked by Chinese forces. The marines promptly counterattacked, routing the Chinese and sending them back in full retreat. The next morning a Chinese delegation approached the marines under a flag of truce and delivered a letter for the marine CO before returning back to their own lines. The neatly typewritten letter is said to have read 'To the Marine Commander: Please accept our most humble apologies for attacking you last night. We thought you were the Army.'

If it isn't true, it should be. The point, however, is that this is how one communicates with the enemy before, during and after a fight, to this day. There is a protocol, a routine to it, and Eaton followed it to the letter. So did the commander of the Derne fort, who sent back a note an hour or two later.

> My head or yours.

That, of course, settled that.

Whatever plan Eaton had was joyously postponed at around 2PM, when a sentry reported sails coming over the horizon. The bonfires had helped, but not quite with the desired result: instead of *Argus* and *Hornet,* with their thirty-two guns, it was the schooner *Nautilus,* twelve guns, John Dent commanding. Fresh out of repairs at Livorno, Italy, *Nautilus* had been sent south to see if she could lend a hand in General Eaton's expedition. Her timing was perfect. Her firepower was not that of *Argus* or *Hornet*; *Nautilus* only shipped twelve six-pounder carronades, small smoothbore weapons that had a respectable punch, but not much range. On the other hand, they had a comparatively high rate of fire and could be operated quickly with smaller crews. Alas, Dent had no knowledge of the whereabouts of the other two ships. He did, however, have

A Wide Spot In the Road

something almost as good: artillery. Ready to be loaded in a boat and hauled to the beach was a fieldpiece and as much ammo as they could pack for it. Eaton must have jumped with joy when he found out. The odds had gotten just a bit shorter, even with just one gun.

Clearly now no attack could come off, at least not on the 26th. It would take a great deal of time and manpower to bring everything ashore. It is unlikely anyone had any genuine sadness about that, and even less so when they realized that since the fort had left them completely unmolested up to this point that likely meant that Yusuf's army had not in fact reached Derne yet. On the other hand, they were coming and time was finally of the essence.

Eaton called a council of war with his officers and Lieutenant Dent. There could be no more delay, no more waiting; a point with which, for once, found everyone in total agreement. It was far too late in the day now to even think of bringing the cannon ashore from *Nautilus*. Without it, an attack was still far too dangerous. Accordingly, Eaton directed that the cannon be brought ashore at first light and assembled in battery so as to cover an attack in the early morning of Saturday, April 27th. The marines would get one more night's sleep. Whether or not it would be a good one was a debatable point.

At 0530AM on Saturday morning, a shout of victory and triumph went rippling through the little army. With the timing of stage heroes, *Argus* and *Hornet* came lancing over the horizon into the bay. The cheers would have easily been heard in the fort but, like the gunners in the Tripoli harbor batteries when *Intrepid* came calling on the *Philadelphia* so long ago, they remained oddly silent. Based on the more recent performances of Middle Eastern militaries, there is at least the possibility that the fort's garrison simply assumed that if they just kept their heads down, perhaps the infidels would simply go away. In any event, it became clear very quickly that the infidels were not, in fact, going away and were planning to stay.

Hornet also had a fieldpiece that they intended to send ashore. This gave Eaton a total of two cannon that he could bring to bear on the fort. Better than nothing at all, of course. Before they could even get them to bear they had to get them ashore, and that was one more challenge. With the exception of the harbor of Derne itself, the entire shoreline was made of low but sheer cliffs that came up about twenty feet above the water. It was well after breakfast before the fieldpieces even got into the boats. At more than half a ton each, it was going to be a struggle to get them ashore. Said struggle lasted almost three hours. Only one of the guns, though with a good deal of powder and ammo, was landed before Eaton said the hell with it and sent the other one back to the fleet. It was now just before two, and Eaton was about to put his plan into action.

Eaton's men were divided into two groups. First, the bulk of the men, under the direct command of Hamet Bey himself with Colonel Leitensdorfer advising,

To Barbary's Far Shore

comprised the southern flank of the force. Their job was to swing around the town to the south and west and take a building described as a 'castle' that commanded the approaches to the city. The second group, under Eaton and O'Bannon, with Farquhar and Aletti along for the ride, was the marines, the midshipmen, twenty artillerymen to handle that one lonely cannon, and fifty mercenaries whose job was to charge the governor's house with its symbolic value and place that howitzer on the roof.

Eaton and his detachment would have to charge across the second wadi that was the eastern edge of the 1805 town. That was far easier said than done. From the eastern edge of the wadi to the fort's walls was just about half a mile, twenty-six hundred and forty feet of open space with no cover, no hiding place, no nothing. Once that was accomplished, Hamet's men would sortie from the castle and meet at the governor's. Once that had been accomplished, they would have the fort flanked to the south and west, while *Argus*, *Hornet*, and *Nautilus* pounded the daylights out of the water battery and its eight cannon. Once it was silenced, Hamet and Eaton's recombined force would charge the fort while the Navy provided fire support. Simple as simple can be. All things considered, it was probably the best idea anybody could have come up with, and it did play to their strengths. Now all it had to do was work.

They were lined up on a hill southeast of town. It is hard to say today exactly which one, but likely at about 32'45'N, 22'38'W from the description. They couldn't see a damned thing; the breeze blew grit and God alone knew what else all over them. All Presley O'Bannon knew was that he was not at all happy to be on this miserable little hilltop in this miserable place, walking for once and waiting for General Eaton to get around to sending them forward.

For his part, Will Eaton, his massive black stallion safely to the rear, was pacing to and fro. He swung his sword back and forth in front of him as he looked at the fort with undisputed glee. The fort looked deceptively peaceful, but the red Tripolitian flag that fluttered above it belied that. This was enemy territory now, without question or ambiguity. O'Bannon took the sight in for a moment, then looked up to see Aletti watching as well, then turn to him, shake his head sadly, and roll his eyes heavenwards.

Eaton few more steps, turned and swung, a few pebbles smote by Eaton's steel, then another couple steps. He stopped and yanked his watch out of his vest pocket. On reading it, Eaton's face broke into a grin from ear to ear. He turned to the men lined up below and behind him.

"Oh, bloody hell," Paddy Duffy muttered to himself, "it's the damned witching hour . . . "

"MEN OF AMERICA!" Eaton roared loudly enough for the men in the fort to hear him. Everyone else in the wadi just looked around and tried to figure out

A Wide Spot In the Road

how many actual Americans were in that Tripolitian hole. "THE WAY HOME IS THROUGH THOSE GATES!!" With a flourish Eaton waved his sword in the direction of the fort.

He didn't hear Bradshaw speak. "I don't suppose it ever occurred to His Nibs to just put us on the boats and let us go home that way?"

"YOU HAVE SACRIFICED MUCH," Eaton bellowed, "AND THE END IS NEAR!"

"Could have phrased that differently," Pritchard muttered.

"FOR THE MEN OF THE *PHILADELPHIA*, FOR OUR NATION, FORRRRRRRRRRRRWARRRRRRD!!!"

The shout started from Eaton's men. It rippled south and around. Muskets were lowered. Everyone fired in one long, ragged volley that spat and sputtered. A curtain of gray/white smoke appeared and raced south, like a spectral arrow as it curved around Derne.

"Captain Hull!" Isaac Hull spun on his heel to look where Josh Blake was watching the hilltop with a telescope. "They're at it!"

Hull smiled, the first time that day. "In that case," he said, "I suppose we'd better get to it; bad form to not attend a party you've been invited to. Leftenant Blake, you may do the honors."

Blake raised one eyebrow and grinned from ear to ear. "Thank you, sir, your kindness is truly appreciated. MASTER GUNNER, OPEN FIRE!"

"AYE AYE, SIR, OPEN FIRE! READY TO PORT, STAND BY TO STARBOARD, COMMENCE FIRING!"

With that, *Argus*' port battery let go a beautifully synchronized broadside. Flame, smoke and horror lanced out at the fort ashore. The helmsman nimbly spun the wheel and she pivoted sharply to starboard. Before a single heartbeat had passed, *Hornet*, in trail behind *Argus*, let go her smaller but just as deadly salvo. As she turned, John Dent in *Nautilus* gave hard to port. She headed directly for the shoreline, every sail straining.

The sound of the musketry hadn't yet echoed away. Eaton was loping forward, roaring unintelligibly, O'Bannon right behind him, looking around to make sure he wasn't the only one to have actually been insane enough to charge. He wasn't; Bancroft was behind him, waving his musket over his head and bellowing.

"COME ON, YE SONS O' BITCHES! DO YE WANT TO LIVE FOREVER?"

To Barbary's Far Shore

The marines came up out of the smoke behind him, a line of blue and white and screaming, muskets at port arms and fury in their eyes. Behind all of them was Farquhar and the mercenaries. What had once been a brilliantly colored line was now all varying shades of brown, but yelling to wake the dead. They were running headlong down the hillside now; gravity making sure they could do what courage alone might not be enough for. Green banners, the holy color of Islam, snapped back behind the charging force, giving lookouts on the ships a pretty good idea of where the line actually was.

Ahead of them, the defenders of the Derne fort realized that they were not going to get out of this without a fight. The fortifications erupted in geysers of dirt, rock and Berber. The wounded's cries echoed forward, letting the Americans know that the Navy was at the top of its game today. Give them this, though, as soon as the first salvos buried themselves in the fort's hide, they popped up and opened fire on the advancing troops.

Now, keep in mind that these are not highly trained and skilled soldiers, or marines, facing a ferocious attack. These are men who are not more than a step or two past banditry and who understand only that when they fire, the enemy is supposed to do the decent thing and die. They have had little, if any training on what to do if the enemy keeps coming. Given the speed with which, understandably, the attackers were coming, and the amount of time that it takes to reload a musket, the defenders will have enough time to get off one shot, assuming they can see anything through the haze and smoke and dust.

The next salvo of gunfire from the fleet arrives with that rustling-leaves sound and the THUD of impact.

Nautilus is dancing inshore as close as John Dent dares, and then closer. Her gunners get their aim, and then let fly with a sound that is a cross between thunder and a banshee's wail. Six carronades hurl fire and brimstone at the heathens in the fort. At this range, their lighter projectiles are just as hideously effective as the heavier ship killers aboard *Argus* and *Hornet*. They are doing their job. The cannon crews at the fort are discovering the meaning of the phrase "discretion is the better part of valor" as they keep their heads most securely down.

Eaton cannot see Hamet's men as they charge to the southwest. He hasn't been able to since just a few moments after the festivities began. The marines, the mercenaries, and that lonely cannon with its twenty-something man crew are now in a small cluster of homes just at the eastern edge of the second wadi, trying to get themselves reorganized. O'Bannon looks behind him. A quick count satisfies him that everyone, at least all his marines and the midshipmen, is still there, panting like winded racehorses. They take advantage of the brief pause to reload their muskets.

A Wide Spot In the Road

Rounds are pelting the eternal dust around them, little geysers of sand grit. Bancroft, half bent over, dashes the few feet to O'Bannon's side. He yells over the din. "Sar, I'm not sure our friends," the Sergeant gestures with his left hand towards the mercenaries, or at least where they're supposed to be, "are as eager as we thought! We might need to settle here for a few minutes and try to get formed up again!"

Before O'Bannon could answer, Eaton, just a few feet in front of the Leftenant, turns and snarls. "Wait here?? For what, Sergeant? One good push and . . . "

Eaton was interrupted by a crackling BOOM from a hundred or so yards away. The artillery finally got into action, followed by the rolling thunder of the Navy throwing everything they had at the fort.

About three thousand feet away, Hamet Bey, trying to look as if he is leading his force while doing everything possible to actually stay out of the lead, has actually gotten his men across the wadi. Leitensdorfer and some of the other leaders were behind Hamet, as much to keep him going in the right direction as in support. The Tripolitian troops in the wadi were outflanked by a force of unknown size but most precisely known ferocity. They scrambled back up its west wall and scattered in the town, far more interested in taking cover than anything else right now.

Leitensdorfer reined his horse to the right, and charged with his sword extended. Bellowing in three different languages, he swung his sword in a circle over his head. He managed to convince enough men to follow him that everyone is eventually running northwards. One of the other officers rode up to him. Leitensdorfer needed to think for a second about which language to use. "Tell His Matchesty zat he needts to follow his troops, ziz vey, ziz vey!"

The Navy has been doing its job quite well. The water battery hasn't gotten off more than a few rounds. They've been harmless, except for one that blazes through *Hornet*'s rigging and nearly drops her ensign. Richard Zacks relates the marvelous story. One of her junior lieutenants grabbed the bunting in his arms and raced up the maze of rigging to get it flying once more. The occasional musket ball whistles past. One of them struck his watch fob. What would have been a fatal wound was deflected by a piece of brass randomly hanging on his waistcoat. Aside from what would have surely been an unpleasant bruise, the lieutenant returns to the deck unharmed.

Hornet, in the meantime, proceeds to make her own special contribution to the gunfire. She is firing broadside after broadside of grapeshot at the water battery. Essentially a giant buckshot round, with lethality to match its size, the tennis-ball size grape cut through the air and just about anything else they meet,

To Barbary's Far Shore

including humans. The gun crews at the fort are staying low because of it, but they are now starting to get a sustained fire going on the flotilla off shore.

The Tripolitians who faced Eaton are now getting their bearings back. Perhaps Eaton started his charge just a bit farther out than he should have. Else he would be closer to the fort by now and perhaps even there. Gunfire is now lancing from the fort in steady volleys. The marines, along with their allies, can answer the fire but nowhere near quickly or heavily enough to make a difference. The fieldpiece is banging away, sufficiently hidden by smoke and flame that the fort's defenders can't see them. The artillerymen can't really see what they're shooting at either, but that isn't the point. The artillery was to make the Tripolitians keep their heads down, and it's becoming less and less effective.

It becomes even less effective. The artillerymen, eager to keep firing as fast as they can, forgot the basic rule of loading and firing their piece. There is the chest-thumping THUD of the fieldpiece firing. That sound is simultaneous with a terrifying BUZZWHIRRRRRRRRRRRRrrrrrrrrrrrrrrrr. It comes from behind the marines and rushes over their head. The noise is so thoroughly bizarre, even in battle, that the blue-jacketed Leathernecks throw themselves to the ground.

Joe Pritchard turns to Paddy Duffy. "What in the bloody hell was that?"

Eli Danielson, his face whiter than it's been on the entire trip so far, speaks with wide eyes. "I don't rightly know, gentlemen, but I think it was on our side!"

Obie O'Brien, eyes just as wide, answers. "For our sake, it better be!"

It was, as it turned out, the rammer. The one and only rammer for the fieldpiece was left in the muzzle by the eager gun crew. The fieldpiece is now out of action for the rest of the fight. The artillerymen, now unemployed, seem singularly unwilling to pick up muskets or swords or join the fight.

On the southwest side of the city, Hamet and Leitensdorfer have reached the 'castle.' No better description has come down to us, so likely it was no more than a particularly large home that had seen better days. They secured it through the simple means of surrounding it and running wildly through it. It was empty. Hamet's men have not lost a single soul during their run.

Leitensdorfer is on the ground, trying to get things sorted out. The only thing more disorganizing to a non-professional nineteenth century infantry force than a defeat is a victory. Everyone relaxes for a few moments and hopes that the leaders don't want to go any further either. After a few minutes, however, the Bavarian realizes that he is the only one making any effort at all to get the men back into something resembling a formation before preparing to charge the rest of the way to the fort.

A Wide Spot In the Road

Charging up the steps to the roof of the house, Leitensdorfer grabs a telescope to see if he can locate any sign at all of Eaton and his men. The curtain of dust and smoke to the northeast has lifted slightly. The Navy can be seen sailing in lazy racetrack circles offshore, sending regular broadsides into the fort. Ragged, uncoordinated salvos thunder back in reply.

But where is Eaton?

Leitensdorfer does his best to peer through the somewhat murky optics. He sees a few green banners flying in a small cluster of structures, perhaps two hundred yards away from the fort. Eerily silent fans of black/gray smoke come from muskets he cannot see. The fieldpiece is silent. Whatever's happening, it can't be good. They are not moving, and should have been much closer to the fort by now.

Leitensdorfer spots Hamet, surrounded by bodyguards twice his size, and moves quickly to him. "Your Matchesty," he says with firm Bavarian correctness, "ve must at vonce rally our vorces und press on to ze vort! Eaton and his men are pinnt down und . . ."

Hamet turns to the Bavarian, a look of combined fear and anger in his eyes. "NO," the Bey snarls, moving his arm across his waist in a slicing motion. "We stay here. It is dangerous to go any further!"

Leitensdorfer's look is at first one of confusion, followed by stunned disbelief. His anger boils quickly, but the looks he is getting from the four towering bodyguards quickly dispatches any thoughts he had about grabbing the little weasel by the collar and slamming him to the wall. Or, for that matter, any thoughts he may have had about removing Hamet from command in a most severe and final fashion.

Gathering calmness the best he could, Leitensdorfer responded through gritted teeth. "Your Matchesty, I do not zink you haff unterstoot. Eaton may be trapped; he may not be able to charge ze vort."

"Then he will determine the best way out, and he will let us know if he needs help," Hamet snapped. "Until then, we will stay right where we are!"

With that, Hamet folds his arms across his chest and looks into the distance. His bodyguards cast sidelong glances at Leitensdorfer, insuring that their leader's thoughts remain undisturbed. Leitensdorfer stomped back downstairs, wishing he could disturb them anyways.

It is surprisingly early in the fight, only about two forty-five on a warm Mediterranean afternoon. The old saying that no battle plan has ever survived first contact with the enemy has gotten a genuine workout today. Eaton, O'Bannon and Farquhar/Aletti are pinned down with their men just a minute or so's brisk run from the fort. Their own artillery is out of action and the naval

To Barbary's Far Shore

gunfire, though accurate and constant, is losing its effect. To the southwest, Hamet Bey and Leitensdorfer have achieved their objective, the 'castle,' with ridiculous ease and Hamet has refused to proceed any further.

Will Eaton cannot know of Hamet's hesitation, no, let us call it what it is, cowardice, but he has his own worries. The gunfire from the fort is growing more fierce. The men who ran from Hamet's force have now entered the fort from the south. Now they're bringing their weapons to bear on Eaton and his men. The fort's artillery fire is now starting to notably slacken, but that is cold comfort.

Eaton, O'Bannon and Farquhar are gathered in the safe lee of a nondescript little building They crouch over a more-or-less accurate outline of the fort, and the area immediately to its front, scratched in the sand. Their discussion is punctuated by the random crackle of musketry. Their force tries to hit something, anything, at this distance. Eaton looks at his watch; he realizes they will have to do something, and do it quickly.

"I'm open to suggestions, General," Farquhar says affably. "Do we want to send a party out and try to contact Leitensdorfer and Hamet?"

Eaton shook his head. "It's half-past three now. By the time anyone got there and got back, it would easily be five o'clock, and by then it would be too late to do anything, and we can't fall back."

"And I'm not at all sure our men would still be here at five o'clock," O'Bannon pointed out in a dubious tone. The officers looked around. For the first time, they could see their troops were starting to look over their shoulders at the wide, empty spaces behind them, with the expressions of men who desperately wanted to be Anywhere Else But Here. Their fire was slacking off. Whether or not they knew they could fall back was irrelevant. Thoughts of survival often trump mere logic.

The British troops Richard Farquhar had served with would have been most unlikely to break and run, except in the most spectacular extreme. Even that was not a sure thing. A typical line unit would probably be overrun before breaking, and an elite regiment would quite possibly go down to the last man, right where they were ordered to stand. American troops, such as the militia Will Eaton had served with, had a tendency to crack a bit faster, more by lack of training. But, with proper training, they could take quite a pounding. These men, though; they hadn't yet taken a single casualty and were getting ready to show the white feather and try to get out.

If Eaton didn't do something, and quickly, it was all over. For William Eaton, then, this was it. The moment he seemed to have live his entire life for; the one he had been building up to for years now, the one that would define his

A Wide Spot In the Road

mission, define him, in the eyes of his nation and of history. In that heartbeat where he realized all of that, Will Eaton saw what he had to do.

Straightening up to his full height, Eaton looked towards the fort. He made sure his hat was properly pointed fore and aft. Then he turned to O'Bannon and Farquhar. In quiet yet firm tone, he announced his intentions. "Gentlemen, I intend to order a charge against that fort in exactly five minutes. Please have your men load their weapons and fix bayonets."

There was dead silence. O'Bannon and Farquhar looked back at Eaton, seemingly not understanding what they'd been told.

Eaton took this apparent confusion serenely, repeating his order with a clarification. "Gentlemen, I intend for us to charge that fort with cold steel. If we hit them fast enough, they will be able to get off, at best, one volley. We should spread out the best we can, so that one volley does as little damage as possible. Aim for that point right there." Eaton pointed with his sword just behind the water battery. "Their artillerymen will run for it and that will cause confusion among the rest. I am sure of it, gentlemen. Sure of it!"

O'Bannon and Farquhar could do nothing but nod mutely and return to their respective troops.

The response they got, especially with the marines, can best be described as incredulous, but marines have never been known to shy away from a fight. Without a further sound, they loaded their muskets once again, and lined up without a command. None was needed. This was what they trained for; the moment that they always knew would come, if the gods so decreed.

And the gods failed them not.

About two hundred yards, that's all. An active man can run that in a minute, assuming good health. The marines were healthy enough, but they did have their packs. They also had the fact that they were United States Marines on their side. That can cover a great many drawbacks. They lined up once more, just the way they had on the hillside; O'Bannon stood in front, Bancroft a few steps behind him. There was a crash of drums and a peal of trumpets as the five minutes abruptly ended. O'Bannon looked behind him to see Edmund Bradshaw, his musket at port arms. O'Bannon called over the din. "Ready, Sergeant?"

"Absolutely, sar," Bradshaw replied. "But I feel it's my duty to point out to th' Leftenant that th' General is quite mad, y'know."

O'Bannon had to grin in spite of himself. "A little late to worry about that now, Sergeant. See to it though it gets into the report. STAND TO!"

To Barbary's Far Shore

The rustle and crash of leather, wood, and boots echoed above the drumrolls. Shadows started to lengthen now in the afternoon sun, shadows that will point to places not yet visited, ones called Chapultepec Castle, Tarawa and Iwo Jima.

George Mann swallowed hard. In a still, small voice, he says, "I'm scared."

Bancroft turned his head to look behind him and spoke to Mann. "Good lad. Be worried if you weren't."

Eaton raises his sword.

The gunfire from the fort seems to slacken for a moment, just a moment. Then it starts rising. Balls spatter into the dirt.

On Board Argus, Off Derne

"CAPTAIN HULL!"

Josh Blake's voice is high and worried for the first time since Isaac Hull has known him. That gets his attention immediately. *Argus* is turning to starboard and unmasking her port broadside for another pass at the fort, *Hornet* behind her. *Nautilus* lies in front of them both, getting ready to turn away after another crack at the water battery.

Hull leaped into the rigging. He looked through his own glass and saw Eaton's men lined up. Ragged and dirty banners flew once more. Hull needed no detailed message from the General to tell him what was about to happen; a charge worthy of legend. One, however, that, if Hull does not support it, will end in an unmitigated defeat, no, disaster, for the liberators.

Exactly one way to fix this, Hull thought. "Helm, hard a starboard, NOW!"

The helmsman spun the wheel hard. *Argus* turned tight, tighter.

Hull roars. "MASTER GUNNER! PUT THEM IN THE FORT, THE HELL WITH THE BATTERY!!"

This will take some doing, and in damned little time. The master gunner knows exactly what he is doing. His crews quickly adjust their elevation as *Argus* reaches the apex of her turn.

The Fort At Derne.

O'Bannon, Farquhar and Aletti raised their swords. They were rewarded with a bellow of devotion and courage, one unlike anything they have heard on the raid so far. They sincerely hoped their men meant it. Why their attitude has changed so quickly, we will never know; quite possibly, it is simply the fact that they're not going to just sit here any longer and get shot at but rather they are

A Wide Spot In the Road

going to take one last swing at the damned heathens. They seem eager enough to get on with it.

Eaton's sword snaps down. Before the tip is lower than his belt, he moves forward. Boots pound the sand; his cape flowes back behind him. In that instant, everyone else is running and yelling too.

On Board Argus, Off Derne

Argus hit the very apogee of her turn. The starboard rail now catches the wave tops. The master gunner earns his pay.

"FIRE!"

The cannon race back across the gundeck. Their restraining ropes snap taut. The crews physically pull them back up the slanting deck. The balls are on their way, arcing up and over little *Nautilus*. Her crew wonders just what in the hell is going on. Her lookouts see Eaton's men moving forward, gaining speed, heading for the battered mounds of dirt that mark Yusef Bey's fort.

John Dent spun *Nautilus* around almost on a dime. Although he is fouling the range of the two larger ships, he does not dare leave. Grapeshot is the only thing keeping the gun crews away from their posts.

The tops of the fort's revetments flash orange and red. The sound is a crackling rattle like balloons popping. The black powder of the muskets sprays sparks from their muzzles. Billows of smoke roil upwards. Balls whirr and buzz and snap through the air like deranged insects looking to feed.

The Fort At Derne.

The marines and their followers thumped across the open space. Their chests are tight, their hearts pounding, and they are breathing like blown horses at the end of a particularly unpleasant race. Every now and then they feel the thwack of a ball hitting.

Close, but not close enough, you heathen bastards.

They raise their heads just high enough to see. The wall of blinking lights starts to weaken, to hesitate. Colossal peals of thunder race up from behind them. Black/brown geysers erupt from behind the walls of the fort.

The Navy's gunfire is telling now. It registers with a precision that would impress future sailors with laser guided rounds. The men along the wall now realize that the ships are firing broadside after broadside; not around the battery, but indiscriminately inside the fort's enclosure. They are terrified. Most will not be able to reload before the little force gets there. They have no bayonets and only short daggers to defend themselves with.

To Barbary's Far Shore

They are now under no illusions whatsoever about what will happen next. They know of the Marines. They know that the only thing marines do as well as shoot is use their bayonets. They have no intention of being trapped in what is certainly about to become a killing pen. In ones and twos, then more, the defenders fall back from the fort's east wall and run for the west. They can take cover in the town and stay there till things die down. Perhaps they will be back, *inshallah*. For now, pure, simple survival took first consideration.

The ground started to rise in front of the marines, slowly at first, then a bit more steeply. The end, one way or another, is in sight, but there will be, at last, a price. The last defenders, as would be expected, are hard men who have no intention of letting mere kafir, infidels, drive them out of what is their property. A few heartbeats from the walls of the fort the last remaining defenders rose up. They fired their last organized volley. Their aim was not good, but it was good enough. The price is paid.

Several of the Greeks go down. Presley O'Bannon hears a strangled wheeze. John Whitten suddenly spun to his left. His musket flew away and he crumpled motionless to the ground. Edward Stewart stopped. He hesitated for a heartbeat, less than that, as he sees his friend go down. That is all it takes for some damned Musselman to line up a shot that knocks him down as well.

Eaton turns and roars. "FOLLOW ME!"

His command turned into a scream of agony. A ball neatly shatters his left wrist. A gout of blood and bone sprayed across the space between him and O'Bannon. The lieutenant simply kept waving his sword and bellowing.

"GO! GO! MOVE, YE DAMNED FOOLS, MOVE!"

In that moment, O'Bannon is at the top of the fort. He looked down as his men pour over, screaming and yelling and, God help them, smiling as they pursue the fleeing Tripolitians out of the enclosure. Most of them, anyways. More than a few are kneeling on the hard-packed ground, waving their hands above their heads in time-honored gesture. The troops can handle them. O'Bannon leads his surviving men down into the fort on the run. They are going to end this, once and for all. The marines level their muskets to fire. The enemy fires first. Edmund Bancroft goes down with a ball to his right arm and right thigh, bellowing like a wounded ox. The volley from the marines knocks the enemy down.

The bayonets finish them.

Isaac Hull and Josh Blake looked through their telescopes at the fort. Not a single soul could be seen in the dirt and smoke. For a long moment, their hearts stop, wondering what has happened. They saw Yusuf's red banner go slipping

A Wide Spot In the Road

down the flagpole, and, in that instant, they knew that somehow the impossible has just happened.

Artist's impression of the fall of the Derne Fortress, 27 April 1805. Source: Naval Documents related to the United States Wars with the Barbary Powers, Vol. 5, Naval Operations Including Diplomatic Background, From September 7, 1804 Through April 1805, (Washington, DC: 1944)

To Barbary's Far Shore

TWELVE: A BIT OF COLORED CLOTH

Saturday, April Twenty Seventh, Eighteen Hundred and Five

O'Bannon looked around at the fort's square. It was now filled with some of its former defenders, kneeling in submission with their hands on their heads. Some were quietly watchful. Some wailed in unintelligible supplication. Some were pale and simply terrified. That was understandable. In that part of the world, then as now, defeated soldiers looked forward to slavery at best, execution and a mass grave at worst. Curtz was next to him, looking for all the world like a grim-faced statue. He stayed one step behind O'Bannon, his job making sure that the Herr Leutnant stayed safe.

"SAR!"

O'Bannon turned to see Pritchard and Duffy trot up behind him, sweating, filthy and grinning from ear to ear. Coming to attention, they saluted. "Sar, Private Pritchard reports; the fort is secure!"

Presley O'Bannon had wanted to cheer so loudly that it crumpled the walls, but right now all he could feel was a sense of utter and profound relief. "Right, then," O'Bannon replied. "Get something white, or even close to white, and get up there to let the fleet know we're all right. Off with you!"

As the two marines ran off, O'Bannon heard cheering, shooting and ululating coming from the other side of the fort. Borne on the shoulders of a dozen screaming mercenaries whose enthusiasm was infecting the rest of the men was General William Eaton, his hand wrapped in a bloody bandage and laughing his fool head off. O'Bannon looked up to Curtz. "If anyone here so much as moves, make him regret it."

"'Make him regret it' . . . jawohl, Herr Leutnant!" Curtz snapped his heels with a crack that would have gladdened the heart of Baron von Steuben himself. He turned to keep an eye on things as O'Bannon headed to meet Eaton.

A Bit Of Colored Cloth

"O'Bannon!" Eaton roared with laughter. "We did it!! We did it!!" With that, Eaton jumped down and embraced the lieutenant with a joy traditionally missing among Marine officers. That in turn moved the troops to even greater heights of ecstasy.

Disentangling himself from the raving Eaton, O'Bannon pointed to the rampart and shouted. "I'm sending word to the fleet that it's safe!"

Eaton was still laughing with an almost hysterical edge. "Safe? Of course it's safe, Leftenant O'Bannon! We have made this place as safe as our own homes!"

O'Bannon was, to put it politely, skeptical of that particular claim, but he held his tongue. "General, I truly think we should look at getting the fort in order to repel a counterattack. I do not believe that the Berbers will sit still for this for long!"

"Tut tut, O'Bannon! You'd think . . ."

Before Eaton could finish, there was a shrill scream. One of the prisoners, infuriated by the joy of his conqueror, leapt to his feet. Hands outstretched, he headed for Eaton at a dead run. Unfortunately, his fury blinded him to the sight of six feet, three inches of Swiss marine within arms' reach. Otherwise, the Berber would have foreseen the possibility of Curtz flipping his musket, taking the barrel in one massive paw like a club and smacking him squarely in the face. The Berber did a most impressive backflip and lay still near what looked to be a half-buried cannon ramrod. The others looked studiously away.

For his part, Curtz looked at the astonished O'Bannon. "I belief he regrets it, Herr Leutnant." Curtz gave a little bow, flipped his musket back and went back to watching the prisoners, now remarkably quiet.

The ships offshore saw the piece of linen Duffy and Pritchard were waving and came to the correct conclusion. Within minutes, boats were headed ashore with more men, doctors, and whatever supplies could be spared. Surgeon Richard Cable was in the first boat ashore, knowing he'd be needed and wishing he wouldn't be. That hope was dashed when his boat pulled onto the sand.

Corporal Campbell came running out, practically pulling Cable up the beach. "Ye must come quick, Doctor. Edmund, John and Ned are in bad shape; very bad shape indeed!"

They ran up the beach and into the fort. Campbell guided them through the chaos to a room that had probably been a storeroom of some kind. That was where he found O'Bannon and O'Brien tending to the three marines.

"Doctor . . . Richard . . . please . . . "

To Barbary's Far Shore

Cable opened his bag and pulled out every tool, device, and supply he had, trying to decide who to go to first. He knew that, if he chose wrong, one more man wouldn't go home that day. He was determined not to let that happen. Bancroft was clearly in pain, but breathing and alert. He could wait a moment. Kneeling next to Edward Stewart, he touched his fingers to the marine's carotid artery. Cable felt just a faint, thready fluttering as Stewart's eyes rolled back in his head and he took painful, gasping breaths.

"Doctor," O'Bannon said quietly, "John Whitten?"

"Is dead," Cable said quietly as he examined Stewart. "Has been for at least a few minutes." Whitten's eyes were still half open, gazing at the ceiling as if he was awakening from some intriguing dream. Cable looked up to O'Bannon and spoke, softly, but with as much authority as he could muster. "And Stewart is dying. I'm sorry, Presley. Had I gotten here sooner . . . "

O'Bannon swallowed hard and simply shook his head. "Is there anything at all?"

"Make him comfortable," Cable answered as he moved to Edmund Bancroft. He examined where the balls had impacted his arm and thigh, at close enough range that they'd gone cleanly through. "Sergeant," Cable asked loudly, "can you hear me?"

"Of course I can hear ye," Bancroft snarled back with a grimace of pain. "This damned room isn't th' size of one of the heads, and you're yelling." Bancroft's head fell back to the ground and he hissed in pain.

Cable had to probe the wounds to find out just how bad they were. That brought another roar of pain. "Looks like no major blood vessels, but that's only by good luck than anything else," Cable said with professional detachment. "I can suture him, but he's going to need to stay here; far too much risk getting him back to the ship."

Silence.

"Presley?" Cable looked beside him to see O'Bannon carefully closing Stewart's eyes.

Private John Whitten , Regiment of Marines.

Private Edward Stewart, Regiment of Marines.

Honor and praise unto them. They were not the first marines to be lost in combat, and sadly, they wouldn't be the last. All of this nation's wars are hard on marines; as the smallest service, casualties were always proportionally greater. In the First World War, 2,461. The Second World War would take 36,950. Korea, 4,266. Vietnam 13,091. Desert Storm, just 24. The war on those

A Bit Of Colored Cloth

who took up the mantle of Yusuf Bey and his ilk has claimed 1,100 marines, as of Spring 2012. That will climb until we furl the last flag and declare victory sometime in 2014, whether we are victorious or not.

But it is the proportion that is striking. Whitten and Stewart were, to be precise, more than twenty percent of the original American detachment who charged the wall at Derna that day. Military leaders tell us that a force that takes ten percent of its strength in casualties, 'decimated' in its original sense, not the current usage that simply describes a unit that has been badly battered, is no longer combat capable. Yet Eaton's detachment kept going, kept moving, never stopped attacking. If, for no other reason, than they had no idea they were supposed to stop. Quite likely, they never considered doing anything else.

There was no time right now to mourn. O'Bannon and Eaton would see to it later that they were properly attended to. For right now, there was still a battle to fight. The Tripolitans were now outside the fort, disorganized by their retreat. But, once they realized how badly organized the victors were, they'd be back, and with a vengeance.

Sure enough, some courageous souls among the Tripolitans began to quickly organize the defenders among the houses and souks that so closely surrounded the port. A counterattack now might well have been thoroughly final. But the Tripolitans, in their eagerness to get clear of the fort, had left almost all of those cannon on the rampart primed and loaded. They were now quickly turned on the Tripolitans and rapidly fired down into the town.

The niceties we observe today about keeping artillery fire clear of civilians had no place in early nineteenth century military tactics. Indeed, it was simply considered something you might try to do, emphasis on the 'try,' if your own civilians were in the way. Eaton knew only that the enemy was there, at point-blank range, and he would have to be driven back.

After a few salvoes, Isaac Hull knew it too. He maneuvered *Argus*, *Nautilus* and *Hornet* into position a little east of the fort to join in the general revelry. Yet another superb demonstration of what later generations would call NGS, or naval gunfire support, ended just after four o'clock, when the Tripolitans decided that things were far too hot and pulled back just beyond the edge of town.

The governor was not, interestingly, among them. He seems to have believed that, not only were his forces going to completely crack, but they would probably hand him over to Eaton and his forces. One would think that this would indicate at the very least a guilty conscience on the part of the governor regarding his behavior towards the good people and garrison of Derna, but never mind. Instead of leading, even in retreat, he headed for one of the many mosques that dotted the city and claimed sanctuary, at least for the time

To Barbary's Far Shore

being. His next refuge was somewhat more intriguing, but we shall get to that shortly.

It is just before five o'clock post meridian, Saturday, April Twenty-Seventh, the year of our Lord Eighteen Hundred and Five. Smoke is still pouring out of the blasted homes and other structures around the fort, as well as some fires within the fort itself. The flagpole at the center of the fort is poking skyward, like a finger, demanding attention and silence. Attention it shall have, but silence is at this point beyond the capabilities of the four-hundred odd mercenaries who are now in and around the fort.

First Leftenant Presley O'Bannon gives a sharp, precise command. He comes to attention, followed by Pritchard, Curtz, Duffy and O'Brien. The midshipmen, no longer just teenagers but combat veterans now, snap to as well, with Eaton at their head. His left wrist still seeps crimson, further staining the filthy sling that binds it to him. It now sends fiery tendrils of pain up his arm and into his hand. But his right arm is unscathed, and it obeys a reflex from far into his past as it whistles up into a salute.

"Forr-arrd . . . HARCH!"

O'Bannon steps off on the left, as men have done since Babylonian times. His men follow precisely behind him, at order arms. He carries no weapon, save for a discharged pistol shoved into his belt. In his arms, instead of a musket, is the most precious thing he has ever borne.

To an objective observer, as if one could be found that day, it is no more and no less a tightly folded triangle of cloth; Rich blue, with a constellation of five-pointed, white-silver stars spread across it, as if God himself had simply decided to scatter them across a velvet night. But, to the five marines who escort it to that lonely wooden flagpole, it is far, far more than that. It is a symbol of home, a symbol of devotion, a symbol of all that they have tried to do and stand for since the day in Washington or Philadelphia or whatever exotic spot they were in when they raised their hands to swear the Oath. The Oath that they will bear true allegiance to the United States.

The Oath says, without saying, that this flag is the symbol of your allegiance. Not to kings and princes, but to a constitution of men and ideas; the men who worked so hard to build the young republic and the idea that the Constitution, and the natural rights it acknowledges and guarantees, are worth defending. Worth dying for, if need be. The Oath says, without spoken words, that this flag is a symbol of freedom, where you have the right to say what you will and believe what you will, and worship what you will, in peace and safety. That any man who seeks its protection is worth defending.

Worth dying for, if need be.

A Bit Of Colored Cloth

"P'ltoon, HALT!"

No parade ground halt was ever as sharp and precise as that the marines executed at that moment. O'Bannon stood two steps from the flagpole and its frayed halyards. He steps forward, once, twice. He transfers the flag under his left arm as he undoes the rough lines. Small broken strands prick his hands like needles. But he'd been good at this in Washington; it was, perhaps, the only thing he had been able to do well. A little discomfort was quite all right. One reinforced edge slides smoothly from the folded triangle. O'Bannon deftly runs the halyards through the grommets at either corner. The fort is going mad now, but the lieutenant and his men cannot hear a sound. They are marines. For them, this task is as important as the march or the attack, and they address it with the same cool skill and precision.

With his right hand, O'Bannon starts hauling back on the lanyards. He keeps the flag from touching the ground as it neatly unfolds and begins to climb the flagpole. No command is necessary for his marines to present arms in perfect, sublime unison.

Were we able to stand on that parapet, the flag we see would look slightly odd to us. At that time, the tradition, all twenty-nine years of it, was that with every star added, another stripe would be added as well. Instead of the trim, rectangular banner we know today, there was a slightly stocky flag with fifteen stripes, eight red, seven white, and, of course, the stars would have been much, much bigger, in five staggered rows of three stars each. But it is still the flag the Marines would carry to other places with names like Chapultepec, Bois de Belleau, Guadalcanal, Iwo Jima and Fallujah. No one from any time in our history would have any problem understanding what was happening.

But to be precise, what was happening was this. For the first time in history, the flag of the United States of America was being raised after a conquest in the Old World. The Union Jack, the French and Spanish tricolors, the crimson with white crescent and star of the Ottoman Empire, all were common sights in North Africa. After all, for all their criminal organizational skills, the Berbers were not at all good at actually running countries. Parts of the Barbary Coast, as well as other areas of Africa frequently changed hands and flags at a moment's notice. Conquest didn't necessarily mean that the leadership changed hands; far from it. Quite often, the old Bey or Pasha, or whatever he might be called, simply promised fealty to the new imperial overlords, wherever and whomsoever they might be. For the most part, things went on as usual.

This time, it was different. That flag had been born in revolution, but as a symbol of freedom, not of conquest. Those who held it high did not, could not, believe there was any need to conquer anything, not with the millions of square miles of trackless frontier in North America. Americans had everything they could ever want or need; all that was required for it was to work hard. The hell with the rest of the world. We could ignore them.

It didn't take long for Americans to be disabused of their notions as to how to deal with the other nations that made up the planet. Notions of peace and brotherhood melted very quickly in the face of simple, brutal fact. At the beginning of the nineteenth century, nations took what they wanted and the rest of the world be damned. Americans learned very fast that, without the Royal Navy to look after their ships, others could take those ships and hold their crews for ransom. When we finally had enough, we went in and made sure those others understood that there was a limit beyond which we would not be pushed. And when it comes down to that point, the only thing that will get an enemy's attention is the sight of your flag flying high and proud over what used to be theirs.

The flag went up the pole, buoyed even higher by the wild screams and ululations of the mercenaries. They knew nothing of what it all meant, and cared even less. They probably assumed only that they had won and would therefore be paid, but that was all right.

The flag reached the top of the halyards with a sharp, crisp SNAP. O'Bannon tied the lines off with a practiced motion, then took two steps back and brought up his hand in the best salute he'd ever given. He held it for a perfectly timed three-count, then whipped it back down. Then and only then did the marines finally cheer their accomplishment.

There was one other duty to address, taken care of the next morning. Isaac Hull himself came ashore in the best uniform he had, all ruffles, gold braid and massive dignity. He was accompanied by the rest of the marine detachment and two sail makers. They brought with them a roll of stiff, white canvas.

The first part of the ceremony was performed in a small, bare room away from the rest of the world. The marines gathered under O'Bannon's supervision. They carefully laid Whitten and Stewart into the rough canvas. Each of them said a quick, silent goodbye before filing out and letting the sail makers do their work. They in turn wrapped them gently but securely in the sections of canvas, then sewed it tightly together with the heavy, curved needles that pierce the canvas with ease. One last stitch, by tradition, is respectfully slipped through the noses of the deceased. A practice we find at best queasy, but men of the sea then understood that this was the way things were done. They held no grudge against the men who would clothe them for their final journey. When they are done, one of the sail makers opens the door, and the marines file in, and lift Whitten and Stewart to take them to their rest.

A slow march outside; the marines move in precise unison. This is the last thing they will do for their friends, and they want it to be right, as much as they

A Bit Of Colored Cloth

hope that, no matter what the circumstances, someone will take that care for them. O'Bannon is at their head, Curtz, Pritchard, O'Brien, and Duffy behind. Just a little behind them, Edmund Bancroft, sweating in pain, is assisted through the procession. Eaton, Hull, and the midshipmen bring up the rear. A little off to the side are Farquhar, Aletti and Leitensdorfer, even now slightly odd men out. The mercenaries watch quietly, for the most part remaining silent and avoiding been seen to stare too closely. Muslims tend to be superstitious enough about dealing with the dead, but even more so when it comes to the dead of other faiths.

They eventually arrive at a palm tree a few yards away from the east wall. It is a fairly empty area, well enough away from dwellings, where Whitten and Stewart can rest quietly. Two locals have dug more-or-less rectangular holes for the donation of a few pieces of gold. Beside each of them are three pieces of rope. The marines lay Whitten and Stewart on the ropes, then fall in.

Isaac Hull places his hat squarely on his head, and steps forward to lead the service. There is a prayer to be said during a burial at sea, but they are on land. No matter; the words apply equally well.

"We therefore commit the earthly remains of Privates John Whitten and John Stewart, Regiment of Marines, to the deep, looking for the general Resurrection in the last day, and the life of the world to come, through our Lord Jesus Christ; at whose second coming in glorious majesty to judge the world, the sea shall give up her dead; and the corruptible bodies of those who sleep in him shall be changed, and made like unto his glorious body; according to the mighty working whereby he is able to subdue all things unto himself.

"Amen."

All hear these words; but, in their heads and hearts, they come out differently. *Lord, we are sending you two good men. They did well by us, please take care of them.*

The marines of *Argus* fired a salute into the air. The Americans raised their hands in final honors. A flock of gulls, startled, wheeled into the sky and off to the horizon.

For reasons that should not be at all too hard to fathom, the Tripolitians were distinctly unwilling to allow matters to stand as they were. The sight of that bit of colored cloth waving above their fortress could only have enhanced that attitude. Quite understandably, they wanted to liberate the city of Derna and return it to the gentle, enlightened rule of Yusef Bey. But, above and beyond that, there was a very real concern, from private soldier to general, that failure to get the Bey's city back would lead to consequences far more unpleasant, and, indeed, terminal, than a mere encounter with the bastinado.

To Barbary's Far Shore

Keeping in mind that the Tripolitian army wasn't an army in the Western sense, their response was quite speedy. Yusuf Bey knew within a day or so what had happened. He responded with alacrity, sending the bulk of his remaining forces down the coast road to Derne. By May 1st, they had the town quite neatly surrounded, though they did not seem to be in any great hurry to make a brave charge into glory. That day *Hornet* shoved off for Syracuse with reports from Eaton, O'Bannon and Hull intended for Commodore Barron, who they most sincerely hoped would be healthy and coherent enough to read them; he was.

But over the next seven days, the Tripolitians slowly, very slowly, crept back towards the center of town. In fairness, the commander of the Bey's army understood that American tactics and the United States Navy's artillery had routed the Derne garrison with surprising ease. On the other hand, he understood that another thing the Americans had going for them was the attitude of the locals. The people of Derne were flexible, one had to be, and the Americans were the 'strong horse' at that point. But, turn enough of them against Hamet Bey, and things might be much easier.

So the Tripolitian commander spent the next few days doing his damnedest to cajole and, where needed, bribe the population to turn against Hamet. They were aided in their efforts by the former governor; as one will remember, he was hiding out in one of the larger mosques. The governor made a break for it, but only got as far as the home of a local notable who was a backer of Hamet's. But, for reasons unfathomable, he decided that the Muslim laws of hospitality and sanctuary required him to put the governor up for an indefinite stay. To further complicate matters, it appears that the governor was put up in the notable's harem. One can at least compliment the governor for finding the most pleasant place one could to sit out a war.

Eaton got wind of this. He wanted to take an all-Christian force to the notable's home and put a stop to things once and for all. But, as so often happens in situations like this, he was convinced by Hamet, of all people, that such an action to defend himself and protect those who had laid their reputations and lives on the line to support the rightful Bey would be offensive and insulting. Why Hamet had such a break with reality at this point is unknown, but it is more than possible the timid streak that lay within him had come back with a roar, or, perhaps more properly, a squeak. It is also quite possible that Hamet had understandable qualms about the depth and strength of his support among the citizens of Derne, and wanted to do nothing that could risk it to even the smallest extent. In any event, word got back to the governor. Before Eaton could come up with Plan B, he made one last escape. He slipped right under the nose of Eaton's men and into the Tripolitian lines.

The next day, Monday, May 13th, the Tripolitians made their first full-dress counterattack. They had been whipped into a respectable furor by the

A Bit Of Colored Cloth

governor. He assured them that the good people of Derne were just waiting for the Dey's men to appear to rise and throw the infidel devils out. The Tripolitian efforts were inadvertently aided by one General William Eaton, who had managed to convince himself that the Dey's men didn't have the courage to come after them.

It was a near-run thing. The Dey's forces, about twelve hundred of them, lined up on the low hills that rose south of Derne, ready for a fight. They made some careful probing attacks first. Then they decided to go after Hamet's cavalry, which was billeted a mile or so south of town. The cavalry fought well despite being outnumbered, but eventually it began to tell. Cavalry, regardless of nation or tradition, isn't equipped, supplied, or trained to fight a prolonged defensive battle under any conditions. Think of General John Buford's magnificent stand on the first day of Gettysburg, up against the spearheads of the legendary Army of Northern Virginia. Or, and perhaps more appropriately, the quick, horrifying annihilation of George Custer and his men at the Little Bighorn. Whichever example one prefers to go with, the results are depressingly similar; after the initial stand comes a weakening, a wavering and then the command to fall back.

Although *Nautilus* and *Argus* were still offshore, the Dey's commanders had chosen their approach well. They were able to push the cavalry back, while being in such a position that the fort's artillery and that of the ships could not touch them. Within a few hours, the Dey's army was back in the town proper, sending Hamet fleeing from his new capitol at the former governor's residence back into the fort. Success, however, has its own drawbacks. The Dey's men were now in the maze around the fort once more, which drastically slowed down their advance, and, not incidentally, finally opened them up to the ships' guns.

Eaton was now truly fearing for his safety, from the hands of his own troops. Their loyalty had been, at best, less than rock-solid even during the assault on the fort, but mercenaries are practical men; one does not get paid unless one wins. Now however, they had not been paid; or at least, not what they thought they should be paid. That combined with the growingly worrisome supply shortage to create a distinct lack of enthusiasm about going on the offensive. Eaton solved the problem by taking the men he could rely on, for all practical purposes, the Christians, and turned his artillery upon the town once more. Even then, despite impressive efforts, the initial salvoes were ineffective until a lucky shot killed two of the Dey's own riders. Riders who were busily engaged in making sure the Dey's men stayed focused on capturing the fort and Hamet, instead of engaging in poorly controlled self preservation.

As has been seen in recent wars with Arab armies, when the men who know what they are doing are lost, or the enforcers killed, cohesion and discipline take a back seat to a quick retreat. What was left of Hamet's cavalry,

To Barbary's Far Shore

backed up by naval gunfire, drove the Dey's men back to their original camps south of town. There they dug in, apparently expecting the mercenaries to follow.

A pen and ink sketch of Argus *as she appeared in 1803.*
Source: U.S. Navy

That was unlikely. Eaton was now discovering just how bad their supply situation was. They weren't going to be on starvation rations, yet, and the fort's water supply would be adequate for a while yet, but only just so. The ships could ferry over some, and only some, supplies, but that would be small comfort. *Argus* and *Nautilus* were small ships to begin with, and relied on regular replenishments from victuallers, the larger frigates like *Constitution* or *Congress* or closely scheduled stops in port. If they had to remain at Derne for any prolonged period of time, they'd eventually have to leave to replenish. And sure enough, on May 18th, *Nautilus* sailed for Syracuse with more dispatches. This reduced the entire USN presence on the largest foreign expedition the United States had to date mounted to exactly one vessel of great courage and skill, but middling size.

A Bit Of Colored Cloth

The bottom line was that any dreams Eaton had of pressing on to Tripoli itself were at best going to be delayed, and, more likely, would be cancelled entirely; not that the General had any intention of actually admitting that. And not going to Tripoli endangered the entire point of the damned raid in the first place, freeing Will Bainbridge and his men. On the other hand, going back to Alexandria wasn't much of an option either. Trying to march out by land was impossible, obviously; insufficient supplies. In any event, doing so would lead to a quick and final end. The Dey's men would simply overrun whatever defense Eaton and his men could put up out in the open. An evacuation by sea was a credible option, but it would require a good part of the Mediterranean Squadron to pull it off. That would not seem to be happening any time soon.

So, unless anyone could come up with something better, Eaton, the marines, the midshipmen and four-hundred something heavily armed and ill-tempered mercenaries were going to be cooped up together at the Old Fort of Derne until either victory, retreat, or surrender. The first two could be dealt with.

The last did not bear thinking of.

To Barbary's Far Shore

THIRTEEN: THE PRESENCE OF POLITICS

The Occupied Fort, Derne, Tripoli. Saturday, June First. Eighteen Hundred And Five

D awn that morning had been cool, but the sun wouldn't take long to turn the Old Fort into an oven, and a crowded one at that. O'Bannon mounted the steps up to the south wall, where Pat Duffy and several of the newer marines stood sentry duty. With a glass, one could see the Tripolitans' camp, all smoke and haze against the horizon. The marines, along with a smattering of Arab troops, were keeping an eye on things. The truth be told, the marines were keeping an eye on the Arabs, the Arabs were keeping an eye on the marines, and both of them were making sure that they wouldn't be surprised by any of Hamet's men wanting to make things even more unpleasant.

Duffy rattled to attention and saluted. As O'Bannon returned it, the breeze off the ocean took with it a whiff of alcohol, and not particularly good alcohol, at that, from Duffy. The lieutenant's nose wrinkled slightly. For his part, Duffy merely grinned. "The Leftenant has to admit . . . it's a damned sight more pleasant than the way our allies smell."

O'Bannon had to hide a smile at that. "Private Duffy, the only thing I'll admit to is an unquenchable desire to find out where you're finding that rotgut. Have you seen any activity from our friends?"

Duffy shook his head. "Nothin' this mornin, sar. Not even scouts. Wonderin' if we might have gotten their attention the other night . . . "

The 'other night' Duffy referred to was May 28th, when the Tripolitans decided that they would take another crack at the Old Fort. This one showed slightly more skill and planning than the previous one, sending a strike and a foraging party up the ravine to the fort. The problem this time was that they only sent about seventy-five men, which meant they were outnumbered about seven to one by the fort's defenders. Not to mention, they were funneled quite nicely

The Presence of Politics

towards the defenders' weapons by the terrain they were using to mask their approach. A short fight ensued, and the Tripolitians decided that now was not a particularly good day to enter Paradise.

O'Bannon was about to comment on the likelihood of them trying again when there was a shout from behind them, up on the north wall. The Arab sentries were gesturing frantically at the ocean, where *Argus* sat as watchful as her namesake. Far off on the horizon, a speck of white that could have been mistaken for a seabird quickly resolved itself with a moment's gaze into a sail.

It was *Hornet,* inbound from Tripoli, with news that was at once heartening and disturbing. All of it boiled down to two things. Through "negotiations of a Most Sensitive and Delicate Nature," Lear's toiling seemed to be bearing fruit; cheering enough, if it meant Will Bainbridge and his men might be heading home. The other thing was that they should start looking at the possibility that they were going to have to leave their hard-won prize behind. There was a disconnect between the two things that made little sense to Hull, O'Bannon and Eaton. If nothing was sure, then why should they prepare to leave?

Of course, no matter what was happening, or about to, they now faced the very real problem of what to do next. Eaton had no intention of listening to anyone or anything regarding his plans. The general opinion among his subordinates was that he was, indeed, as mad as a hatter. On the other hand, everyone had to admit that the raid had gone far better than anyone had a right to expect. Maybe, just maybe, between Eaton's insanity and the United States Navy, he might have a chance of accomplishing something once he got to Tripoli. Without question, that was still Eaton's plan.

This writer, no matter how hard he tries, simply cannot believe, even for a moment, that Eaton actually expected to take Tripoli with two hundred and fifty to three hundred men and a handful of light cannon, along with dwindling supplies and increasingly surly troops. Sadly Eaton's thoughts on the matter have not made it down to us. On the other hand, Eaton, no fool, would have taken note of the way Yusef's forces cracked when faced with even a few trained men with fire support. He would have known about the earlier USN bombardments of Tripoli and their next-to-ineffectual counterfire, and he certainly would have known about the slow-motion response to *Intrepid*'s first visit. The traditions of the Barbary militaries called for them to behave like bullies when they initiated combat, and to run as fast as they could when they came up against trained men with firearms who had no intention of rolling over. Eaton had to have been picturing himself sailing through Yusef's terrified men like a wolf through a flock of sheep, but the time for descending on the pasture was running out.

Yusef's men rudely interrupted Eaton's reveries on the 10th with one last attack on the fort. Now reinforced by a number of Arab troops, whose pay must

To Barbary's Far Shore

have been taxing Yusef Bey's generosity to the utmost, they threw a hard swing at the Fort, crashing through the town and heading straight for the gates. Fire from the ramparts stopped them long enough for Hamet's cavalry to ride to the rescue, but it took four long hours and superbly accurate fire support from *Argus* to push the attackers back through town and far enough away to call it a victory for Eaton. There was but one casualty, Paddy Duffy, who broke his leg bounding down a flight of stairs while trying to get into the fight.

The fight, however, reinforced a brutal fact; their supplies were running lower. At this point, a sortie from the fort to Tripoli was out of the question, but it was just as well. The next day, another sail appeared on the horizon. This time, it was Authority Embodied: the frigate *Constellation*, inbound.

After the warship wheeled into place and reefed her sails, a message went out from Captain Hugh Campbell. Eaton, Farquhar and O'Bannon were to report at once; Captain Campbell's compliments to Captain Hull, and he could hie himself aboard as well. Campbell was a decent man, but there was no decent way to tell them what was coming. It was best to be blunt.

Lear had concluded a treaty, in principle, with Yusef Bey on June 3rd.

Constellation's day cabin was cramped at best, but today it was nearly oppressive, even though only five men were present. They sat or stood around Campbell's desk. The only sounds were the muffled noises of the ship's daily routine and the rhythmic slap of waves against her hull.

Campbell, glasses perched upon his nose, first read the dispatches from Tripoli, then the treaty. He most likely paraphrased it to get past the diplomatic and other irrelevancies that stood like manned barricades in the way of the parts they needed to know. In almost sepulchral tones, Campbell read Article Three.

" . . . Article the Third: All the forces of the United States which have been, or may be in hostility against the Bashaw of Tripoli, in the Province of Derne, or elsewhere within the Dominions of the said Bashaw shall be withdrawn therefrom, and no supplies shall be given by or in behalf of the said United States, during the continuance of this peace, to any of the Subjects of the said Bashaw, who may be in hostility against him in any part of his Dominions; And the Americans will use all means in their power to persuade the Brother of the said Bashaw, who has co-operated with them at Derne &c, to withdraw from the Territory of the said Bashaw of Tripoli; but they will not use any force or improper means to effect that object; and in case he should withdraw himself as aforesaid, the Bashaw engages to deliver up to him, his Wife and Children now in his powers . . . "

Well. That was about as straightforward as it could get. The forces at Derna were to disband and disperse immediately. They would not be resupplied, they would not be reinforced and it was, at best, unlikely that, if any pay was still

The Presence of Politics

coming to them, that they would receive it. But the worst part of it was the clear, final, and complete abandonment of those 'subjects of the said Bashaw' who had taken up arms against him.

One remembers the Vietnamese who fought and worked and served alongside the US military for years, only to find themselves watching in disbelief as the last helicopters cleared the embassy. The final act here would be nowhere nearly as dramatic, but it left precious little time for those affected to leave as quickly as possible. Worse still for all of them, but most of all for Eaton, was the fact that the man they had dragged out of the Nile reeds and marched to within an inch of his rightful throne was now going to be tossed aside like some bad penny.

The treaty said "persuade the Brother of the said Bashaw," but no one in that room had any illusions as to what the treaty and its negotiators meant. Hamet Karamanli was to be removed from the scene with all haste, indecent or otherwise, if the crew of *Philadelphia* was ever to return home. Willingly, if possible; unwillingly, if need be, but removed he would be. If he cooperated and if he kept his mouth shut and if he stayed right wherever the United States Navy was going to drop him off, then there was a reasonable chance that Yusef would restore his family to him.

But all of it rested on the unpleasant fact that the crew of the *Philadelphia* was still in durance vile. Yusef Bey was playing all the tricks at his command to drag things out, including at least one semi-veiled threat to kill all of his hostages. This was enough to give everyone present pause, insofar as, and has been noted, one simply did not kill one's hostages. They were money in the bank, and that was that.

Yusef, quite frankly, was beginning to feel just the tiniest bit nervous. One of his most important outposts had fallen, and apparently quite easily. The US Navy no longer seemed to be the comic-opera force that he had played cat-and-mouse with for some years now. Worst of all, the United States itself, peace loving, slow to anger, but always quick to pull out its checkbook, was now beginning to behave in ways that echoed the behavior of the British. The British were people Yusef left most severely alone, because they had shown both the ability and the intent to hurt him.

The Americans now appeared willing to hurt him. That was a change, a violation, the truth be told, of the rules of the game. Yusef Bey Karamanli was not going to stand for that. Of course, being the realist he was, he knew he would have to. That wouldn't stop him from yanking a few American chains, and the first one was the one needed to get them to back off on their military advance. If the Americans thought that the crew of the *Philadelphia* were goners, then they might, would have to be, inclined to put General Eaton and his men onto a very short leash. That is exactly what they did.

To Barbary's Far Shore

Campbell threw down the treaty copy as if it were leprous and leaned back in his chair, rubbing the bridge of his nose with his fingertips. He took a deep breath, pausing for a moment to phrase his words. "Gentlemen, I . . . for God's sake, every officer from Commodore Rodgers on down are simply at a loss . . . but Colonel Lear has given his orders and we have to obey."

Eaton pursed his lips. "Let him say what he pleases. I have more time in the privy than Tobias Lear has in planning or decreeing military matters. It will take days at best to arrange . . . "

"Consul," Campbell said quietly, but firmly, "it will take you . . . " Campbell shot a look at the clock on the mantelpiece. " . . . approximately thirty-six hours. Those are my instructions. They allow for no discussion."

"But . . . "

"But nothing, Consul. My orders are to be underway no later than tomorrow night, to transport you directly to Syracuse, and then see to it you return to the United States aboard the *Franklin*. There are apparently some questions regarding the funding for your mission that the Secretaries of War and Navy wish to address to you."

Eaton's jaw tightened at that, but he remained silent. There are times when being sans phrase is best, and, for once, William Eaton may have recognized one.

There was quiet for a moment before O'Bannon spoke, a quiet icy tone to his voice. "Captain Campbell, I must respectfully but firmly protest. We hold," he quickly corrected himself, -"held every advantage, and some . . . some . . . secretary signs it all away? We have allies here, we can go all the way to Tripoli and . . . "

Campbell put down his glasses and leaned forward. In the little cabin, the distance seemed much narrower than it was as Campbell spoke in a tone that was little more than a growl.

"Now you listen to me, Leftenant, and listen well, for I am not going to repeat myself. You may be right. We may have been just a hair's breadth from toppling Yusef and that band of cutthroats he calls a government. We may be just hours away from releasing Will Bainbridge and his men, and it matters not! People in fancy suits that cost more than we make in a year have decided that the game is not worth the candle, and that is that!

"Do you not think that the captains, and the commodore himself, didn't make every effort to stop this? Don't you think that we would have liked nothing more than to sail into Tripoli and chase Yusef out? We tried, and we have been told no. We shall do it another way, and, by God, whether we like it

The Presence of Politics

or not we shall do it because we swore to obey the civilians . . . even if they are wrong!

"I am not happy, Leftenant; I am God damned furious that Americans have died and their treasure has been squandered, only to leave us in a place where we shall be back here ere long to do it all over again, but we have been ordered by our betters to pack up and get out!

"Now, Leftenant O'Bannon of *Argus* and victor of Derne, is there anything about 'pack up and get out' that your thick Marine brain does not understand?"

Campbell punctuated that last with a slap of his hand against the desktop, a noise that rattled every object in the compartment.

O'Bannon's clenched teeth allowed, "Sir, no sir," to pass.

The silence returned again, uncomfortably so, until Hull discreetly cleared his throat. Campbell looked up, the beginning of a snarl on his lips. Hull ignored the implied threat. "Captain Campbell, in that case, we have a great deal of work to do and far too little time to do it. With that in mind, I should point out that, right now, our most serious problem is not the orders we have, but the four hundred or so mercenaries we've hired who are going to be most annoyed once they find out we're leaving."

Campbell reflected upon that for a moment. "Besides those of us here, how many are at the fort?"

"My men and the marine reinforcements," O'Bannon replied, "plus Hamet and his entourage, Aletti, and Colonel Leitensdorfer. A total of about seventy-five men."

"Don't count us in," Farquhar said. "Aletti and I would do far better if those men believed we were betrayed as well. We have to go back to Alexandria and resume our activities on behalf of His Majesty. A great many men out there would cheerfully assist us in shuffling off this mortal coil if they thought we'd abandoned them."

"We've not betrayed anyone," Eaton said, almost to himself.

"I'm reasonably sure we won't be paying them," Farquhar pointed out most affably. "Believe me when I say they will feel, and act, quite betrayed."

"In the meantime," Hull interjected, "the marines and Hamet?"

"We could just quietly rotate the marines back to the ships," O'Bannon said, his voice tight. "It would take a few hours, but if it's done with a bit of care we can pull it off. The Mohammedans aren't likely to take notice of one more American walking down to the shore."

Campbell nodded. "And Hamet?"

To Barbary's Far Shore

Eaton looked up sharply at that, eyes aflame. "I shall deal with that. He is here by my pleading, and I shall insure his safety."

"No games, Consul. No delays, no last minute plans," Campbell warned. "He has a choice; sail with us or remain on his own. Do you understand?"

"Perfectly."

Campbell's expression was skeptical, but he simply did not have the time to be more doubtful. "Right, then. We will be sailing at approximately forty minutes past seven. I strongly suggest no one miss the last liberty boat. Get to it."

Hull, Eaton, and Farquhar stood and headed for the hatch. O'Bannon stood at ramrod attention, looking directly at Campbell. He took a moment to realize the young officer was still there. "Can I assist you, Leftenant?" Campbell's voice was just one notch below a snarl.

"I would respectfully remind Captain Campbell that the remains of two of my men are here at the fort. Would the Captain kindly instruct me on how I am to retrieve them?"

Campbell said nothing, looking down for a moment, before O'Bannon's voice filled the little space.

"WOULD CAPTAIN CAMPBELL PLEASE TELL ME HOW I AM TO RETRIEVE THE BODIES OF MY DEAD MARINES??"

"Leftenant," Hull said quietly; far too quietly. "You are, without question, thoroughly and completely over the line."

"No, Hull," Campbell said, still looking at his desk. "O'Bannon, I shall give you this one time. We cannot return their remains home. There is no time to disinter them. If there was, we have no way to get them back to the squadron, and, if there was, we have no way to get them decently back to the United States. The best you could hope for is a burial at sea, and there is simply no way to get them even that far. I am sorry. I am sorry beyond words.

"Now do your job, Leftenant. Dismissed."

There were approximately four hundred men, more or less, at the Derne fort. Roughly six seventh of them were none too enchanted with the remaining seventh. the fact that the six-sevenths were willing to take the money of the remainder notwithstanding. The victors of Derne had to face some very hard arithmetic, they were outnumbered by their putative allies.

O'Bannon, Eaton and Farquhar would have made the ride back to the fort in silence, but Farquhar was far from pensive. He was looking at the fort's seawall, judging lines of sight, estimating distances, and he was thinking. As

The Presence of Politics

they bumped ashore, the Briton turned to Eaton and O'Bannon and discreetly motioned for them to step a few yards down the beach with him.

"Gentlemen," he said, "when it comes to our force of noble warriors, when are they at their happiest?"

O'Bannon folded his arms and looked around while Eaton looked back out to sea. "When they're killing, looting, and pillaging," the marine replied, "or contemplating a happy day thereof."

"But of course," Farquhar smiled. "So, if we want to keep them amused and occupied, I suggest we tell them to prepare for a pleasant evening's violence."

There was still a sizable force of Yusef Bey's lions within sight of the fort. Since they'd had their heads handed to them more than once, they were certainly in no great hurry to throw themselves at the fort again, especially with the United States Navy hovering protectively within long cannon range. On the other hand, returning to their garrisons without the explicit permission and grace of Yusef Bey Karamanli, peace be unto him, would have resulted in most unpleasant homecomings. Since they had plenty of food and supplies, and the pay wagons were still rolling in from Tripoli, it was felt prudent to remain in bivouac around Derne, not too closely, of course, and keep an eye on things.

Farquhar proposed that they tell the force to prepare for an attack on the Bey's men just after sunset, around 8:15 on the evening of the 12th. It would not be hard to whip them into a frenzy; especially if a few hints were dropped about possible riches just delivered to the enemy. Some judicious deployments within the town itself would help as well. In town, they couldn't see the quiet withdrawal. It was now roughly noon. They had about seven hours. If they started quietly ferrying men out to the squadron within the next hour or so, the last boats would be leaving just as it started getting dark. It was going to take some quick thinking and no small amount of doubletalk, but it would, hopefully, work. If it didn't, the best result they could hope for was to be handed over to the Bey's men as a peace offering. The worst didn't bear thinking of.

In the meantime, Eaton was going to have a quiet talk with Hamet Bey.

The United States Marines have faced only one clear-cut, unalloyed defeat in their entire storied history, the surrender of the Fourth Marines during the Japanese invasion of the Philippines in World War II. They went down fighting during that grim spring of 1942; but, for all that, they were wiped out almost to the last man, with only a few bedraggled survivors making it out.

They have faced reverse and retreat, without question; most notably, at that God-forsaken little body of water in north central Korea called Chosin Reservoir. There, a little bantam rooster of a man named Lewis Puller, known

To Barbary's Far Shore

throughout the Corps as 'Chesty' for his habitual stance, commanded the First Marine Regiment through a frozen Gehenna that has since become epic legend in the Corps. They were surrounded by nearly seventy thousand howling Chinese; to a great extent, because of the incomprehensible ego and attitude of Douglas MacArthur, the same general whose errors doomed the Fourth Regiment. Puller looked at his men. "We've been looking for the enemy for some time now, and we've finally found him. We're surrounded. This simplifies things."

The First Regiment cut its way out down icy mountain roads under constant, never-ending attack. They walked all the way to the Pacific coast with Puller at their head; in keeping with the honored tradition that no lost marine is ever left on the battlefield, only the dead rode in the few surviving trucks. It was a deliverance so perfect, so well executed, that, for this and an encyclopedia of other exploits, to this day, marines still say, "Good night Chesty, wherever you are."

A decade later, at another once-anonymous village called Khe Sanh, the iron-willed North Vietnamese felt that they could subdue the United States Marines the same way they had once wiped out France's finest warriors in the very small place that became Hell at Dien Bien Phu. For seven months of ferocious siege, the two sides grappled. But, in the end, it was not the marines who broke and withdrew into the green shadows. Unfortunately, after a 'decent interval,' the marines were told to evacuate the base they had so bravely held with their blood, sweat, tears, and two thousand, eight hundred and fifteen casualties, two hundred and seventy-five of them lost. The North Vietnamese took possession of the blasted firebase's remains, once they were reasonably sure that the Americans weren't going to bomb them out of existence for doing so, and took a picture of the base they had so valiantly captured. Sadly, the American media called it a defeat as well, though it was nothing of the sort. The marines of Khe Sanh would have understood, only too well, what Presley O'Bannon's marines were about to face.

The night is quiet. As far as anyone knows, Yusef's men stay quietly and obligingly in their encampment. By ten, most of the troops are asleep, or almost so. The four leaders are very quietly running down the plan once, twice, three more. The march and the attack might have been planned on the fly, but, by God, this would not be. Every movement, every walk to the beach, would be planned to the minute. If anyone, anyone, made a mistake, the troops would take no more than a few moments to put two and two together and end the Derne expedition in a most abrupt and terminal manner.

It is just after five thirty in the morning, Wednesday, June Twelfth, 1805. The marines were quietly awakened at five. They ate their breakfast in shifts over the next hour or so, while the Muslims observe their sunrise prayers just

The Presence of Politics

after 6. They are, at least nominally, occupied for about ten minutes, so now is the time. The word would go out, quietly, without any fuss; first, down to the sergeants who would have tightened their jaws, somehow gotten out a strangled "yes, sir," and then went off to tell the men. Their briefings, in ones or twos, would have been quite simple; we're leaving, we're keeping it damned quiet, and if you know what's good for you you'll keep your mouths shut and your eyes and ears open. There are widened eyes and stunned expressions, but no words. Everyone just puts their heads down and keeps eating.

Hamet Bey Karamanli, rightful ruler of the nation of Tripoli, is in a dank, smelly room with a single weak sunbeam to illuminate it. He'd sunk into a rickety chair and stared at the flaking walls as William Eaton tells him it is all over. Eaton has not pulled any punches.

His report to Hamet is direct and to the point. The American government has abandoned you; abandoned us. At the end of an expository speech that, for William Eaton, must have been breathtaking in its brevity, he looks to Hamet. "The squadron will leave in around seven hours. There will be room for you and your most trusted advisors if you wish, but you must decide now if you wish to leave."

The silence in the room is almost palpable. "Why did we come here? Why did your government lie, Eaton?" Hamet weakly asks.

"They did not lie," Eaton said with more confidence than he felt. "The plan simply ... changed."

Hamet took a deep, shuddering breath as his eyes began to harden. "Perhaps I should change my mind. Perhaps I should fight on and try to save my family. Perhaps I should tell my people that the Americans plan to kill me and abandon them, and we should kill you first." That last came out with a sibilant hiss, like a snake declaring its intention to strike.

The William Eaton who had sailed down the Nile to find Hamet would have stood as tall and proud as a warrior statue and proclaimed his innocence. The William Eaton who marched four hundred-odd miles across the desert would have reared up and struck anyone making such a threat. The William Eaton who marched up to a fortress and demanded its surrender would have demanded satisfaction, and gotten it. The William Eaton who, a few hours ago, was told the greatest adventure of his life was ending in failure no longer had the courage to fight back. Eaton simply met Hamet's gaze with his. "Your Majesty, if you wish to strike me down, do so. I am unarmed and unguarded, and, right now, it would be a mercy unto me."

Hamet only lowered his head. "I shall take my advisors with me to your ships, and you shall not say a single word about it. We will go with you to Sicily. We will be prepared to leave in one hour. And, may the Prophet mark my

To Barbary's Far Shore

words, if my wife or my children are harmed in any way, I shall find you, even if you dwell in high towers. In the meantime, get out of my sight, kafir." A harsh clearing of Hamet's throat, followed by his spitting on the dirt floor, told Eaton that the interview was fully and most thoroughly over. Eaton did not reply; indeed, how could he? Instead, he bowed slightly and backed through the door, careful not to turn his back upon Hamet; not from distrust, but rather from respect.

After all, he was a king.

One Thirty.

Boats had been arriving and leaving ever since the *Constellation*'s arrival the day before and again since dawn. Now the pace was going to pick up a little bit, and there'd be no way to hide that. Fortunately, Colonel Farquhar's plan allowed for that. At about one-fifteen, Farquhar and Leitensdorfer started rounding up the troops and telling them there would be an attack that night that would require just about everyone. The men would be divided into three groups. One would go out ahead on either flank, while the third group would go down the ravine. As the flank groups engaged Yusef's troops, the third would come charging down the ravine and right up the middle of the enemy camp, which, according to a nonexistent defector, had just been paid.

The mercenaries reflected on this for a moment. This put Yusef's men a step ahead of them, and, therefore, this was a situation that could not be tolerated. The troops began getting very excited about the possibility of recompense that night. They behaved accordingly, aggressive and hard to handle, which was exactly what Farquhar had in mind. It would take hours to get them sorted out and settled down so he could lead them out of the fortress.

In the meantime, the Americans and Hamet would quietly scuttle away.

Farquhar, Aletti and Leitensdorfer watched as the troops milled about in the fortress. "Colonel, when planning to deceive a great many people at once, never underestimate the power of human nature," Farquhar said.

Leitensdorfer smiled tightly but said nothing.

Farquhar looked at him quizzically. "You're leaving, aren't you?"

Leitensdorfer thought for a moment, the small smile still on his lips. "Vell, I'm not going to be popular mit zese nize pipple vonce zey realisse zere not gettink pait. Und, der trut be tolt, I vasn't doing zat vell in Alexandria. Not much opportunity for a man like me. Perhaps I vill go to America; see if, perhaps, I can confince zem to pay me, iff no von else does."

The Presence of Politics

Farquhar nodded in agreement. "Eugene, if you convince the Americans to pay you a single shilling, you owe me half." Farquhar extended his hand. "It's been a pleasure, Colonel Leitensdorfer."

For his part, Leitensdorfer shook hands firmly. "Ze pleasure, und honor, haff been all mine, Colonel Farquhar." Heels click softly, one last grin, and Leitensdorfer disappears into the mob. Farquhar looks to Aletti with a questioning eyebrow. The massive Turk simply shakes his head and shrugs.

Two Thirty.

The first squads have been back aboard *Argus, Constellation*, and *Hornet* for some hours now. The next groups are slowly returning to their ships. One of *Argus'* boats is ashore now. O'Bannon and his men are gathered below the north wall of the fortress, with only a few dozen yards of beach between them and home. O'Bannon looks to all of them and tries to find the words to send them off.

"Right," he finally says. "As of now, Sergeant Bancroft is in charge, and I need you to . . ."

"No," says Bancroft, quietly but firmly.

"I beg your pardon?"

"My apologies, Leftenant O'Bannon. No, sar. I will not take charge. I will stay here and evacuate with you and Mister Eaton."

"Bancroft, be reasonable. You're injured . . ."

"I truly don't care a whit about it, Leftenant. I marched all this way to do a job that now appears was unneeded at best, unnecessary at worst, saved yer bloody backside and lost my wife and my retirement in the process? No, sar; I'll be stayin' to the bitter end."

O'Bannon's expression was one of stunned disbelief. "Sergeant Bancroft, this is not a matter for discussion, you will get on that boat . . ."

Bancroft leaned in close, his voice menacing. "Or what, Leftenant O'Bannon? Ye'll throw me out and send me home? I wasn't supposed to be here in the first damned place! And believe me when I tell ye that none of these fine lads is gong to lift a finger to get me aboard!"

O'Bannon looked up at his men. They all looked everywhere but at the confrontation before them. O'Brien whistled a tuneless little ditty as he watched a gull soar noiselessly over the commotion.

"Leftenant," Bancroft finished, "I'll not leave this miserable place unless it's in the last boat. Ye owe me that."

To Barbary's Far Shore

Insubordination, true enough. But Bancroft had a point. He'd been told to give up a great deal for a mission that, by rights, he shouldn't have even been on. O'Bannon shook his head at the futility of continuing the argument any further. "Fine. Campbell, you take charge. Get everyone back. If we do not make it back, you are to report to Captain Hull, and quite likely, the Commodore, as to the events of the mission. Am I understood?"

Campbell came to attention and saluted. "Perfectly, sar. I'm bettin' though you'll get back whole."

O'Bannon simply shot him a sour look. "I might get back whole, but I have doubts how long I'll stay that way. Go."

Curtz spoke up quickly. "I should shtay mit der Herr Leutnant. It might be . . . um . . . dancheruss." The giant Swiss spoke that last word with positive anticipation.

O'Bannon shook his head. "No. Everyone goes, and goes now. Off with you, then." The marines turned and headed for the boat.

After a single step, O'Brien turned around. "Leftenant?"

"What is it, O'Brien?"

Obie shook his head, as if trying to get something to make sense. "I'm sorry, sir. It's just that . . . well, I thought we'd won."

O'Bannon and Bancroft looked at one another with knowing smiles. O'Bannon spoke. "Private O'Brien, you will learn that there a great many definitions of the word 'won,' Now, off with you."

With that, O'Brien saluted once more and trotted off down the beach.

Four Thirty.

Farquhar is giving double-talk laden instructions to the mercenaries, hoping that the confusion he's causing will be written off as typical kafir stupidity instead of deliberate obfuscation. Aletti is, as always, beside him, a monstrous scimitar in one hand. Farquhar casually surveys the mess as one of the mercenaries carefully walks up to them, his hand on the hilt of his own sword.

"Farquhar Bey," he calls in heavily accented English. "I have words!"

Oh dear, thought Farquhar, *here we go.* "What is it?" he asked impatiently. "I have business!"

"You will tell me business! Where Eaton Bey? Where Obanion Bey? Where kafir Amrikans? They lead, not you!"

The Presence of Politics

"They are meeting to decide battle; not your matter, fahima, understand? Go back to your post!"

The mercenary shot Farquhar a wicked look and obeyed. As he left, he pulled aside another soldier. At that moment, Farquhar knew, beyond any shadow of a doubt, that everything was about to go, without any delay, straight to hell.

Aletti looked to his master and raised a questioning eyebrow.

Farquhar shook his head. "No, not now. Too many witnesses. Let's just keep playing the game and see how long we can hold out."

Seven Thirty.

There is a constant roar from inside the fort now, and it cuts through William Eaton straight to his soul. He has not only been defeated, he now fully expects to be torn limb from limb by the men he commanded so confidently just a few days before. Presley O'Bannon hears it too. With the confidence of youth, he is certain he'll get out of this; though, right now, he will be damned if he knows exactly how. Edmund Bancroft has seen more than a few moments like this in thirty years, and he's under no illusions at all about his mortality. With that, he touches O'Bannon on one shoulder, startling him out of his grim reverie.

"Presley," Bancroft says, "I do believe it's time to catch the last boat for home."

There is a single boat waiting for them on the beach, *Argus*' whaleboat, manned by Isaac Hull's best oarsmen. Not only will it be able to get back to the ship in a few minutes' time, they have a great deal of incentive to set a speed record on the way.

O'Bannon, in turn, leans toward Eaton and speaks more quietly than he means to. "William . . . General Eaton . . . we must go. Now."

Eaton rubs his eyes for a moment and nods quietly.

They are in a small portal that leads into the fort from the north wall. It is a walk of perhaps a hundred yards to the boat. With nonchalance none of them feel, and thinking that at any moment they will feel a ball between their shoulder blades, they start the stroll down the beach. O'Bannon's heart is pounding but he keeps his head up, his tread even over the crunching sand until he hears Bancroft growl.

"F'r God's sake Presley, try not to look like a bloody toy soldier. Yer goin' fer a walk and comin' right back, not goin' up the hill with the bloody Duke of York."

To Barbary's Far Shore

It was all O'Bannon could do not to positively stroll the rest of the way to the boat.

Farquhar and Aletti came bounding down to the shore, the pulsing cries of the troops in the fort behind them. Eaton, who had been about to give the order to shove off, instead cried out.

"Wait!"

The rowers halt obediently, if a little reluctantly. The mercenaries are now screaming at the top of their lungs, and something seems to be going on in the fort's square behind them. The heavy wooden walls of the *Argus* wait for them. If it were indeed up to those six sailors they would be leaving a roostertail behind them right now as they headed for home.

Eaton and O'Bannon jumped back out once more, splashing through a few feet of surf to meet the two men they had to leave behind.

"I hope you'll forgive this last little bit of theater," Farquhar said loudly. He gestured angrily for the benefit of the furious men a few hundred feet away. "But every little bit helps."

A loud roar went up from the fort. The men reflexively looked up to see the flag ripple one last time, then sail downwards like a tired bird seeking rest. "That is most definitely your cue, General Eaton," Farquhar said. "I'm afraid they've caught on. I must confess that this was one of the more fascinating missions I've carried out for King and Country. My report will be a challenging one to write."

Eaton looked at the flag disappearing behind the walls of the fort, now glowing with the sunset. Tears beginning to well up in his eyes as he turned to Farquhar. "I'm only sorry it didn't end the way we planned for."

For his part, Farquhar gave a wolfish grin. "Wouldn't have missed it for the world, General. Fair winds and following seas. To you as well, O'Bannon."

"Thank you, Colonel," O'Bannon replied, gesturing as wildly as he could to keep up with the masque. "Till we meet again."

"Indeed," Farquhar said and began to turn back. Before he could, Eaton looked up to Aletti. "SIR, IT HAS BEEN AN HONOR AND A PLEASURE TO SERVE WITH YOU. MAY YOUR GOD KEEP AND SAVE YOU."

Aletti replied with perfectly phrased and accented Oxford English. "It has been my pleasure as well, General, but I do wish you would speak to me in a more reasonable tone. I am not deaf."

Eaton's jaw nearly bounced off the sand, and O'Bannon's was not at all far behind it. It took a few heartbeats before Eaton found his voice. "Aletti . . . you . . . you can speak! Why have you not done so before?"

The Presence of Politics

The Turk smiled. "You have never annoyed me sufficiently before, General. Be that as it may. For your own safety, you must now depart. Please allow me to say farewell in the words of my people."

With that, Aletti raised his massive arms over Eaton's head. From the walls, it could have been a threat as well as a blessing. He intoned in a noble, dignified voice that belonged in a house of worship, not a battleground.

"Be'koach, Be'shalom, Hazak Ve'ematz, G'mar Chatima Tova. Shalom."

Eaton looked up at Aletti. Now he was openly and unashamedly crying at the sound of those soaring, ancient, rhythmic words ascending over the sounds of barbarity and violence. "Sir, I am touched beyond words . . . that was truly beautiful, truly a thing of joy," Eaton said, gulping back a sob.

And with that, Eaton stopped for a moment. A thought grew in the back of his head. Sounds he hadn't heard since his days at Dartmouth formed in his brain, grew familiar and translated themselves. His eyes grew wide and he spoke haltingly. " . . . truly . . . Hebrew . . . ?"

For his part, Aletti simply rested his index finger alongside his nose. With a smile that would have intrigued La Gianconda, he simply whispered as he walked away. "Shhhh . . . ,".

O'Bannon was doing his best to keep a straight face. Eaton watched the Turk walk away for a brief second, then turned to Farquhar with face that betrayed utter and complete befuddlement.

Farquhar simply grinned. "Who knew?" He spun on his heel and stormed back up the beach towards the fort, screaming something in Arabic and gesticulating furiously.

Eaton and O'Bannon watched for a moment longer, then heard Bancroft discreetly clear his throat. "Gentlemen," the sergeant called, "I do hate to be interrupting fond farewells, but I suspect we're all going to be in mortal peril in a few moments if we don't leave, and I've cheated the Reaper more than enough times on this trip."

"Wisdom, as always, Sergeant Bancroft," O'Bannon said as he took Eaton by the elbow and guided him back to the boat as the rowers prepared to push off.

Aboard *Argus*, Joshua Blake peered through his glass at the shore. "Cap'n Hull; they're moving, sir!"

Hull strode over to the rail as he watched the whaleboat start to move over the water towards them. The length was so short in terms of mere distance, but so very, very long in terms of the danger that still lurked over their shoulders. He nodding grimly. "About bloody time. I was beginning to think they'd want to

To Barbary's Far Shore

send wreaths and chocolates to those damned heathens before they'd leave. Are we at departure stations?"

Blake lowered the glass and took one more look around. "That we are, sir. As soon as the whaleboat's aboard we can get underway."

"Mmm. Were up to me, we'd throw down a rope and they could hang on till we got out of range."

O'Bannon reflected that he'd never seen a boat crew row quite as fast, or as steadily, as the six members of *Argus'* crew were now. *Argus* was the closest of the three ships to the shoreline, but still three or four hundred yards away. That was an eternity right now; one that suddenly got a great deal longer.

Bancroft's eyes narrowed. "Beggin' the Leftenant's pardon?"

"Yes, Edmund?"

Bancroft tilted his head towards the fort. "I do believe our friends back there are firin' a salute in our honor . . . only horizontally."

O'Bannon and Eaton whipped their heads around. Small puffs of white/gray smoke started to appear on the ramparts. Right on cue, small splashes erupted a few feet to either side of them. The rowers, of course, understood and pulled even harder.

"CAPTAIN CAMPBELL!"

Hugh Campbell bounded across *Constellation's* quarterdeck to the rail. He snatched the glass from the lookout, to see a few dozen Arabs firing their muskets as quickly as they could. Others crowded in between them. "Some days," Campbell snarled, half to himself, "it simply does not pay to put on the uniform. MASTER GUNNER!!!"

"SAR!"

Campbell lowered the glass and bellowed. "RUN 'EM OUT!!"

"AYE AYE, SIR. RUN 'EM OUT!!"

With a rumble that vibrated through the ship, *Constellation*'s main battery of twenty-eight eighteen-pounder long guns crashed out through the gun ports, fully shotted and ready for action. The bellows of the gunners and their commands echoed through the gun deck, slow matches at the ready. "The main battery is at yer disposal, Cap'n!"

"Damned right, Master Gunner," Campbell called back. "Stand by for my orders!"

"Aye, aye, sar!"

The Presence of Politics

Aboard *Argus*, Isaac Hull had seen the same thing as Campbell. Since great minds think alike Hull ordered his main battery out as well. *Argus* couldn't field quite as many, or as heavy, weapons as the much larger *Constellation*, but, at this distance, it would have the same brutal effect.

"SIR!" Josh Blake's voice was urgent. "I believe the damned heathens have a long gun on the wall!"

Oh, NO.

Hull took the glass from Blake and scanned the fort's parapet. Sure enough, there was one, no, two, cannon poking their ugly snouts over the pale wall. It looked like they'd remembered just enough of their training to be loading the damned things. *Constellation* had her long guns run out. For that matter, so did *Hornet*. They were either hesitating for some reason, or simply hadn't seen the cannon coming up.

By rights, Campbell, as the senior officer present, has the responsibility to give the order to open fire. But, Hull thought to himself, *by that point, the whaleboat's been blown out of the water. The hell with it; perfect end to a perfect bloody day.*

"MASTER GUNNER," Hull called, "RUN 'EM OUT AND PREPARE TO FIRE!"

The rumble and crash of cannon on their heavy carriages thundered through every fiber of the ship for a moment before the Master Gunner called back.

"THE MAIN BATTERY AWAITS YOUR ORDERS, SIR; AND WE WERE A STEP AHEAD OF YOU ON THE 'READY TO FIRE' PART!"

Hull wanted to smile at that. He decided to wait a moment. His hesitation was rewarded with a thick gout of black smoke lancing out from the parapet. A hollow THUD followed that almost rattled his teeth. Almost before his brain could register that, a fountain of dirty water exploded upwards perhaps a dozen yards from the whaleboat. A shockwave rolled it to one side for a moment, before it settled back down.

Hull idly noted with professional pride that the crew never once stopped rowing in perfect synchronization. He rolled his eyes Heavenwards, to a God who had so clearly decided to use them for His sport this otherwise beautiful evening.

"OPEN FIRE!!"

"OPEN FIRE!" Josh Blake repeated the command. There was the sound of a dozen commands echoing up from the gun deck. Eight twenty-four pounders and a single twelve pounder opened up; their target, the wall of the Fortress at Derna.

To Barbary's Far Shore

The flash and roar got Hugh Campbell's attention most quickly. He made a mental note to have Isaac Hull keel-hauled later. "MASTER GUNNER, COMMENCE FIRING!!!"

"AYE AYE, SAR, COMMENCE FIRING!!"

Argus' main battery, though small, was more than sufficient to get an enemy's attention. *Constellation* was meant for business; business she delivered behind a sheet of flame and a wreath of smoke that erupted into the gathering darkness. The cannonballs, as were sometimes the case, moved slow enough that one could see them.

And see them Presley O'Bannon did. "DOWN!!"

There was the sound of rustling leaves overhead followed by a whirrrrr like some berserk toy. First *Argus'*, then *Constellation'*s, broadsides went thundering overhead to land on the beach they had just left. Fans of sand and dirt and masonry erupted into the evening sky.

Eaton sat back up to look back at the beach. Bancroft's crutch landed on his shoulder and forced him back down. "Not quite yet, General Eaton!"

Eaton almost snapped back up to take a swing at Bancroft. He heard the leaves come up from behind him once again, then the watery thud of the ball impacting a few yards past them. This time, the boat heaved upwards, almost standing on its stern. It fell back with a sickeningly weightless feeling and a loud SLAP. They lay motionless for a long heartbeat. The oars caught once more and then they were moving towards *Argus* again.

"Master Gunner, keep firing!!"

"Aye, sir!"

Argus' gunners had the range now, and the next broadside was as well-timed and precise as Hull could have wished for. *Constellation's* second broadside hit the beach with a heavy WHOOMP! *Argus'* shots sailed cleanly through the sand and mud and rock to impact the fort's wall. This time there was no mistaking the effect.

The crew of one of the cannon had already headed back into the shelter of the fort with quite reasonable assumptions of discretion and valor. The second crew had stuck around a bit longer, hoping to get just one more shot in at the lying infidels. Devotion is one thing; pushing one's luck something else entirely. In this case, "pushing one's luck" applied. *Argus'* broadside tore through the fort's walls like the Furies enraged. Wooden braces cracked, collapsing the rampart for a good thirty feet. Razor-edged splinters flew through the falling bodies of the gun crew as they and their weapon landed in the dirt with heavy, hollow THUMPS.

The Presence of Politics

It took a few moments for the freshening breeze to blow the smoke and dirt away. When it did, Isaac Hull could clearly see in the gloom that the cannon were gone. So was just about everyone else.

"CEASE FIRE!"

Then he turned to Josh Blake. "First, get us moving. Then, bring us about so we intercept the whaleboat. I really do not feel like staying within range of those savages any longer than I must."

"Aye, sir!"

Constellation, the Yankee Racehorse, shook her sails out and moved nimbly ahead. Campbell kept her port side pointed directly at the fort. *Hornet* started to move out for the open ocean, making sure nothing was going to impede them from that corner. It was getting dark now, but the sky was clear and it was still easy to see. *Argus* glided noiselessly forward as her sails filled, aiming directly for the whaleboat that still splashed its way through the surf.

Campbell turned to his first officer. "First, keep an eye on those bastards. If they so much as twitch, blow them to Kingdom Come."

"With pleasure, sir!"

With that, Campbell went back to the quarterdeck and began the process of getting *Constellation* out to the open sea. By all accounts, she should have been far enough out that grounding wasn't a threat. But, for all Hugh Campbell knew, *Philadelphia's* crew still wasn't out of their captivity and he had no desire to join them. In any event, *Argus* could still put sufficient iron onto the fort to make the Musselmen wish they'd never put their heads back up.

Argus herself was picking up speed and heading straight for the whaleboat. Hull stood by the helm, ready to take it himself if he had to. The helmsman was doing just fine, threading the brigantine through an invisible needle that, if they succeeded, would get them and their returning lambs nicely out of harm's way. Of course, not doing so meant at best turning around and trying again or running the whaleboat down into the turbid waters off Derne. *Professionally embarrassing and personally mortifying,* Hull thought to himself, *but there was nothing else for it.*

The oarsmen were still sweeping them for all they were worth. Eaton looked at the approaching *Argus*, rapidly turning into a dark, menacing bulk in the darkness. "O'Bannon . . . am I wrong in believing that they're headed directly for us?"

O'Bannon squinted into the gloom. "Your eyesight fails you not, General. They're headed directly for us. My guess is they are going to try and throw us a

To Barbary's Far Shore

line and tow us out of range of the fort. On the other hand, they could be planning to just run us down and forget this entire damned affair."

Eaton's face went slack and so pale in the full moon that was rising over the velvet darkness of the ocean, O'Bannon had to pat him on the shoulder. "Just joking, General. We'll be out of here in just a few minutes."

Hull looked to the bow, where a group of marines and sailors stood, preparing to throw a line to the whaleboat. The boat crew would have a few seconds to secure the line and then work it so they would be towed alongside the *Argus*, now picking up even more speed.

"Bosun!," Hull called, "do your work!"

"Aye, sir!" The bosun picked up his hailing trumpet and called out. "Ahoy the whaleboat . . . all STOP!"

The whaleboat's crew was nothing if not well trained. They raised their oars as one. As they glided to a stop, they could hear Hull give a command on *Argus*' quarterdeck. The oncoming ship, seemingly close enough to touch now, pivoted sharply to starboard. A weighted line came sailing towards them, whistling as it flew through the air. It landed neatly across the whaleboat's middle. The crew got it secured to the stanchion at the bow in record time, just as the line started to curl, then stretch, then go taut. With a jerk and a splash, the whaleboat started obediently following *Argus*, now headed due north behind *Hornet* and *Constellation*. The men in the whaleboat looked around and realized they were moving out of danger, and smartly at that. With a collective sigh of relief, they leaned back in the little boat and enjoyed the ride. It took a minute or so, but finally, Sergeant Bancroft spoke up.

"Sar?"

O'Bannon had closed his eyes for a moment in relief, but opened then again to look at Bancroft. "What is it, Sergeant?"

Bancroft tilted his head towards the smoking ruin they'd left, now being illuminated by pinpoints of torches. "Worst damned liberty port I've ever seen."

They stopped a few miles further on and hauled the men and the whaleboat aboard. The men staggered up a rope ladder, Bancroft going first. He went up *Argus*' rough, splintered flank, every step an agony. A blue-clad arm reached over the side and grabed his wrist. The extra effort was enough. Bancroft threw himself over the gunwale, collapsed against it, panting for breath. He looked up to see Sergeant William Williams still holding his wrist and grinning like a fiend.

"About damned time, Edmund. Thought you'd decided to go back to Baltimore on your own."

The Presence of Politics

Bancroft couldn't help but smile. "Be thankful I don't throw you into the bloody boat myself. Now, if you're done making jokes at my expense, please escort me to the sickbay."

O'Bannon was next, followed by Eaton. He came up slowly, hesitantly, waving away any offers of help. Footsteps on the deck turned out to be Isaac Hull, who paused and crossed his arms.

O'Bannon snapped to attention, saluted the quarterdeck and then saluted the captain. "Leftenant O'Bannon reporting for duty, sir."

Hull returned it after a moment or two. "I am truly glad to see you, Leftenant. You did your best."

"Thank you, Captain. I gather we're heading for Syracuse?"

Hull nodded. "We'll catch up with Commodore Barron there. Hopefully we'll have some better news about Will Bainbridge and his men by the time we get there. How is . . . ?" Hull inclined his head towards Eaton.

O'Bannon thought for a moment. "Truly, sir, I think he wishes he was dead right now."

Hull nodded. "Depending on how well he explains himself, that may well be the better option. Get some rest, O'Bannon. The adventure's over and we're back to work in the morning. I expect to see your report ready to submit to the Squadron by the time we get to Syracuse." With that, Isaac Hull strode aft towards his quarterdeck.

O'Bannon remained for a moment while the whaleboat's recovery wound down around him. William Eaton, Captain, Consul, General, and Conqueror, stood at the rail, quite severely alone. He watched the fires of Derna disappearing steadily into the darkness. The sea hissed fluidly along *Argus'* hull, the sounds of the ship and her men in counterpoint, as he felt his body adjust to the roll of the ship. Then, and only then, did Presley O'Bannon realize how tired he was and how hungry he felt.

He decided at that moment to do two things. First, see if Blodge was still awake and get him to cook up a little bit of decent food, even if he had to bribe him to do it. Second, to crawl into his rack, that uncomfortable, claustrophobic, itchy, creaking, wonderful rack, and sleep until they got to Syracuse.

They came into Syracuse very, very early in the morning a week or so later, the winds fighting them every step of the way. Hull had asked the officer of the watch to wake him when they entered the harbor, and he kept his promise. Hull came topside to a cool, breezy morning with the sun just beginning to lighten the eastern horizon of the Ionian sea. A few pinpoints of light marked the villages of Isola and Carrozzierre off to port. There were other lights as well; *Hornet*, astern

To Barbary's Far Shore

to starboard, *Constellation* straight ahead. In the port, flickering lanterns in a tight formation amid the scattered lights of the merchantmen awaiting their turn at dockside; the United States Mediterranean Squadron, at rest. There was at least one big frigate there, along with several other ships. Hull guessed that someone had finally decided to hang the expense and send out another frigate, which looked to be carrying a commodore's pennant, just in time for the festivities to end. *Lovely,* Hull thought. He had been hoping to put off reporting to the Commodore for a few days yet at least. Apparently, even that small mercy was to be denied him.

Frigate Constellation aka 'The Yankee Racehorse'. Source: U.S. Navy

A warship never truly goes to sleep. When the sun came up, the activity on *Argus'* deck only increased, with more noise. By the time Presley O'Bannon came up on deck, there were men working rigging, holystoning the deck and doing all the many things that kept a ship in fighting trim; trim enough to satisfy Isaac Hull's eye. The marines were at their posts; some on watch, some below, all back in their daily routine. Everyone, the sailors and marines alike, was at their job, and the whole thing was moving as smoothly as a precision pocket watch.

All except the one man aboard who had no job, William Eaton. Eaton had barely stepped from his cabin over the last week. He'd hardly eaten, but had drank quite heartily; to the extent that Captain Hull had raised a concerned eyebrow over the matter. Not that Eaton had been a problem, far from it. When he did emerge, he was quiet, polite, and seemingly most rational. But all of them had been in long enough to know that when a man took to his quarters and

303

The Presence of Politics

began pickling himself, especially on the less than high quality local spirits to be found aboard ships of the US Navy abroad, there was concern to be had.

Right now, however, Eaton was seated by the rope ladder, a light jacket across his shoulders. The mountain of baggage he had transferred to the *Argus* so long ago had dwindled down to two battered bags. Both of them carried only a few items he had left aboard in the first place. The rest was gone, probably draped across some barbarous Musselman back in Derna who would never appreciate what he'd found.

O'Bannon quietly walked up to Eaton and stood aside for just a moment until the diplomat looked up. "Good morning, O'Bannon. Beautiful morning."

O'Bannon nodded. "That it is." Pause. "Not long now, I gather?"

Eaton shook his head slowly, looking a few hundred yards away. Another ship lay swinging gently at anchor, a boat being lowered over her stern. "The *Franklin* is sending her boat now. Commodore Barron and Captain Hull have made it quite clear to me that I should be ready to board her boat immediately upon its arrival. And so I am."

O'Bannon squinted across the water. Sure enough, a small boat was now inching away from the *Franklin*. Her crew was not putting a great deal of effort into it, but steady enough for all that. Ten, fifteen minutes at most. O'Bannon could only stand there, not at all sure what to say.

Eaton suddenly spoke up. "It was indeed an honor serving with you, Presley. You alone made this possible. You and your marines."

O'Bannon smiled gently and shook his head. "Well, as I remember, Colonels Farquhar and Leitensdorfer had considerably more military experience, and your refusal to give in kept us going But, if you'd like to credit me with all of that, please be decent enough to put it in your final report. It may just convince my superiors that I am not beyond hope."

Eaton smiled, the first one O'Bannon had seen in days. "If that is your wish, my friend, I shall cheerfully do so. But I fear anyone named in my final report may end up in the dock beside me."

O'Bannon said, "Not likely, William. And if they do, rest assured that at least one Leftenant and his marines will spring to your defense at a moment's notice. Not to mention the crew of the *Philadelphia*."

Eaton looked up, his eyes red. "I hope," he whispered. There was silence for a few moments.

O'Bannon looked thoughtful and put his hand on Eaton's shoulder. "William, I shall return in a moment. Even if the boat gets here, please wait for my return."

To Barbary's Far Shore

Eaton hesitated for a moment. "Well . . . of course, Presley, but why?"

By that time, O'Bannon was down the deck talking to Willie Williams, who could be seen to smile broadly as they both headed belowdecks. Eaton turned back to watching the boat draw closer. Each sailor was clearly visible now. The faint sounds of the bosun calling the oars drifted across the water. There would be no escape this time; no fast talk or noble speech. He would go to the *Franklin* and he would stay closely watched in his cabin until they arrived back in Washington. Then he would answer for his sins.

Can't be helped, Eaton thought. His mind ran silently for a moment then said, quite simply, worth every last bit of it. *No matter what happens back in Washington, it would be worth. . . every last bit.*

His mind wandered for a bit. He was wrenched back to the present by first, the sound of *Franklin's* bosun commanding the little boat alongside *Argus*. Before his mind could even process that, there was the thump of boots and the rattle of gear. Heels snapped to attention on either side of him.

Eaton looked up to see O'Bannon at attention, and his marines lined up on either side of him at present arms. All of them. Williams. Bancroft; barely standing, but proudly so. Curtz towered over the rest. Duffy with one leg in a cast, wobbled unsteadily. Campbell stared straight ahead. Pritchard grinned slightly, with his shako askew. Even Obie O'Brien stood as tall as he could, given the circumstances.

Memorial to the sailors lost on board the U.S. Ketch Intrepid

Eaton stood a bit unevenly, surveying the remains of the force he had led from Alexandria a lifetime ago. He realized that they had, indeed, won.

For his part, Presley O'Bannon, resplendent in his best uniform, pulled his right hand up in a salute so sharp it fairly whistled.

"General Eaton . . . mission accomplished."

Eaton reached back in his soul. To the sergeant, who had fought his way through the Revolution those many years ago. To the captain, who had once

The Presence of Politics

dared the whole Northwest Territory to defy him. He sent back a salute just as sharp.

"God bless you all," Eaton whispered. He lowered his hand, picked up his bags, and went carefully over the side, down the ladder to his future.

The boat wobbled just a bit as Eaton settled down. As soon as he had, the crew lowered their oars into the water, carefully turning back towards the *Franklin*. At first, there was just the sounds of the oars through the water, birds crying, and waves slapping the sides of ships. Then, just for a few moments, Eaton heard a fiddle play a gentle lament, softly and carefully.

To Barbary's Far Shore

FOURTEEN: L'ENVOI

As it turned out, no one on the expedition knew about its ultimate success until they got to Syracuse. As has been previously mentioned, Yusuf Bey wasn't at all happy about having to give in. He was doing his best to convince the Americans that he was actually crazy enough to do something as suicidally dangerous as kill the *Philadelphia*'s crew. He wouldn't have, of course; that would have been a disastrous changing of the rules. Not to mention, if anything would convince the Americans to come directly after Yusef, the execution of nearly three hundred Americans would have done it. Every other nation or satrapy in range would have stood aside and looked the other way. But no one was willing to take that chance; nor could they be blamed.

On May 26th, Lear and Commodore Sam Barron arrived off Tripoli aboard *Essex*, and Barron formally resumed command of the squadron from John Rodgers. That meeting must have simply overflowed with mutual respect and politeness. Barron was enjoying a brief moment of good health, and he probably hadn't been completely up to date as to what was going on regarding the expedition. Bringing him up to speed would have been just as motivational as the fact that John Rodgers was still there. Once the formalities were out of the way, *Essex* raised a white flag of truce and stood into the harbor that had seen *Philadelphia*'s captivity and destruction and *Intrepid*'s last tragic ride.

Yusef apparently decided, at that point, that enough was enough. A white flag climbed the staff above his palace. The Spanish consul, having offered his services in the name of peace, came out to officially welcome the American delegation.

There were five days of back and forth and bad weather until the 29th, when the Spanish consul met Lear aboard *Constitution*, which had more room to host the two delegations. At this point, via the Spaniard, Yusef laid his cards on the table. He demanded $130,000 for the crew of the *Philadelphia*, and the immediate return of any Tripolitanians in American hands. To his credit, Lear politely but firmly told the Dey no. An immediate prisoner exchange was the only thing they were interested in. Obviously, that wasn't going to happen, but the Dey had to have understood now that the game was well and truly up. He replied with his final demand; sixty thousand dollars and a full prisoner swap.

Done.

L'Envoi

A moment, please. Having agreed to said terms, the Dey realized he could not, in good conscience, release any Americans until his subjects, so dear to his heart, were released first. Lear dug in his heels, and laid down the law. Either accept the US offer, or turn it down; with, by the way, the Squadron in long gun range of your palace.

The next morning, Lear got his answer. A freshly scrubbed and shaven Will Bainbridge was rowed out to *Constitution*, escorted by Danish consul Nissen. Officially on parole, Bainbridge was both grateful to be out of the pestilential hole that had been his home and thoroughly frustrated with the pace of negotiations. One cannot be angry with him for it.

What must have been a spirited discussion followed in *Constitution*'s day cabin. Lear decided to hold his ground; and, by the by, thank the Spanish consul for his efforts and inform him that his assistance is no longer required. There seems to be the possibility that the Spanish consul was taking advantage of the situation, for either national and/or personal gain. On the morning of June 2nd, William Nissen came back to the *Constitution*, now empowered by Yusef Bey to wind things up.

The treaty was, from the diplomats' point of view, fair and reasonable, but Yusef insisted on the clauses throwing Hamet to the wolves. Having one's relatives try to take one's throne was an occupational hazard in this part of the world; indeed, it was one of those things a ruler simply expected. But Hamet had gotten far too close for success and Yusef had no intention of letting anyone get that close again.

The final treaty was signed on June 3rd. On June 4th, while O'Bannon, Eaton, and the rest of the Derne expedition were still holding on to their prize, the American consulate at Tripoli re-opened. Crowds of Tripolitians sang the praises of the new Consul, Colonel Lear. Or, at least, they did if they knew what was good for them.

A few minutes after that, Will Bainbridge led his men dockside. Ships of the Squadron waited for them with food, clothes, and medical care. They marched out of that miserable hole on the Tripoli waterfront with their heads held high and their pride and dignity intact; examples to the others who, most surely and sadly, would follow.

As has been mentioned earlier, Americans had suffered captivity before and would continue to do so again, and in situations far, far worse than the comparatively benign conditions of the Tripolitian captivity. Anyone who has suffered imprisonment as part of their honorable service is regarded as someone who has served far above and beyond what was required of them. But the men of *Philadelphia* were the first Americans to fall en masse into the hands of a culture that did not regard them as even human. They were kafirs, unbelievers,

To Barbary's Far Shore

infidels; men who were not worthy of the slightest respect or decency because of how and who they worshipped. All else about them was subordinated to that one simple detail.

It would be nearly two centuries before it happened again, and we would be shocked when it happened again. Indeed, for some reason, we are always shocked when this sort of thing happens in that part of the world, but never mind.

But for now, they were going home as heroes. Will Bainbridge would always get skeptical looks from his peers and his superiors for what had happened, but no one would ever be able to question the nobility and bravery that he and his men had shown.

There was one other event that should be recounted here. When *Constellation* arrived at Derne, she carried a passenger who went ashore in the same boat with the message from Captain Campbell for the American officers to report to him. That passenger was an envoy from Yusef Bey, come to gather the good people of Derna into his fold once again. He was waiting in the governor's palace, such as it was, for the Americans to make their getaway.

It turned out that, after the final repulse of Yusef's troops on the 10th, those worthies had decided that perhaps it would be wise to withdraw out of range and wait for reinforcements. Because, of course, the huge American frigate had brought more troops to defend the fort. They were, of course, mistaken, and most seriously so; but, in any event, they were effectively out of the fight. Troops were on their way down from Tripoli, but it would be a day or so before they arrived, and they would be under the command of Yusef's emissary.

After things calmed down on the night of the 12th, events which included a thorough and complete pillage of the American base inside the fort, the emissary ordered the people of the town rounded up and brought to the main square. There, he made a short speech, explaining Yusef Bey's willingness to understand that mistakes had been made. Of course, they had found themselves facing overwhelming force. All would be forgiven. He closed his speech by waving letters of pardon before the assembled subjects. He lovingly assured one and all that Yusef's magnanimity knew no bounds.

For some reason, the good people of Derne were, shall we say, skeptical. Instead of raising three cheers for Yusef Bey, they ignored the emissary and prepared to defend their town once more. The mercenaries took advantage of the confusion to scatter. Almost assuredly, Colonel of Volunteers Richard Farquhar was ahead of the pack for the long ride back to Alexandria. One may safely assume that the return of Yusef Bey Karamanli's authority to the beknighted city of Derna was neither a peaceful nor pleasant occasion.

L'Envoi

Syracuse Harbor, Sicily, Thursday, August First, Eighteen Hundred And Five

Without question, the previous evening had been the most riotous Presley O'Bannon could ever remember, and that was saying something indeed. Unfortunately, one must pay the piper eventually. Eventually came for O'Bannon at three bells on the mid-watch. The thudding of boots and muffled laughs, curses, and other sounds of *Argus* coming awake got his attention. He suddenly snapped awake; or, at least as awake as one can be without opening one's eyes.

The last thing O'Bannon could remember was the sight of Edmund, Sergeant Edmund Bloody Bancroft, United States Marines, dancing a jig atop a table at the tavern they'd selected for his retirement celebration. *Actually, O'Bannon thought, it wasn't so much that we'd selected that particular one as we'd sort of stumbled into it on their way from the previous place we'd selected. Oh, well,* O'Bannon thought. *At least we made it home.*

We had made it home, hadn't we?

Slowly, carefully, O'Bannon winched one eyelid open and looked around. *Let's see now. Low overhead, a guttering candle, thoroughly uncomfortable bedding, and the muffled slap of waves against wood.* Admittedly, it could have been any one of the dozens of taverns that lined the Syracuse waterfront, but O'Bannon was reasonably certain that he was, indeed, home and dry aboard *Argus*. The confirmation came a second later. A peal of thunder rolled through the compartment, emanating from the thin door and reverberating off O'Bannon's eardrums to set off a headache that went from nonexistent to excruciating in a heartbeat. O'Bannon had intended to roar a suitable response to the miscreant outside his door, but all that came out was a dry, sandy croak that vaguely resembled the word "What?"

The door swung open with a piercing squeak that seemed to travel right down his backbone. Campbell poked his head in. "G'marnin to ye, Sar; three bells and all that."

"MMph" was the only thing O'Bannon could say in response as he closed his eyelid again. *Good Lord,* he thought, *what DID we drink last night?*

"Sar?"

O'Bannon opened his eye again and tried to fix his gaze on Campbell.

"Sar, we do need ye to be getting' up." Campbell said it with all proper deference, but with the smile of an asp. "Beggin' yer pardon, sar, but did ye overindulge a bit last night?"

"You ought to know, Corporal; you were there."

To Barbary's Far Shore

"Now that I was, sar, that I was. But I have to admit that, when the Leftenant gets to celebratin', it's truly difficult indeed to tell when he's crossed the line."

"Campbell, get out of here," O'Bannon said with a spastic wave of his hand. "Or I shall take the utmost pleasure in shooting you."

"Assumin' ye could find yer pistol this marnin, sar."

"GO."

"Aye aye, sar." Campbell closed the door with a rattling bang that set O'Bannon's teeth on edge. He was sure he could see the corporal's grin through closed eyes, bulkhead and all. There was nothing for it then but to try and get up, with an emphasis on 'try,' O'Bannon slowly sat up, making sure that he didn't crack his skull on the overhead.

"Surgeon Cable!"

Staggering out with one hand on the rack to steady himself, O'Bannon peered carefully into the upper bunk. Surgeon Cable was there all right and, by the looks of things, still breathing. *Well, good for that*, O'Bannon thought. *Hate to have to explain to Captain Hull how we lost the ship's surgeon.* Getting dressed without moving too quickly was a challenge, but one Presley O'Bannon was up to. A better way to put it was that he had regained sufficient coordination to insure that he didn't try to pull his jacket on over his feet or his boots on over his hands. The only interruption to his efforts was an occasional strangled snoring sound from Cable. He also seemed to be dreaming about a young woman named Contstanza, who was apparently none too happy with him.

By the time he was fully shaved, dressed and reasonably presentable, it was close to five bells. O'Bannon left his stateroom and took a seat at the wardroom table. Blodge had a pretty solid breakfast laid out this morning, but O'Bannon's stomach was rebelling at the mere thought of food. A cup of tea would have to do, and in any event was far more practical. O'Bannon poured himself a steaming mug with a little honey and sat down heavily at the table, leaning forward with his eyes closed.

Pascal Peck walked in, far happier than any human should have been that ungodly time of the morning. "G'morning, Mister O'Bannon," the midshipman piped, pouring himself a cup of tea. "How was Sergeant Bancroft's fling last night? Enjoyable, I trust?"

O'Bannon winced as he took a sip of tea. "I'm sure it was," he said wearily, "but I'll be damned if I can remember much of it."

"Including Sergeant Bancroft dancing the jig?" Peck was grinning like a fiend around his mug.

L'Envoi

O'Bannon squinted at Peck with as much firmness as he could muster just then, which wasn't much. "And, may I ask, how you happen to know about that?"

Peck was still grinning as he downed the last of his tea with a deeply contented sigh. "It's a small ship, Mister O'Bannon. A very small ship indeed."

Peck pushed his chair back with a scraping noise that went right through O'Bannon's lacerated senses. "In any event, I've got to get back topside. Captain Hull wants me to arrange for Sergeant Bancroft and Private Duffy's gear to get over to *Congress*. I'll see you later."

"Mmph," was all O'Bannon could reply.

Peck stuck his head back into the wardroom with a wicked smile. "By the by, sir, how did things turn out with Surgeon Cable and Constanza?"

The look on O'Bannon's face was sufficiently irate that Peck ducked quickly out of the way, lest a teacup come flying at his head.

Little brat, O'Bannon thought as he sipped his tea, the warm taste of the honey gliding down his throat. Around him, the crew was starting to get things secured for their departure that morning. The Squadron, now heavily laden with supplies, ammunition, and anything else they could think of, would stand out in a few hours, hopefully around eight. Someone had told him yesterday where they were headed for but, for the life of him, he couldn't remember it just now.

Another sip of tea, and there was the creak of an opening door. Richard Cable staggered out and collapsed into a chair. O'Bannon shot him a sideways glance. "Dear God, Richard, do you . . . "

Cable's eyes were tightly closed and he was breathing heavily. "Yes, Presley, I feel every bit as bad as I look. I would indeed hate to feel this awful without showing it." With that, Cable slowly began to lean forward, until his head touched the table with a gentle klonk.

O'Bannon blinked once, then leaned over. "Richard, are you all right?"

Cable's reply was muffled by the tabletop. "No, Leftenant O'Bannon, I am not all right. I consider myself dead, passed away, no longer of this earth, for I am suffering the tortures of the damned with every breath."

"Ah," O'Bannon nodded sympathetically. "Shall I call for the surgeon?"

There was a pause, and then Cable replied with a truly impressive string of obscenities, most of which questioned O'Bannon's parentage, legitimacy and other highly personal matters. O'Bannon listened politely as he sipped his tea, then answered in a tone of mock hurt. "Tsk, tsk, Surgeon Cable. I was merely trying to help."

To Barbary's Far Shore

"You can help best by leaving me alone."

"Suit yourself." O'Bannon took another sip of tea. "By the way, how did things turn out with you and . . . what was her name? Constanza?"

For the first time, Cable raised his head, but only to squint painfully through barely open eyes. "How did you know about Constanza?" he asked before lowering his head once more.

O'Bannon smiled. "It's a small ship, Richard. A very small ship."

"Far too small to have to deal with the likes of . . . " Cable was interrupted by Josh Blake thumping into the wardroom.

O'Bannon toasted him with his teacup. "G'morning, Leftenant Blake." Cable could only give a barely audible moan.

Blake went to pour himself a cup of tea and fix a small plate of beef and potatoes as he looked curiously at O'Bannon and Cable. "Good morning, gentlemen, and I use the term loosely." Sitting down next to Cable, Blake asked, "what's wrong with him?" before tucking into his food.

O'Bannon swallowed the rest of his tea. "Hard to say, really. Probably the plague."

Blake froze in mid-bite, his eyes locked on Cable, who remained moaning but motionless. "The plague?"

"Oh yes," O'Bannon nodded cheerfully. "First signs are a deep lassitude and fatigue, much like this." O'Bannon picked Cable's head up by the hair, held it for a moment, then released it to fall back on the table with a marvelous wooden thump. "Then, of course, the buboes start to swell up. It's a most revolting sight; great purple-black lumps that . . . "

Blake swallowed hard, his mouth suddenly dry as he grabbed his tea and plate and pushed back quickly from the table. "Pardon me, O'Bannon, but I'd better take this back on deck, keep an eye on the midshipmen, y'know." With that, Joshua Blake disappeared back out of the wardroom.

O'Bannon smiled wickedly as he finished his tea. The muffled voice once more rose sepulchurally from the table beside him.

"I'll get you for that, O'Bannon. If it is the last thing I ever do . . . "

O'Bannon rose and patted Cable on the shoulder. "That's the spirit, Richard. Set a goal and stick to it, that's what I always say." Heading aft, O'Bannon decided that now was as good a time as any to check on the marines and make sure they'd be ready for formation in a few minutes. *Argus'* decks were rapidly becoming a maze of crates, barrels, piles of rope and masses of everything else that a ship of war needed to survive at sea. It was small

L'Envoi

consolation that, as crowded as it was here, it was infinitely worse aboard the big frigates like *Congress* and *Constellation*, which had to carry enough to replenish the smaller ships as well.

Sure enough, when he got there, the marines were just tightening up their uniforms. Sergeant Williams keeping a keen eye on each of them until he saw O'Bannon walk in. Williams snapped to attention, no easy feat on *Argus'* cramped decks. "Atten-shun!"

The bustle of activity came to a sudden and abrupt halt. "G'morning men, how are . . . " O'Bannon stopped midsentence.

The marines had been securing their hammocks in the overhead. Each of them was as tightly bound and perfectly stowed as anyone could have asked for. But one of the hammocks was badly rolled, lumpy, poorly tied . . . and moving.

There wasn't a sound as O'Bannon slowly walked up to the animated bedroll, just a few inches away from Williams' head. Muffled shouts and imprecations were issuing forth from it. A pair of infuriated eyes bulging out from a gap in the bedding were the only clue as to what lay inside. O'Bannon examined the apparition for a moment before saying a word.

"G'morning, Sergeant Williams."

"G'morning, Leftenant O'Bannon."

O'Bannon slowly, deliberately walked around the bedroll, which was now twisting violently. He turned back to Williams. "Sergeant, who in God's name is in there?"

Williams smiled broadly. "Private Duffy, sir!"

"Ah." There was another pause while Duffy continued to scream . . . well, something, then O'Bannon looked at Williams again. "And . . . why exactly is Private Duffy secured in the overhead?"

Williams shot a thoroughly disapproving glance at Duffy, then looked apologetically at O'Bannon "Well, Leftenant . . . this is one of those things . . . that is . . . it's possible that you might not entirely understand the circumstances."

O'Bannon took another mildly disbelieving look at Duffy, who was now thumping his heels against the overhead timbers. He looked back at Williams. "Indulge me. Please."

Williams swallowed and then began to explain. "Well, Leftenant, you know how Surgeon Cable ordered him confined to the sickbay until he transferred to *Congress* this morning?"

"Yes?"

To Barbary's Far Shore

"Well, Paddy, that is, Private Duffy, didn't want to miss Sergeant Bancroft's going-away party last night, so . . . he snuck off the ship."

O'Bannon looked at Duffy. "Private Duffy, is that true?"

Duffy bellowed, or, at least, tried to bellow. "MMph! Imp jsutph wnthpht tooph saph ggophbyph . . . " Williams clapped his hand over the bedroll approximately where Duffy's mouth should be.

O'Bannon considered this for a second. "I'll take that as a yes. But, setting aside the question of just how he got off the ship, he wasn't at Sergeant Bancroft's party. I was, and though I remember damned little of it, he most definitely was not there."

Williams shook his head. "No, sir, he wasn't. Apparently he ran across several other affairs first and graced them with his presence instead. He never actually did make it to Sergeant Bancroft's, though I gather a couple of times he just missed you."

"I see. That still doesn't explain why he's in the overhead."

"No, sir, it doesn't, but I was getting to that. When he got back and we caught him. Well, you know how frisky he is when he's had a bit."

"That I do. Go on."

"Well, sir, we tried to get him to calm down, but he wouldn't. I could have put a guard on him, but then whoever got that detail would miss Sergeant Bancroft this morning, and Lord knows that no matter who we put on him and how we do it, Private Duffy is nothing if not inventive. So . . . we came up with this."

"Ah."

Duffy's eyes went wide, and he shouted. "Bluphy serphent wllimphs, oll geph yuph furp . . . "

Williams slapped Duffy alongside the head. The bellowing stopped, though Duffy's eyes were literally rolling from the blow. O'Bannon considered all this for a moment. "Right. You dealt with the problem in a most . . . unique way, Sergeant. I'm sure Sergeant Bancroft would have approved."

Williams smiled brightly. "Thank you for the compliment, sir. I truly appreciate it."

"Quite all right. But, in the future, please notify me first when you feel it necessary to hogtie one of my marines."

"Of course, sir."

L'Envoi

O'Bannon nodded. "Good. Carry on, and get the lads topside." Looking up at Duffy, he patted him on the head. "As you were, Duffy."

As soon as he was gone, Duffy started thrashing again. Williams stood before him, arms folded and looking for the entire world like a disapproving father. "Now, Patrick," he said sorrowfully, "is that any way to behave in front of Leftenant O'Bannon?"

"Gphnyhlf!!"

Williams' eyes narrowed. "Paddy, I'm truly doing my level best to be understanding here . . . "

"Fulfhernft!"

At that, Williams stepped back, eyes wide with indignation, then looked around. "You men clear out, now. I need to have a chat with Private Duffy."

It was unlikely that the marine berths had ever emptied that quickly before. In a blink of an eye it was just Williams and Duffy. Without preamble, Williams went nose to nose with Duffy, and spoke quietly but very firmly. "Patrick Duffy, I consider myself a reasonable man in most things, and though pride be a sin, I do pride myself on being slow to anger . . . "

"Bupmh . . . "

"'Bupmh,' nothing! Now I am warning you, Private of Marines Duffy, if you make one more sound, just one, besides 'Yes, Sergeant Williams,' I will personally see to it you never get off this damned ship. Do you understand me?"

Duffy's eyes were wide with terror, and his voice plaintive. "Npphr?"

Williams jaw was set, his voice low and hard. "Never, Private Duffy. Now, you behave yourself and we'll cut you loose in a few minutes and get you topside. But, so help me God, if you don't, I'll see to it you're here until my enlistment is up. You will be on this ship so long that twenty years from now, new marines not yet born will report here, see you, and say 'There sits Grandfather Duffy, he once annoyed his sergeant. Behold, and profit by his example!' Now, have I made myself quite clear?"

Duffy's eyes were downcast in resentment and resignation. "Yph, Serphent Wllmphs."

Williams beamed at him like a proud parent. "That's a good lad. I'm going up top, and I'll send some of the lads back in a few minutes to get you."

With that, Williams turned and strode briskly away, leaving a chastened Paddy Duffy to reflect on the error of his ways. Duffy gave a muffled, mournful sigh, and then realized that he was in a position to get at least a few minutes more sleep . . . and this wasn't really all that uncomfortable.

To Barbary's Far Shore

O'Bannon climbed the ladder up to *Argus*' deck, carefully because his balance was still a bit off. It was already sunny and warm, one of those days where you looked forward to getting underway. Of course, the sea air didn't hurt either. Helped keep one alert and reasonably conscious.

The harbor was already bustling as the squadron prepared to get underway, and even O'Bannon had to admit it was a magnificent sight. *Argus* was moored to the outside of the formation, while *Congress* was a hundred or so yards away to starboard, her decks pulsing with activity. It would be an especially pleasant day for her; she would be on her way back home in a few days, carrying Bancroft and Duffy with her. To her starboard was *Constitution*, whose crew was already heading up the rigging, preparing to unfurl the acres of linen that would pull her out of the harbor. Ahead of *Congress* was *Constellation*, whose crew was moving at a much more dignified pace, as befitted the Commodore's flagship. And, just visible by their masts and rigging, moored around the big frigates like watchful sentries were *Argus'* consorts, *Hornet*, *Nautilus*, and *Enterprise*. Off to port, O'Bannon looked for a moment at a good-sized merchantman that had had moored alongside them during the night. *It had to have been during the night*, he thought with a rueful smile, *else we would have run into it with the liberty boat.*

The merchantman's crew was moving much more slowly than those on the warships, but that was reasonable enough. It would be a while yet before the anchorage was clear enough to move through and get alongside the wharf, and there was no sense in hurrying up just to wait. *That is a military trait,* O'Bannon reflected. But they would move smartly once they did get there; time, after all, was money, and there were other practical reasons for moving out as quickly as one could. *Most masters regard the present lull in pirate activity as only a temporary respite,* O'Bannon mused, *and they are probably right.*

Admittedly with Hamet now in Sicily, and by all appearances permanently, Yusef Bey had been keeping his corsairs on a very tight leash. But there was no doubt in anyone's mind that eventually Yusef would forget the very close call he'd had and he'd let Murad Reis go a-raiding again, and they'd be right back where they started. Thank you, Consul Lear . . . and President Jefferson, and everyone else who had contributed to this nightmare.

An American flag, larger and brighter than the one he had raised at Derna, but otherwise the same, began to stir fitfully from her jackstaff with an occasional multi-colored flash, each one a heartbeat longer than the first. *A little bit of home*, he thought. Someplace where people sat in spacious homes and ate decent food and didn't have to worry about finding their head on some Berber's pike; someplace where people went about their business utterly unmindful of danger or the men they sent out to deal with it. Well, if nothing else, Captain

L'Envoi

Bainbridge and the crew of *Philadelphia* were on their way home right now, and they could at least take some comfort in that.

Home, thought O'Bannon. *A hell of a word. Either someplace to be safe or someplace to leave.* The flag on the merchant swirled again; in his mind's eye, O'Bannon could see the flag atop the Derna fortress again, its colors unlike anything else in the dirty, dusty pile of rock and mud. But yet there was something fitting about it being there at that moment. Something that said, this is where we have drawn the line, and we will not be denied. Something that said, as long as I fly, this is now a little piece of home, just the same as a white farmhouse in the Blue Ridge, a row house in Baltimore or a shack in New York. It mattered not. It was home. And so too were the two little squares of Tripoli soil where Whitten and Stewart rested eternally in the shade of the palm tree.

Maybe I really should go home when they got back, O'Bannon thought. *After everything we've been through the last few months, a little peace and quiet might not be an entirely bad thing, not a bad thing at all. Father would be his usual self and complain about how tough it was to run the farm without me there, and Mother would fuss over everything when she wasn't cooking enough to feed the entire detachment . . . and it would be so reassuringly normal.* In his heart, though, Presley O'Bannon knew it wouldn't be for long. *Normal is out here, always waiting to see what was over the next stretch of horizon.*

O'Bannon turned as he heard thumping steps coming down the deck and saw Richard Cable walking towards him. Well, in truth it wasn't so much a walk as a stagger, but the surgeon was at least upright and moving as he approached O'Bannon, thrusting a hand out to grab the rail so he could stop and steady himself. O'Bannon watched in bemusement as Cable looked up at him with one eye tightly shut and the other wide open.

"Don't say a damned word."

O'Bannon shook his head. "Wouldn't think of it, Richard. But he might." He tilted his head towards the approaching figure of Isaac Hull.

"Oh, God," Cable moaned as he tried to haul himself to attention. It was a bit easier for O'Bannon to do so.

"Good morning, Captain Hull."

Hull stood for a moment, looking at the two of them as he shook his head in disbelief. "Gentlemen, I declare that I have never seen two more disreputable looking specimens on one of my ships. What did you two get up to last evening?"

O'Bannon smiled as best he could. "Oh, nothing much, sir. Just Sergeant Bancroft's going away festivities."

To Barbary's Far Shore

At that, Hull's face brightened into a wicked smile. "Is that so?" he asked sweetly. "Well, at least you made it back to the ship, which seems to be more than I can say for the good doctor here." Leaning down almost nose-to-nose with Cable, Hull spoke in his best quarterdeck bellow. "And how are you this morning, Surgeon Cable?"

Give him this, Cable tried. Swaying slightly, he saluted, or gave something that looked like a salute. "Bully, sir. Just need a breath of . . . " At that, Cable started to go over like a white pine being felled before he caught himself. " . . . fresh air."

Hull's expression was still one of utter kindness. "Of course you do, Cable. Followed by a decent Christian burial at sea. In any event," Hull clapped his hands together loudly enough to make them both wince, "I'm assuming that your respective departments are ready to get underway?"

O'Bannon nodded. "That they are, sir." He motioned to activity amidships. "Mister Danielson is getting Sergeant Bancroft's and Private Duffy's gear sent over to *Congress*, and the lads are forming up. As soon as we have the formal ceremony, we'll be ready."

"Hmm," Hull nodded. "And you, Surgeon?"

O'Bannon was sure he could see just a hint of green entering the surgeon's face as Cable nodded quickly. "Ready for . . . burp . . . " His eyes crossed for a moment. " . . . anything, Captain."

Hull's expression was dubious. "I'll take your word for it. But speaking of the ceremony, where is the guest of honor?"

It struck O'Bannon just then that he hadn't yet seen Bancroft that morning. "I'm sure he's just getting his things together, Captain. He'll be on his way shortly."

Hull raised an eyebrow, but accepted it. "Well, do what you can to hurry him along. We'll be getting underway shortly, and since Commodore Rodgers doesn't seem to have eaten any of his captains for breakfast this morning, I'd hate to be the first. Let me know when . . . " Hull stopped and his eyes widened slightly as if beholding something he knew had to be a mirage behind O'Bannon and Cable.

Turning, O'Bannon saw Edmund Bancroft stepping up onto the deck from the midships ladder. He wasn't in his familiar Marine blue. instead, he wore a beautifully cut suit of crimson velvet, with brass buttons that caught the rising sunlight. Even from that distance, O'Bannon could tell that the shirt was high quality linen, with lace ruffles that cascaded from the collar and flowered discreetly from each cuff. The shoes were brand new, with gleaming silver buckles and polished within an inch of their lives. The whole outfit was finished

L'Envoi

with a matching tricorn hat, a narrow, tasteful white band around the upper edge. But, for all that splendor, Bancroft stood there with a finger in his collar looking for all the world like a child dressed up by his parents and sent out unwillingly to perform for guests.

"Good heavens," said Hull in a tone of quiet amazement. "I didn't realize that Sergeant Bancroft had such excellent taste in civilian clothing."

O'Bannon blinked once at the apparition before him. "I didn't realize he had any civilian clothing."

The three of them stepped over to Bancroft. Seeing them coming, Bancroft reflexively snapped to attention, only then remembering he was out of uniform. He relaxed, slightly. Hands were warmly shaken all the way around.

"Edmund," O'Bannon smiled, "you are an absolute fashion plate. Where did you get this made?"

"Me wife had it done up the last time I was home. Rose always did have the best taste in these matters. But frankly, Leftenant, I'd rather I was back in the uniform," Bancroft said with a scowl as he tugged at his collar again. "I feel like a bloody idiot in this monkey suit."

Hull grinned. "Now Sergeant Bancroft, you're going to have to get used to it. You get back to Baltimore and you're going to be a man of means and substance. You'll need to look the part."

Bancroft drew himself up to his full height. "With all due respect, Cap'n, the only reason I'll ever be needin' a suit like this 'twould be if someone were gettin' baptized, married, or buried. And I should be pointin' out that those are not activities one usually associates with a tavern."

Hull shook his head. "You've never met my pastor back in Boston. In any event, we do need to be getting underway, so . . . " Hull motioned aft and they headed for the quarterdeck, Cable weaving along slightly behind them.

The marines were in formation, with Sergeant Williams at their head, a chastened Paddy Duffy on crutches to one side, and the midshipmen as tall and proud as they could be. Even O'Bannon had to admit they'd never looked better. Across the deck from them, Will Allen had as many men as he could spare from *Argus*' crew, and under his watchful eye even they looked downright respectable.

Bancroft stopped short at the sight and looked at O'Bannon with mystification. "What's this then, sar?"

"Only what you've earned, Edmund," O'Bannon said as they fell into line alongside each other, with the easy grace of men who have done something so many times it has become a reflex.

To Barbary's Far Shore

Hull took one last look around to make sure all was well, and then came to attention. With a roar that could have been heard back home, Hull called out.

"SHIP'S COMPANY . . . "

Williams snapped to, and with a parade-ground thunder that could rattle windows, called "MARINES . . . "

Both men snapped out the command "ATTEN-SHUN!"

With a crash of metal, wood, and boots, the marines and sailors came to attention. The marines stood perfectly straight and aligned in blue and white uniforms with perfectly polished muskets, the sailors infinitely more relaxed in a grab bag of trousers, hats, and scarves.

"SIR!" Williams barked, while bringing up a salute. "THE MARINE DETACHMENT IS ALL PRESENT AND ACCOUNTED FOR!"

O'Bannon returned the salute. Pascal Peck stepped forward, carrying a tightly rolled parchment scroll, bound with red and gold silk cords. O'Bannon took the scroll and deftly untied the cords. With a flick of his wrist, it unrolled with the dry rustle of fine paper, a heavy red wax seal at its bottom. Still at attention, O'Bannon began to read.

"Attention to orders! To all who see these presents, greetings; whereas Sergeant of Marines Edmund Raphael . . . " At that, every eyebrow on the deck raised a notch. Even O'Bannon shot Bancroft a sidelong glance while silently mouthing "Raphael?"

Bancroft's only reply was a dark scowl, and O'Bannon thought better of any further questions.

Clearing his throat, O'Bannon began once more.

"Whereas, Sergeant of Marines Edmund Raphael Bancroft has served his country with distinction and credit since the Tenth of November, Seventeen Hundred and Seventy Five,

"And whereas he was seriously wounded at Derna, the Kingdom of Tripoli, on the Twenty-Seventh of May, Eighteen Hundred and Five, incurring those wounds as a result of brave, heroic, and most noble actions,

"And whereas those wounds have left him unable to execute his duties in a safe and reasonable fashion,

"By order of the Honourable Robert Smith, Secretary of the Navy, Sergeant Bancroft is hereby discharged from any further obligation to his service, effective the First of September, Eighteen Hundred and Five. Further, by request of Secretary Smith and by Act of Congress assembled, Sergeant Bancroft is hereby granted a pension in the amount of five dollars per month, effective the

L'Envoi

First of October, Eighteen Hundred and Five, and to continue at said rate for the remainder of his natural life.

"Sergeant Bancroft is to be commended on a long and honorable career, in which he has brought honor and credit to himself and the United States Marines. Signed, Franklin Wharton, Commandant Lieutenant Colonel, United States Marines."

O'Bannon rolled up the scroll, faced left and respectfully handed it to Bancroft. "Sergeant Bancroft, the detachment is yours."

Bancroft took the scroll in his hands carefully, looking at it with a curiously far-away expression, as if he was unable to quite accept that the moment was, indeed, finally here. Stepping off, he went to inspect his marines one final time. Bancroft made the facing moves with a precision born of decades of practice; decades that time could never wear away. Anyone who saw him, even as dandified as he was that morning, would know that he had been in service to his country for a very long, very proud time.

Starting with the last rank, Bancroft slowly moved down the row. He briefly inspected each man, with a murmur of encouragement or farewell at each one. Halfway down the last rank was Bernard Curtz, ramrod straight, every button glittering, and his face as sharp and chiseled as a block of ice. Bancroft looked him over quickly, and as usual, found nothing. With a gentle grin, he looked up at Curtz. "Well, ye damned Kraut . . . ye just may make a decent marine o'yerself some day."

Curtz smiled from ear to ear like a trained wolf that had just gotten a compliment from his master. "Danke, Herr Feldwebel. I mean, sank you, Herr Feldwebel . . . Serchent Benkroft . . .

Bancroft rolled his eyes heavenward. "Curtz?"

"Ja, Herr Feldwebel?"

"Shut up, lad." Bancroft's smile was a wide as Curtz', and as he started to step off, he stopped and stepped back for just an instant.

"Curtz?"

"Ja, Herr Feldwebel?"

"For heaven's sake, learn some bloody English."

Curtz' heels snapped together like a gunshot. "It vill be mein pleasure, Herr Feldwebel!"

"Knew it would, lad; knew it would."

To Barbary's Far Shore

A few steps further down and Bancroft was facing Joe Pritchard. Bancroft reached up to adjust his shako, then stepped back with a questioning look on his face.

Pritchard looked at him in mystification. "Something wrong, Sergeant?"

Bancroft shook his head sadly. "I'm disappointed in ye, lad. Me last inspection and ye don't have a bloody word to say."

Pritchard smiled. "Struck dumb with emotion, Sergeant."

Bancroft gave a derisive snort. "That'll be the day ye're struck dumb by anythin', Private Pritchard."

Pritchard still grinned, but it softened as he replied, "'Tis true this time, Sergeant. May the wind always be at your back."

Bancroft gave a gentle nod in acknowledgement. "Fer once, Joseph . . . I believe ye. Take care o' yerself."

"I will, Sergeant."

Two sharp corners, a few more steps, and there was Bernard O'Brien, even in his leather collar and shako barely coming up to Bancroft's head. A quick look up and down; not a thread out of place.

"Ye'll make a fine marine, Private O'Brien."

O'Brien could barely keep his voice from cracking, with a smile that belonged opening presents on Christmas morning, not on the deck of a man-o'-war. "Thank you, Sergeant Bancroft."

"Assumin' ye don't get yer damned fool head blown off first." The smile collapsed as quickly as it had appeared, as O'Brien swallowed hard and said nothing. Bancroft leaned over and said gently, "Now ye listen to me, lad. You pay attention to Sergeant Williams and Leftenant O'Bannon, ye hear? One good fight does not make ye an old soldier, an' ye still have a very, very great deal to learn about this life. Keep yer ears open and yer mouth shut . . . and once ye've learned to do that, come see me and I'll teach ye everything I know. Agreed?"

The smile came back, more subdued this time, but with an understanding behind it. "Agreed, Sergeant Bancroft."

"Good lad." Bancroft patted O'Brien on the shoulder and finished the inspection before stepping before Williams and coming to ramrod attention. The two old friends simply looked at one another for a moment, unsure what to say, each one wanting to say the right thing if it turned out to be the last thing. Finally, Bancroft spoke.

L'Envoi

"William. The lads look fine . . . just fine. Keep an eye on Pritchard, though. He still hasn't learned how to wear his damned cover."

"I shall, Edmund. And they look fine through your work, not mine."

Bancroft gave a short wave of his hand, both acknowledging and dismissing Williams' compliment. "Well . . . in any event . . . they're all yours now, William. Take good care of 'em."

"I shall, Edmund. You have my word of honor on that."

"Never a doubt, Willie . . . never a moment's doubt." Bancroft extended his hand, and Williams grasped it firmly with both of his.

"May God go with you, Edmund."

"And with you, William . . . and with you." Bancroft stepped away, paused, and turned back as if he had forgotten something. "By the way, Willie . . . "

"Yes, Edmund?"

Bancroft leaned in with a conspiratorial air as he whispered, "Don't forget to stay close to O'Bannon. He's a good man, Willie; the finest officer I've ever known. But there are days the lad couldn't get his boots on without a little help." Both of them turned to look and shoot quick smiles at O'Bannon, who stood there looking suspicious and knowing that something was going on but damned if he could figure out just what it was.

Marching back across the deck to face O'Bannon, Bancroft stood there for just a moment, not at all sure what to say or do next. Finally, Bancroft spoke quietly. "That'll be it, then?"

"That'll be it, " O'Bannon nodded.

"Right, then." Bancroft looked around one last time, taking in the sights and sounds and smells and people, burning them into his memory forever. Sniffing once, Bancroft sighed, then smiled gently. "Well, sar, it wasn't much of a living, but it was truly one hell of a life."

Bancroft saluted O'Bannon one last time and O'Bannon returned it, a salute that was more than the final honors accorded each other by two old friends and comrades. This would be the last one, the final acknowledgement that one career was over, and that another, still young, would now be on its own.

Isaac Hull stepped forward and bellowed "THREE CHEERS FOR SERGEANT BANCROFT! HIP, HIP . . . "

He interrupted the reverie, and then the entire ships' company, sailor and marine alike, erupted in a "HOORAY!!" that flew out to sea like a frigate under full sail, echoing off the towering minarets of Alexandria, where the Egyptians

To Barbary's Far Shore

looked out to sea at the ruckus in utter mystification. Infidels, they thought. Mad as well as unholy and barbaric.

"HIP, HIP . . ."

"HOORAY!!"

John Rodgers looked up from his breakfast aboard *Constellation* at the sound of the cheers coming across the water. *Argus*, he thought with a snarl. *Damned undisciplined rabble over there, that's what they are.*

"HIP, HIP . . . "

"HOORAY!!"

Richard Cable had been able to hold on through everything so far, but the roaring cheer finally did him in. He staggered back against the rail, stumbling down its length far enough away from the formation so that when he became ill over the rail, no one would notice. Hopefully.

Every last man on *Argus'* crew broke into applause and cheers, with the sad exception of Surgeon Cable. Even Joshua Blake was able to overlook the fact that they were all supposed to be at attention and applauded lustily, if a little self-consciously.

"Be damned," Bancroft said quietly.

Hull stepped forward to shake Bancroft's hand. "Sergeant, it has been a genuine pleasure. You've got an honored place on any ship I ever command."

"Thank ye, Cap'n," Bancroft said with genuine humility. "Though, if it's all the same to you, I think me wife might be preferrin' that I stay home from now on."

"It'll be my pleasure to get you there, Edmund," O'Bannon said. "We need to get you over to *Congress.*"

Bancroft discreetly cleared his throat. "Beggin' the Leftenant's pardon, but there is one matter we still need to take care of."

O'Bannon looked at him with a mystified expression. Bancroft tilted his head towards Arthur Campbell, still at attention in the ranks, and then it hit him.

"Oh no, Edmund, no; not now. Why ruin a perfectly good day?"

Bancroft's grin was enough to strike terror into O'Bannon's heart. "Regulations are regulations, sar."

O'Bannon gave a resigned sigh as he held up an index finger to Hull, wanting him to wait just another minute. O'Bannon came to attention himself. "Corporal Campbell and Private O'Brien, front and center!"

L'Envoi

With a thud of boots, Campbell and O'Brien sprang forward and snapped to attention in front of O'Bannon. O'Bannon looked back at Bancroft, who made a brief shooing motion with one hand, and O'Bannon looked as if he was tasting something particularly unpleasant. Taking a deep breath, O'Bannon spit it out as quickly as he could:

"InaccordancewiththeregulationsoftheDepartmentoftheNavyandhavingbeen approvedbyCommandantLieutenantColonelWhartonCorporalArthurCampbellish erebypromotedtotherankofSergeantUnitedStatesMarineseffectivethisdatewithallt hepayprivlegesandentitlementsthereof . . . and may God have mercy on my soul."

O'Bannon turned on one heel and stalked off as Bancroft stepped up to Campbell, whose jaw would have been on the deck had he not been at attention. "Weren't expectin' that, were ye?," Bancroft asked with a wry grin. All Campbell could do was blink in surprise as Bancroft took a pair of beautifully embroidered sergeants' stripes from his jacket pocket, their metallic threads catching the sunlight, and handed them to Campbell.

Campbell swallowed hard, still not sure what to say, but finally getting out a stunned "thank you, Sergeant Bancroft."

"Don't fall in love with 'em, lad," Bancroft replied. "And thank Leftenant O'Bannon and Sergeant Williams. They picked ye, and over a fair amount of protest, I might add."

Campbell looked at the stripes in his hand, still not knowing what to make of them. "Sergeant Bancroft . . . I'm grateful, sergeant, truly . . . but you had said . . ."

"I know what I said, Sergeant Campbell. The first prerogative of a non-commissioned officer is to change his bloody mind." Bancroft paused for a moment, then leaned in close to Campbell. "Ye are a royal pain in the arse, Arthur. Ye are obnoxious, arrogant, and not just merely annoying but ye know everything and the bloody price of it. But for all of that . . . Sergeant Williams and the Leftenant know they can count on ye. Do not let them down, d'ye understand?"

"I won't, Sergeant. I promise . . . " Campbell's hand closed around the stripes, as if he'd just latched onto the treasure of the Stuarts.

"Ye'd better not. Otherwise, I'll come back here and fix the problem meself. Now, ye'll be needin' a corporal . . . "

At that, Campbell brightened noticeably. "I know exactly who I'll be choosin', Sergeant; Private Curtz. He's . . . " Campbell hesitated at the sight of Bancroft shaking his head lugubriously. "Not Private Curtz, Sergeant?"

To Barbary's Far Shore

"Not Curtz. He's a good man and he'll be a fine marine once he learns the damned language. I was thinking more along the lines of . . ." At that, Bancroft placed his hand on O'Brien's shoulder, and O'Brien almost fainted from the surprise. "Private Bernard O'Brien, here."

"Oh, no, no, no, " Campbell declared. "Obie's a good lad, but that's all he is, a lad . . ."

"And a smart and brave one who needs to be taken in hand and taught the ropes. . ."

"No, Sergeant! I am indeed grateful for these. . . " Campbell shook the stripes he was still gripping, "but 'tis a sergeant's prerogative to choose his corporal."

Bancroft remained calm, but one eyebrow started to ascend upwards. "And it was felt you should choose Private O'Brien."

Campbell leaned forward, jaw out, as pugnacious as he dared be on deck. "No Sergeant! I have spoken on this and 'tis all I'm saying! My corporal will be Private Curtz!"

Bancroft stood there a moment considering this, his arms folded and his right shoe gently tapping the deck. Then he leaned forward again. "Arthur . . . how would ye be liking to set a record for going from corporal to sergeant, and back?"

Campbell was about to say something thoroughly unpleasant, but then decided on something simpler. "You . . . wouldn't . . . dare . . ."

There was a twinkle Campbell found very upsetting in Bancroft's eye. "Try me."

Campbell and Bancroft were eye to eye for a moment until Campbell realized that there was an exceptionally good chance that Bancroft meant every word. Well, one must learn what battles to fight and which ones to walk away from, and Campbell came to the reluctant conclusion that this one needed to be left alone. Coming back to attention, Campbell gritted his teeth and looked straight through Bancroft. "As always . . . Sergeant Bancroft's years of wisdom and experience win out."

Bancroft's smile was peace and light itself. "Knew ye'd see it my way, Arthur." Turning to O'Brien, Bancroft snarled, "Ye got any questions?"

"NO SIR, SERGEANT BANCROFT!"

That happy grin once more. "That's a good lad. Pay attention now and ye'll do just fine." As Bancroft turned and headed back down the deck, he was sure he heard Campbell mutter, "my arse ye will," but figured it was Willie's

327

L'Envoi

problem now. Passing O'Bannon with a brisk stride, Bancroft never even paused.

"Now I can get off this damned tub."

O'Bannon and Hull shrugged and fell in behind him. As they strode towards where the boat crew had rigged the ladder over the side, they passed Surgeon Cable. He was finally back in something like an upright position and slightly less green, though walking in a straight line seemed to be just a bit beyond him for the time being.

Peck and Danielson were overseeing the boat crews when they got there and snapped to attention. Hull peered over the rail. "Are we ready to get Sergeant Bancroft started home?"

Before either of the midshipmen could reply, a white-hot torrent of Irish obscenities came blasting upwards. They all looked over to see Paddy Duffy trying to hang onto a rope ladder with one hand and keep his broken leg off the ladder with the other. He accomplished nothing other than to get the ladder twisting from side to side like a berserk pendulum, always staying just out of reach of the two sailors in the whaleboat who were trying to get him in.

Hull gave O'Bannon a doubtful glance. "Are you sure Duffy's no longer fit for duty? He sounds terribly healthy to me."

There was a solid, ringing THUMP from below, followed by an even stronger volley of malediction. O'Bannon shook his head. "Well, at this rate Captain, if he isn't unfit now he will be soon."

One of the boat crew looked up. "Beggin' the Captain's pardon, but we'll have this gentleman secured in just a moment, and then you can send Sergeant Bancroft down."

Hull looked about the anchorage and then shot an anxious glance at his watch. "Well, look alive down there, but please try not to kill Private Duffy. We don't have the time to do the reports before departure." Turning back to Bancroft, Isaac Hull shook his hand once more. "Sergeant, it has been a pleasure. I don't know which I've appreciated more; your bravery and courage, or your ability to keep Leftenant O'Bannon from harming himself and others."

"'Twas nothing really, Cap'n. Once ye get 'em trained, they pretty much take care o' themselves."

"Truly? You will have to show me your methods some day. Heaven knows I'll need it if I'm to continue working with certain marine officers."

"Actually quite simple, sar; patience, understanding, and the occasional cuff alongside the head. Always did wonders for Leftenant O'Bannon."

To Barbary's Far Shore

"Never would have suspected. I'll have to try it sometime."

O'Bannon cleared his throat, as much to get their attention as to remind them that the subject of their conversation was still standing right next to them.

"Oh, Presley," Hull said with mock graciousness. "You're still here?"

"If that's the way I'm to be treated, I'll go back ashore and offer my services to Yusef Bey," O'Bannon replied with mock hurt. "I would like a moment with Sergeant Bancroft, however."

"Most certainly," Hull answered, pushing Cable and the midshipmen a few steps out of the way before lowering his head to have a quiet but intense conversation with George Mann.

O'Bannon shrugged, then turned to Bancroft. He took a deep breath, not quite sure what to say, but deciding to press ahead anyways. "Edmund . . . I owe . . . "

"Ye owe me nothin', Presley. I did me job. God almighty only knows what would have happened if I'd of come back without ye." Bancroft's tone was gentle, with just the hint of a smile. "If ye owe me anything . . . just take care of the lads. Listen to Willie; he knows what he's doing. And bring yer own backside home in one piece. I can't wait for Rose and the girls to meet ye."

"Edmund . . . about your wife . . . "

Bancroft waved him silent. " 'Twas not yer fault, Presley, d'ye understand? You did what you had to do, and that's all there is to it. She knew, she's always known, that the uniform had to come first. I don't blame her for being angry enough to leave, not a bit. She's put up with things no woman should ever have to, and most o' the time she did it pretty cheerfully. I'm guessin' that this time . . . well, it was just one time too many, that's all."

O'Bannon nodded, really at a loss to do anything else. "Will she still be in Baltimore?"

Bancroft spread his hands. "I don't know, lad. I truly don't know. For all I know, she's burned the bloody tavern down and moved back to Anacostia." They both had to smile at that, but they were uneasy ones.

"Actually, that's damned unfair o' me, Presley. Rose might be angry with me, but she'd never do anything that hateful, never in a thousand years. She's a good woman, Presley, a pearl among women. She held things down for all those times I was gone, and she took care o' me home and looked after me daughters. No man can ever ask any more than that. But I promised her I'd be back, and I broke that promise, lad. It was for a good reason, a damned good reason. Willie Bainbridge and his lads are home and safe, and maybe she'll understand now that we can tell her about it. But, if she's gone . . . then she's gone. I'll always

L'Envoi

love her and I'll miss her with every beat o' me heart. But the world simply doesn't have the decency to stop turnin', lad. I've got to go on, and it's something between Rose and I. I'm guessin' she probably did just lock the tavern and take the girls back to her father's place."

"I could ask Captain Hull to see if Captain Alexander might be able to put you ashore at the Navy Yard . . . "

Bancroft held up a hand and shook his head. "No, Presley. Baltimore's me home, and that's where the rest of me life is, one way or t'other. I'm going home, lad. Once and for all."

O'Bannon swallowed and nodded. "As you wish it, Edmund." O'Bannon looked up for a moment to see Hull ostentatiously checking his watch. "Oh Lord, Edmund, we need to get you to *Congress*. Any last words of wisdom?"

Bancroft looked a bit uncomfortable at that, casting his gaze to the deck and shuffling his feet. "Well, Presley . . . this is a bit difficult. You've always, always been a good officer, lad, always looking out first for the men, but . . . "

"But what, Edmund? Please . . . you and I have been through far too much together for me to take offense at anything you could ever say."

Bancroft looked doubtful at that. "Presley lad, are ye sure about that, now?"

"Of course I am. You've been like a second father to me, Edmund. Whatever you have to say to me, I will always take it in that spirit."

Bancroft had a very dubious look on his face, but shook his head with a sigh. "So be it then, Presley. It's just . . . well . . . "

O'Bannon's expression was starting to move from respect to exasperation. "For God's sake, Edmund, what is it?"

"Presley . . . please . . . get rid o'that damned fiddle!"

O'Bannon's eyes widened as if he had been struck in the head by a plank. He blinked twice before speaking. "I'm sorry, Edmund . . . I don't think I quite caught that."

Bancroft's expression was solid and earnest. "I am truly sorry, Presley, I am, but that tool o'Satan has got to go; d'ye hear? First of all, it looks awful, just bloody awful, lad, in front of the men. Secondly, and just as bad, if not worse . . . " Bancroft shook his head in sadness and sorrow over what he had to say next. "When ye play it, ye sound like a sack o'bloody cats, Presley. I honestly find it one o'God's miracles that grown men don't pull their hair out by its very roots after a few hours of those serenades from Hell."

All O'Bannon could do was blink, his mouth hanging open, trying to frame some sort of response.

To Barbary's Far Shore

Bancroft leaned in. "Lad, close yer mouth in front of the midshipmen. Ye look like a stunned trout."

With an effort, O'Bannon pulled his jaw shut, shook his head to clear his mind, and spoke with as much dignity as he could muster. "Edmund, please . . . don't trouble yourself at all." Tugging on the hem of his jacket and stretching his neck, O'Bannon continued. "I asked for your advice, and you gave it honestly and forthrightly. That is the mark of a true friend, and I am in your debt."

The relief in Bancroft's voice was palpable. "Thank ye for bein' understanding, lad. I hoped you'd take it in the spirit in which it was intended."

At that, Hull, Cable and the midshipmen came back. "Sergeant Bancroft," Hull said, "I do not wish to seem impolite . . . "

"Not a problem at all, Cap'n," Bancroft answered equitably. "Ye got a ship to be runnin' . . . and I've got a voyage to start."

"That you do," Hull said, "but before you go."

Hull placed one massive hand on George Mann's back and gave him a shove towards Bancroft. The midshipman managed to get himself stopped before there was a full-out collision; even so Mann found himself almost face to face with Bancroft's belt buckle. Mann looked up to see Bancroft looking down at him, his expression calm, but definitely unhappy about one more delay.

"Ye wanted somethin', Mister Mann?"

Mann swallowed hard, his mouth suddenly dry. "Y-yes, Sergeant, I did. It's..that is, it's been brought to my attention . . . "

Hull backhanded the back of Mann's neck, and the midshipman caught himself again before he slammed into Bancroft. Swallowing again, Mann continued, "I have realized that I may . . . "

Hull's hand connected once more, this time bringing a yelp of pain from the midshipman. Turning to look at Hull as he rubbed his neck in pain, Mann heard Bancroft rumble, "Go on, lad."

Still rubbing his neck, Mann stepped back a bit, but looked up directly into Bancroft's eyes. "Sergeant, . . . I abused my authority with you. You are a brave and exemplary man, and you did not deserve my treatment of you. I was very, very wrong, sir, and I hope you will accept my most genuine and sincere apology." Mann straightened up as best he could, and thrust his hand forward.

Bancroft considered this for a moment, then looked around at the others before looking down at Mann, and slowly shaking his hand. "Apology accepted, Mister Mann." Mann almost collapsed in relief as he let his breath out.

L'Envoi

Bancroft crossed his arms. "Ye know, Mister Mann, it takes a gentleman, a true gentleman, to apologize like that."

Mann's face brightened at the unexpected compliment. "Why..thank you, Sergeant . . . thank you indeed."

"Yer welcome. Another few apologies and you may earn the title."

Mann's face went bright red and the others stifled laughs.

"'Twas not meant as an insult, Mister Mann. 'Twas meant as a lesson. Ye cannot treat yer men like dogs, sir. Especially when they happen to be the dogs that keep yer backside safe. You pay attention to Captain Hull and Mister Blake; they are true gentlemen in every sense of the word. Learn from them and ye won't have to be beaten into an apology, no matter how genuine it might be. Now, on the other hand, I can't fault yer courage, lad. Not at all. On that account, I'm proud to have served with ye, and always will be."

It was Bancroft's turn to extend his hand. Mann looked doubtful for a moment, but then took it firmly. "From you, Sergeant," Mann said, "that is high praise indeed. Thank you."

Danielson and Peck stepped forward to shake hands with Bancroft as well. Bancroft was grinning as he looked at the three of them. "You know . . . you lads aren't bad, not at all, but God forgive me . . . you all still look like yer only twelve years old."

Danielson gave Mann a nudge in the ribs. "Well, George is; at least, he acts like it." Mann blushed again, but none of them could avoid the laughter, and Peck was about to make a comment when the call came across the harbor.

"Ahoy! Ahoy the warship!" Looking up, O'Bannon could see an officer on the merchantman that had anchored beside them, calling through a massive red trumpet. "Ahoy the warship!"

Hull sighed in utter frustration, "If I ever get this ship underway today . . . " Picking up his brass trumpet, Hull called back, "This is the United States Brigantine *Argus*, eighteen guns, Isaac Hull commanding! Whom have I the honor of addressing?"

The officer turned to a sailor and motioned for him to do something. The sailor broke into a terrified run belowdecks as the officer picked up the trumpet again. "What luck! I am Captain Josiah Downing of the brig *Seahorse,* out of Baltimore to Gibraltar, Alexandria and Constantinople! Have you a crewman named Edmund Bancroft aboard?" Even at that distance, they could sense immense relief in Downing's voice.

O'Bannon, Hull, and the midshipmen looked at Bancroft, and even Cable managed a befuddled squint. For his part, Bancroft looked over to *Seahorse* in mild confusion. "Gentlemen, they've got me there. I don't know anyone aboard any merchantman."

Hull shouted back through the trumpet again. "Captain Downing, Sergeant Bancroft stands here next to me! What can we . . . "

At that moment, the men gathered on the deck of *Argus* watched as something pink, very large, very pink, and unmistakably feminine, suddenly popped up from a hatch aboard *Seahorse*. She pushed Downing aside, faced *Argus* and bellowed. "EDMUND BANCROFT!! YOU GET YER ARSE OVER HERE THIS INSTANT BEFORE I DRAG IT HERE!!"

On *Argus,* they heard every word, quite clearly and precisely, without the aid of Captain Downing's trumpet. At that, Edmund Bancroft's face lit up like a thousand hearths, an absolutely beatific smile coming onto his face as if he was witnessing the Annunciation itself.

"ROSE!!"

As one, O'Bannon, Hull, Cable and the midshipmen repeated Bancroft's word. "'Rose'?"

"DID YE HEAR ME, YE DAMNED FOOL? GET OVER HERE NOW!"

O'Bannon leaned forward and touched the ecstatic Bancroft on the shoulder while the others gaped in disbelief. "Edmund . . . that's your Rose?"

Bancroft nodded slowly, his expression now one of adoration. "It is indeed, Presley . . . a rare and delicate flower, she is." Bancroft suddenly snapped out of his reverie and lunged for Hull's speaking trumpet.

Hull was taken somewhat aback. "Please, Edmund, feel free . . . " He stepped aside to watch.

"Rose, me love! What are ye doin' here?"

"I CAME TO DRAG YOUR WORTHLESS ARSE HOME, YE BLOODY IDIOT, WHAT D'YE THINK I SAILED THREE THOUSAND MILES FOR, ME HEALTH? NOW GET OVER HERE THIS DAMNED INSTANT!!"

"Rose, darlin'," Bancroft said apologetically, "It's not quite that easy."

"YE GIVE ME ONE MORE EXCUSE, EDMUND BANCROFT, JUST ONE, AND I'LL SEE TO IT THAT YER OWN HEAD GETS MOUNTED IN THAT BLOODY TAVERN!" The Amazon aboard the Seahorse pointed at the deck beneath her, and roared. "HERE!! NOW!!"

Heads were now starting to turn on the other ships, and crowds were starting to gather along rails. You could see the Squadron sail anytime; this was entertainment.

L'Envoi

Bancroft called back, "Rose, angel, give me just a moment."

"YE HAVE TWO MINUTES, BANCROFT, AND THEN I SHAN'T BE RESPONSIBLE FOR THE CONSEQUENCES!"

Rose could be seen grabbing Captain Downing's watch from him and holding it skywards to emphasize her point. Captain Downing, it could be seen, had the good sense not to resist.

It was a rapidly panicking Bancroft who called back. "Rose, me dearest, it's not quite that simple. . . ."

"IF I COME OVER THERE, IT'LL BE A SIMPLE BREAKIN' O' YER NECK! ONE MINUTE AND THIRTY SECONDS, MISTER BANCROFT!"

Edmund had gone from enraptured joy to shaking, rattled terror in just a few short seconds. It was without doubt the most awe inspiring display any of them would ever see. Here was a man who had marched, head held high, into the worst the world could throw at him for thirty years. Thirty years later those who were on *Argus'* deck that morning would swear they had never seen anything like it, before or since. O'Bannon himself would later speak in hushed tones of the wrath of Rose Miranda Bancroft, and how, if he'd had her with them at Derna, he could have let Edmund go home without any qualms whatsoever; and, for that matter, everyone else.

Bancroft turned to his shipmates, pale and breathing hard. "Gentlemen, I'm in somethin' of a pickle here . . . ."

"I can see that," O'Bannon said. "The question is, can we do anything about it?"

Bancroft smiled nervously. "Actually sir, once my Rose gets up a good wind, Commodore Rodgers himself couldn't stop her."

They all winced at the mental picture that raised and now realized that their friend was indeed in deep trouble. Danielson looked across to *Seahorse* with his spyglass. "Good Lord," he said, almost to himself, "the woman looks like a schooner under full sail." Realizing what he'd said, he hurriedly apologized. "No offense, Sergeant Bancroft."

"None taken, lad . . . ."

"ONE MINUTE, YE DECKAPE!!"

"She's more of a ship of the line when she's like this. Presley, lad, I am truly open to any ideas ye may have right now."

O'Bannon's mind raced. "Couldn't we just send him over to *Seahorse* and let him go back to Baltimore with them?"

To Barbary's Far Shore

Mann spoke up quickly at that. "Mister O'Bannon, this time I genuinely hate to be a stickler, but the regulations are very clear on this. The only person who can cancel his orders home and book him passage on a civilian vessel is Commodore Rodgers, and I need not point out that the milk of human kindness runs rather thin with him. In any event, we simply don't have the time!"

Bancroft turned to O'Bannon with a pleading expression, but even he was stumped this time. It was just then that Richard Cable spoke up with a weary voice.

"Wait just a moment . . . " Cable rubbed the bridge of his nose, trying to focus his thoughts.

"THIRTY SECONDS!" Rose could be seen casting about on *Seahorse*'s deck for something. She bent over and reappeared with a marlinspike, brandishing it across the open water at the men of *Argus*. At that, Captain Downing and his men removed themselves a respectful, and safe, distance away from her.

"Mister Cable," Hull said warily, "If you have an idea, I suggest you propose it now before I have to tell the crew to stand by to repel boarders."

Cable shook his head once to cast away the cobwebs. "If I remember correctly, Leftenant O'Bannon, I have the authority to immediately, that is, on the spot, discharge any sailor or marine who is unfit for service due to a life threatening condition, do I not, and then have his passage for emergency medical reasons later billed to the Navy?"

O'Bannon thought for a second. "As a matter of fact, I think you do."

Cable nodded slowly. "Good. And as long as you have consulted with the captain . . . "

Hull gave an exaggerated bow. "Consider me consulted, O'Bannon."

"Then I can give him a physical and send him on his way. I will happily note in my report that there was insufficient time to get the necessary paperwork to Commodore Rodgers before sailing."

"TIME'S UP, YE LYIN' DEVIL!" Rose threw her shawl to *Seahorse*'s deck, and looked as if she was ready to strip down to her corsets if she had to in order to make the swim. Captain Downing and his crew, averted their eyes in shock, but wisely kept their distance. O'Bannon and the others looked on at their friend's impending doom.

"Right then," Cable said, squinting at Bancroft. "Whereas I find you in imminent danger of death . . . "

"From what?" Peck whispered to Danielson.

L'Envoi

Danielson handed Peck his spyglass and pointed towards Rose. Peck focused it, observed her for a second, then grimly lowered the spyglass and handed it back. "I see your point."

"I intend to give you a physical examination to determine your fitness for duty," Cable announced.

Mann looked around, stunned that no one else seemed to be noticing this gross breach of regulations; if not in the letter, then in the spirit. "Captain Hull," the midshipman cried, "A full physical examination takes hours, and with all due respect, Mister Cable is in no condition to administer one!"

A commotion across the water on *Seahorse* caught their attention, and they all looked up to see Captain Downing and his crew struggling with Rose, who was fighting back with all the fury and vigor of a Viking berserker. One of the larger crewmen went sailing backwards in a magnificent somersault as Rose's fist connected with his jaw.

O'Bannon turned to Cable and said urgently, "Surgeon, that man's sacrifice just bought us some time, but not much. Do your work!"

"Right," Cable said, wincing from his headache. Taking his watch from his pocket, he held it up. "Sergeant Bancroft, you do not look at all well."

"That's an understatement," Danielson murmured to Peck.

"In order to determine your fitness for duty, I intend to give you a physical test..."

"You can't DO that!!" Mann cried in near agony.

Hull rolled his eyes in exasperation, then swatted the middie about the head, calling back to Cable, "Pray continue, Surgeon."

"Thank you, Captain. Always a pleasure to see a professional officer at work. Now..." Cable paused for a moment, distracted. "Where was I?"

"You were trying to get me off this bloody ship, sar!" Bancroft's anguished response came through clenched teeth.

"Ah, yes," Cable said with relief. Across the water on the good ship *Seahorse*, Downing and his crew were conceding defeat after a valiant, but futile battle. Rose was stomping toward the rail unrestrained by either man or excess clothing, and she obviously meant to come over the side and wreak her black vengeance on Edmund Bancroft.

"Well," Cable said, then looking mystified at his watch until he remembered why he had it out in the first place, "you have one minute to swim from here to the *Seahorse*. If you fail to do so, I shall be forced to discharge you on the spot." All of them looked at Cable, not quite believing what they'd just

To Barbary's Far Shore

heard. Cable didn't realize that for some seconds until he looked up. "You're down to about forty-five seconds, Bancroft."

The realization of what Cable was saying suddenly dawned over Bancroft like a Chesapeake Bay sunrise. Turning to O'Bannon, Bancroft snapped to attention and saluted. "Sar, permission. . . "

"Granted!" O'Bannon threw a salute back, but before he could get his arm back to his side, Bancroft had vaulted over the rail, hat, velvet suit and all, and disappeared into Siracusa Harbor with a gout of foam and spray that flew back into the whaleboat, soaking its crew and Paddy Duffy. Before the men in the boat could realize what had happened, Bancroft came bounding back up to the surface and kicked off towards *Seahorse*. Every crewman and marine on *Argus* was lining the port rail or was perched in the rigging by now, roaring and clapping their approval.

No ship ever built by man, nor any fish created by God Almighty, moved as quick, straight and true as Edmund Bancroft did that morning, heading for the loving arms of his Rose. The lady herself was just about to hike a leg over *Seahorse*'s rail when she saw her husband leap into the harbor waters. She froze for a moment in disbelief, and then started leaping about like a maiden who has just been asked for her hand. Doubtless Captain Downing and his crew were equally overjoyed, but wisely kept their distance until their good fortune could be confirmed.

On *Argus*' deck, the cheering still hadn't stopped forty-five seconds later, and Cable had to shout to be heard. "SERGEANT BANCROFT!"

Bancroft stopped and treaded water for a moment as he turned to face his old ship. "SAR?"

"CONGRATULATIONS, SERGEANT, YOU FAILED! CONSIDER YOURSELF DISCHARGED FROM THE UNITED STATES MARINES AS OF THIS MOMENT!"

The cheers started up again for a moment, even louder than before. Bancroft gave a soggy grin and wave in acknowledgement and continued on towards *Seahorse*. *Argus*' crew started to come down from their vantage points and return to their stations, the morning's diversion done. O'Bannon and Hull watched for a moment longer until Bancroft reached the ship, and bounded up the rope ladder that hung from its rail. It would have been a wet, salty embrace, but it was an honest and loving one. Rose swept him off his feet and they twirled around the deck together, to the everlasting relief and gratitude of Captain Downing and his battered crew.

Hull slowly shook his head as he watched the reunion. "'A rare and delicate flower,' he says. Just when I think I've seen everything . . . "

L'Envoi

"Captain!" The voice came from over the rail.

Hull looked over to see the whaleboat crew still tied alongside, a thoroughly soaked Paddy Duffy strapped to a bench. "Sir, weren't we supposed to take him to *Congress*?"

"Slight change in plans, men. Take Private Duffy over and be quick. I'll set things right later."

At this, Duffy looked up and shouted. "Well, where in the bloody hell is Bancroft going, then?"

O'Bannon smiled. "The good Sergeant is risking life and limb one last time." He waved the boat crew off. With any luck, they could cross the short stretch of water to *Congress* and be back in a few minutes.

Turning back to O'Bannon and Cable, Hull frowned. "I will be setting things right later, I trust?"

"Absolutely, Captain," O'Bannon replied. "We'll handle it, though I'm just not sure how yet."

"You will not," came the teary, cracking voice of George Mann, who stood with his feet planted and fists clenched, while Peck and Danielson unobtrusively took a step back. "I have had . . . just . . . about . . . all . . . this ship is a madhouse! No one, not a single one of you, has the slightest respect for the Navy or its regulations! How could you allow this travesty? How?"

The three officers looked at one another for a moment, then Cable spoke first. "Well," he said reasonably, "I'm the surgeon and Leftenant O'Bannon said it was all right to do so."

"That I did," agreed O'Bannon, "and I did it because I'm the commander of the marine detachment, and Captain Hull said it was all right to do so."

"And," Hull said with a feral grin, "I'm the captain and I can do anything I damned well please on my ship. For instance, like throwing a certain midshipman overboard if he doesn't get to his departure station . . . now."

Danielson and Peck knew their cue when they heard it and gently guided their comrade away as he shook his head in sorrow, simply muttering to himself over and over. "Marines . . . marines . . . "

Turning back to O'Bannon and Cable, Hull looked at the surgeon with newfound respect. "Surgeon, I must confess, that was not only a quick bit of thinking, but a truly decent thing you did. My compliments."

Cable belched once, leading O'Bannon to step back slightly. "The captain is far, far too kind," Cable smiled wanly, "but I must confess myself to practicality rather than decency. The sooner we got him off this tub, the sooner I could go lie

To Barbary's Far Shore

down again. So, with that . . . " Cable came to a swaying attention and saluted, almost jamming a thumb in his eye. Hull shot one back, and watched with a gimlet eye as Cable weaved back towards the hatchway.

"Well, sir, " O'Bannon said pleasantly. "All's well that ends well?"

Hull's gaze was less than gentle. "Not exactly, Mister Leftenant O'Bannon. You and I have some things to discuss." Hull took a step closer, looming over O'Bannon like an ominous shadow.

O'Bannon smiled nervously. "What exactly would the Captain be referring to?"

"The Captain would be referring to the fact that now that our little adventure in aggressive diplomacy is through, we are going to get back to real discipline on this ship. I've let you and that herd you call marines get away with bloody murder, Presley O'Bannon, and it stops this instant, d'you hear?"

O'Bannon smiled with the soul of politeness as he touched his hand to his chest in astonishment. "Captain, I'm shocked, shocked indeed, that you think the lads and I are anything less then the highly trained professionals you have come to rely on . . . "

Isaac Hull turned a marvelous shade of red, and one eyelid started to twitch as if caught in an easterly wind. "'Professional?' Professional?" Hull sputtered. "You and your men cannot do any thing 'professionally' except drink, carouse, and, in general, raise hell! The one and only thing I have come to rely upon you for is that I cannot rely upon you for anything! And furthermore . . . "

"AHOY, *ARGUS*!"

Hull's tirade stopped in mid-rant as he heard the voice of Commodore Barron come whipping across the harbor from *Constellation*. Hull winced with the same sort of pain that one feels from a good, solid punch to the stomach. He picked up his trumpet and faced the flagship like a miscreant about to face a firing squad. "GOOD MORNING, COMMODORE! HOW ARE . . . "

"STOW THE PLEASANTRIES, LEFTENANT COMMANDANT! I WAS WONDERING IF YOU INTENDED TO JOIN THE REST OF THE SQUADRON THIS MORNING AS WE SAIL?"

Hull lowered the trumpet for a moment so that there would be no chance Barron would hear his string of muttered obscenities. Then, with as pleasant a smile as he could manage, he replied. "A THOUSAND APOLOGIES, SIR. JUST A FEW MINOR PERSONNEL MATTERS," at this, Hull shot a baleful glance at O'Bannon, "TO DEAL WITH! WE SHALL BE MAKING SAIL SHORTLY!"

L'Envoi

"WELL, SEE TO IT YE DO!" Barron tossed the trumpet at an inoffensive seaman who had the misfortune of walking past him, and then stalked away. Hull allowed thoughts of mutiny and keelhauling to dance joyfully through his mind for a moment, then turned to see O'Bannon trying to discreetly step away.

"Not so fast, Leftenant!" O'Bannon stopped in his tracks, having the good sense to know when he was caught. Hull caught up with him and this time went nose to nose with his marine commander. "We will finish this later, O'Bannon, you may count on it! Until then, one thing I most assuredly want you to do, and this is an order, O'Bannon, is stop playing that damned fiddle! Bancroft was, as usual, right. It's a disgrace and an utterly miserable display in front of the men, and I won't have it, d'you hear? You will not, repeat, not, play that thing on my deck again. Am I clearly understood?"

To Hull's amazement, O'Bannon snapped to attentio. "Aye aye, Captain! Your word is my lodestar!" At that, O'Bannon smiled, about-faced, and trotted off.

Hull was slightly taken aback by the readiness with which O'Bannon accepted his fate, but there really wasn't time to think about it now. Turning to Josh Blake, who was standing a few feet away supervising a deck crew, Hull bellowed. "Blake, get this damned barge underway!"

"Aye aye, Captain Hull! All right, you damned laggards, get a move on, go, go, go! Anchor detail, weigh that bloody piece of garbage before I kick your backsides all the way to Sicily!"

At that, *Argus'* crew finally got into high gear. Sailors scurried up the rigging to lower the big driving sails. Around them, the rest of the fleet began to move. Anchors were coming up all around the harbor. Streams of brown water poured out of the hawsepipes and back into the sea as the timeless, rhythmic chants of sailors turning the massive windlasses began to get the anchors back into their resting places. *Constellation*, befitting her reputation as the Yankee Racehorse, moved first. Her sails first quivered, and then swelled with the Mediterranean breeze as she literally leaped forward into the azure waters. *Constitution* and *Congress* moved almost as one; officers and men laughing as they urged each other on in a good-natured race to see who would catch up first to the Racehorse. It would be neither of them.

To Barbary's Far Shore

Enterprise knifed out in front of the entire formation, Stephen Decatur laughing his fool head off as the rest of the squadron scrambled to catch up. *Nautilus* and *Hornet* were coming up quickly behind *Enterprise*, cutting through the gathering waves with fans of white spray and blue water flying skywards on either side of their bowsprits. And each of the ships that morning flew the red, white, and blue banner that had flown over Derna, that would fly over Fort McHenry nine years later, and that would still fly over ships named *Enterprise* and *Constellation* two centuries away in places that didn't even exist yet.

Marines At Work. Source: U.S. Navy

But *Argus* would not be denied as she barreled forward. Hull stood serenely behind the helmsman as he gave calm, steady orders that threaded his ship through the formation. He grinned like a schoolboy at play as he watched Stewart and Alexander shake good-natured fists at him. Turning slightly aft, Hull called out.

"Mister Blake!"

"Sir!"

"Get the men singing, I think they may move even better!"

Now that was something of a surprise to Blake, as Hull was not normally an enthusiast of song on his decks. Even Josh had to admit that it could get the crew into a rhythm that could get even the largest ship of the line moving like a

L'Envoi

well-oiled machine. Hull was smiling gently to himself, enjoying every last bit of what might just, after all, turn into a very nice day, when he heard it.

A familiar tune he'd heard in Ireland more than once, as men remembered their ancestors flying from their homeland in defeat. And it was being played on a . . .

FIDDLE.

Before Hull could even move, *Argus*' crew picked up the tune and began singing their hearts out.

"Did ye e'er see a wild goose sailin' o'er the ocean?

Ranzo, ranzo, weigh, heigh!"

Turning slowly around with a look of utter fury on his face and his jaws clenched tightly together, Hull looked Blake dead in the eye, like an archangel come down from Heaven to deliver the judgment of an angry Jehovah. Blake, for his part, could think of nothing else to do than come to attention.

"Just like them pretty girls when they take a notion

Ranzo, ranzo, weigh, - heigh!"

"Mister . . . Blake . . . " Hull snarled, in a low guttural rumble, "Please tell me I do not hear someone playing a fiddle on my deck."

"The other mornin' I was walkin' by the river

Ranzo, ranzo, weigh, heigh!"

Blake's eyes swept around the ship, frantically trying to find the source of the music, then stopped, focused on a spot behind and above Hull. "No, sir!," Blake said with absolute sincerity and authority.

"When I saw a young girl walkin' with her topsails all a-quiver

Ranzo, ranzo, weigh, - heigh!"

"It's coming from the mainmast rigging, sir!"

"I said pretty fair maid, and how are you this mornin'?

Ranzo, ranzo, weigh, heigh!"

Hull's eyes closed in mute surrender to whatever punishment the Gods had decided to saddle him with. He let out a long, slow breath and then suddenly smiled, the sad, pathetic smile of a man who has to face a truly unpleasant reality. "Well, that's different, then, Mister Blake. I'd hate to think my orders weren't being obeyed." With that, Hull walked slowly off, shaking his head while Josh Blake watched in utter and silent confusion.

"She said "None the better for the seeing of you.

Ranzo, ranzo, weigh, - heigh!"

Sergeant Williams leaned over Joe Pritchard's shoulder and whispered. "You're supposed to be at attention, Private, not singing in the choir."

Pritchard's mouth closed with an audible snap. He straightened up and watched out of the corner of his eye as Williams strode down the deck, checking every marine at his post.

Then he simply started humming.

"Did ye e'er see a wild goose sailin' o'er the ocean?

Ranzo, ranzo, weigh, heigh!"

The U.S. Mediterranean Squadron, Source: U.S. Navy

Eli Danielson and Pascal Peck stood at the bow rail, laughing like schoolboys every time *Argus* cut through a wave and splashed them, then rose up again like some giant whale to do it again, Even George Mann decided to join in the fun. They'd be soaked through and through in a few minutes, their wool uniforms would quickly become hot and clammy, and they'd be itchy and uncomfortable all day. They would adore every minute of it, now and for the rest of their lives.

"Did ye e'er see a wild goose sailin' o'er the ocean?

Ranzo, ranzo, weigh, heigh!"

Constellation was out front now and staying that way as Syracuse began to slowly recede. Barron's pennant snapped from her foretop as the wind cracked it

L'Envoi

back and forth. Old Sam Barron himself leapt into the rigging with a spyglass and scanned the seas ahead.

"Just like them pretty girls when they take a notion –

Ranzo, ranzo, weigh, - heigh!"

The sky was a full, glorious blue as Barron came down, as close to smiling as he ever got. "Helm!," he roared, "Set us a course for Siracusa, and be quick about it!"

"Aye aye, sir!,"

The Racehorse's massive wheel began to spin, it's spokes a blur, bringing her about to the northwest. In every other ship of the squadron, captains and helmsmen followed suit. Almost as one, the Mediterranean Squadron turned to the southeast, sails full and taut, as they went forward to see what was beyond the next stretch of the horizon.

To Barbary's Far Shore

EPILOGUE

The Barbary Wars, like so many American entanglements in that part of the world, didn't exactly end with a victorious show of force. Although Yusef Bey became far more prudent, the Dey of Algiers seems to have concluded that, with the Bashaw out of the game and American tensions with the British slowly ratcheting upwards, this left more for him. He therefore went on seizing American vessels before and through the War of 1812.

One of them, the schooner *Mary Anne*, turned out to be slightly more than the Algerians had bargained for. Her skipper led his men in a successful effort to recapture the ship, killing four of the Algerians and putting the surviving four in an open boat and pointing them towards North Africa. The Algerians seem not to have been much for small boat handling, for they did not survive the experience. The Dey, distraught beyond mere solace, sent the United States a bill for $18,000 to comfort him for the loss of his dear subjects. Robert Wallace, writing in Smithsonian, describes what happened next in words far better than mine:

" . . . It is reasonable to think that the United States pointed out to him the way to Hell, but that is not what happened. We paid the bill."

It would be the last time. By 1815, James Madison was in the White House and his patience with the Algerians was distinctly shorter than that of his predecessors. Once the War of 1812 had been settled, Madison sent two full squadrons, under the commands of Stephen Decatur and William Bainbridge, to discuss matters with the Algerians. After one short gunfight, the Algerians finally saw the wisdom of the American position. They signed a permanent treaty of peace and good behavior on June 30th, 1815.

William Bainbridge

William Bainbridge never did quite overcome his reputation as a hard-luck captain, but he went on to make a permanent and lasting name for himself in the history of the US Navy. After his return from Tripoli, he demanded a court of inquiry into the loss of *Philadelphia*. He had to have known it wouldn't be an easy one; after all, in the short history of the USN he had managed to lose two ships and have a third highjacked. He had to have been even more concerned when he saw the board, James Barron, Hugh Campbell, and Stephen Decatur. But, after a review of the circumstances surrounding *Philadelphia*'s loss, the

Epilogue

Board unanimously agreed that no blame could be attached to Bainbridge and his officers. He continued on in various commands until the War of 1812, when he was assigned to command USS *Constitution*. In that capacity, he led her to victory over HMS *Java* in a victory that still rings down through history.

Constitution had been sailing off Brazil on December 29th, 1812, when Bainbridge sighted *Java*. *Java*'s skipper, Captain Henry Lambert, was apparently not expecting combat. He was headed for the West Indies with naval supplies, replacement sailors, several civilian passengers and the new Governor-General of Bombay, who was apparently taking the long way around. Lambert was making for Bahia to water and victual when Bainbridge cut him off. The fight was much like that between the *United States* and HMS *Macedonian*. The *Java* was newly refitted, faster, and slightly more heavily armed than *Constitution*, and Lambert was a highly skilled and respected commander with an experienced, veteran crew. Given his handicaps in passengers though, Lambert decided that discretion was the better part of valor and laid on all the sail he could. Bainbridge did the same, but using the big 'driving' sails that were normally stowed in combat. This eliminated *Java's* edge in speed and maneuverability, and forced Lambert to fight.

Will Bainbridge was determined that this once, just this once – everything was going to go his way; and, by the grace of God and his gunners, it pretty much did. His gunners were in magnificent form that day, catching *Java* with almost every broadside. But Lambert wasn't going down without a fight. In a gutsy move, he had his gunners hold fire as he realized *Constitution* was going to pass close, too close, aboard. When the Royal Navy gunners finally did let go, they inflicted the worst damage Constitution would ever take in a single hit, sweeping her quarterdeck almost clean. One round went directly through *Constitution's* wheel, blasting it to splinters and killing almost everyone around it. Bainbridge was swatted to the deck by a massive iron bolt that was blown from the disintegrating wheel and caught him in the thigh. Although it didn't penetrate his leg or break any bones, he was unable to get back to his feet. Bainbridge, though, was still in full command.

Old Ironsides was starting to careen wildly out of control, away from *Java*. *Constitution's* quick-thinking crew jury-rigged a tiller with chains, beams and tackle, actually steering her from the wardroom pantry. She was hard to control, but she could move, and that's what made the difference. At the last moment, *Constitution* got a broadside off that sheared away most of *Java*'s rigging from her foremast to her bowsprit. The British frigate was still closing fast, obviously planning to close and board.

Java losing her rigging and *Constitution* getting back under control turned the course of the battle. As *Java* shot past, *Constitution* let go two raking broadsides that finished her. Lambert went down, mortally wounded by a marine

To Barbary's Far Shore

sharpshooter. His first officer struck her colors a few moments later. Bainbridge's marines got aboard and were able to capture all of the diplomatic papers aboard *Java*, as well as the Royal Navy's newest signal and code books. Bainbridge himself got aboard a couple hours later; carried more than walking, but determined to see the prize he'd won. Upon boarding, he was told that Captain Lambert wanted to see him. Not knowing that Lambert was dying, he was taken to where Lambert lay in shock and pain as the surgeons tried to at least make him comfortable. Lambert was mostly in delirium but composed himself long enough to sit up. In extreme pain and agony, he offered his sword to the American captain.

Will Bainbridge had been in Lambert's place twice before in his career, and he had known the pain, humiliation, and disgrace that came from losing a ship, even under circumstances where one was blameless. No matter what he did in his life to come, and it would be considerable, he would still be known behind his back as 'Hard Luck Billie.' He knew it, and it would tear at his soul like a knife until the day he died. But, as much as he disliked the enemies of his nation, be they Berber or Briton, by God he would not willingly humiliate another man the way he had been. Squinting through pain-filled eyes, Bainbridge slowly shook his head and placed the sword back into Lambert's bloody, trembling hands.

Java herself was so far gone that Bainbridge ordered her burned, and then headed *Constitution* for San Salvador. Lambert stayed in Bainbridge's cabin under the surgeon's care until their arrival. indeed, all of *Java's* wounded received the best care that the crew of United States Frigate *Constitution* could possibly give them. Captain Lambert died after they arrived at San Salvador, and the crews of both ships stood at attention as one at his funeral. *Constitution*, however, was going to be out of action for a while. The damage to her steering couldn't be repaired away from a full-dress dockyard, and Bainbridge had no choice but to turn her for Boston and home, where they arrived on February 27th, 1813. Bainbridge was unable to get back out to sea again for the rest of the war.

When the war ended he was promoted to Commodore. Along with Stephen Decatur sailed across to the Mediterranean to remind the Dey of Algiers that we still cared. Bainbridge was out of position to assist when Decatur got into the one fight of the brief campaign, but he had the pleasure of watchingthe Dey sign a treaty that essentially put him out of the piracy business, at least as far as the United States was concerned. Returning to the United States, Bainbridge and Decatur served as Navy commissioners in a semi-retired status. When Decatur fought his duel against James Barron, Will Bainbridge was right by his old friend's side as his second. This clouded Bainbridge's reputation with old Navy comrades for the rest of his life, for many of them felt that the duel never should

Epilogue

have been fought in the first place. But to the general public, and most of the Navy, Will Bainbridge was a hero and always would be.

He died, full of years and honors, on July 28th, 1833 and is buried in Philadelphia. His name was commemorated in the first true destroyer of the US Navy (DD-1, 1901-19) and the first nuclear powered –frigate, later redesignated as a cruiser (1962-95). An *Arleigh Burke* class missile destroyer now bears his name.

Presley O'Bannon

Lieutenant Presley Neville O'Bannon, United States Marines, was honored as a hero. His personality was something he would never shake quite that easily, and he was every bit as furious with Lear and the Jefferson Administration as Eaton was. Although Congress was willing to give him a large tract of land, O'Bannon was, admittedly, hoping for something that seemed far more practical, promotion. He did make Captain but never went any farther, quite probably due to the public criticisms he made of the way the expedition had been handled.

His reputation in the Marines had not been the best to begin with, and, even after Derna, O'Bannon had not been the most popular of officers. The old phrase "he was more lucky than good" seems to sum up the attitude of his peers.

Had he held on a bit, though, he could very well have been in the line of succession to eventually become Commandant, although that might have in any event been too long of a wait for Presley O'Bannon; Franklin Wharton held on until his death in September of 1818. He therefore resigned his commission in March 1807 at the age of thirty. In 1809, he married one Miss Matilda Heard in Frederick County, Virginia. Matilda must have been a remarkable woman to capture Presley O'Bannon's attention and affection, but capture it she did. They moved to Kentucky later that year, then on the very edge of the American frontier. In 1812, he ran for and was elected to the Kentucky House, and in 1824 was elected to the Kentucky State Senate and served two years. Sadly, his life with Matilda must have been difficult, for they were divorced in 1826, a most drastic step in that era, but were remarried in 1832. However, in 1843 Matilda O'Bannon was committed to the state Insane Asylum at Lexington, where she spent the rest of her days. O'Bannon never remarried, and spoke with genuine love and adoration of her for the rest of his life.

Whether or not they had any children is surprisingly unclear. There are vague references to two sons, one named Eaton, and there is one written reference by O'Bannon himself to a daughter named Elizabeth, who may have died of cholera in 1835. In any event, O'Bannon went to live with a cousin in Henry County, Kentucky, following Matilda's commitment and died there on 12 September 1850. He would live long enough to see his Marines go from ships'

To Barbary's Far Shore

guards and boarding parties to the first shimmerings of the masters of amphibious warfare that they are today.

First buried in Pleasureville, his remains were moved to Frankfort Cemetery in 1920 by the Daughters of the American Revolution. He rests there today, a simple marker near his tomb stating the bare facts of his life. As a younger man, he might have preferred something more ornate, but at the end of his life he was strangely quiet about his past and this after all might have been more fitting.

There have been three US Navy destroyers named after him, one of which (DD-450, 1942-70) had a reputation as wild as that of its namesake. This included one memorable night where the *O'Bannon* and another US destroyer raced in so close to the Japanese battlewagon *Hiei* that the hulking battleship was unable to lower its massive 14" rifles enough to fire at them. *O'Bannon's* five-inch guns would do no more than scratch the paint on *Hiei's* gray flanks, but her torpedoes were another matter. Without a doubt the *O'Bannon* got at least one in, for she was officially credited thereafter with helping sink *Hiei*. By the end of the war in the Pacific, *O'Bannon's* record was so impressive that Fleet Admiral William Halsey selected her for one of the great honors of those last days. Along with her sister ships *Nicholas* and *Taylor*, the *O'Bannon* escorted USS *Missouri* into Tokyo Bay to accept the surrender of the Japanese Empire.

Only the legendary carrier USS *Enterprise* (CV-6) would have more battle stars than *O'Bannon*, and the destroyer never lost a man in combat. She would serve honorably through Korea and Vietnam, but would be sent to the breakers in 1972. However, the latest ship to carry on the tradition (the *Spruance* class DD-987) would commission just seven years later. Her career lasted until the first decade of the 21st century, when she too was decommissioned with her sisters to make way for the *Burke* class and the forthcoming DDX. On October 6th, 2008, *O'Bannon* was towed off the Virginia coast and sunk as a target. Fittingly, she died hard; it took guns, missiles, and finally aerial bombs to put her down.

But in the end, it is the dashing, cocky and irreverent young officer that history, and the Marines, remembers, and remember him they do. At the Marines' Basic School, the Officers' Field Mess is known as O'Bannon Hall, and the officer graduating with the highest grade point average receives the Lejeune Sword there in recognition of that fact. The Sword, of course, is patterned after the one O'Bannon brought back in Corps' tradition, and the point is clear; lead with the wisdom and skill of John Lejeune, and the courage of Presley O'Bannon. A little of O'Bannon is still passed on to every Marine lieutenant, and if one looks in his or her eyes, he is there in stubborn, daring and eternal spirit. For them, he rests not in the Kentucky countryside but in their hearts, and we are the better for it

Epilogue

Isaac Hull

Isaac Hull continued on in the US Navy through a number of commands, building a reputation as a tough, smart, and caring captain. He eventually took command of USS *Constitution* and led her to one of the most legendary victories in naval history.

Hull got *Constitution* to sea from Annapolis three weeks after the outbreak of war and had headed north to join up with Commodore John Rodgers, who had headed out from Boston a few days earlier. Hull never got there. Instead he found himself up against a full squadron of the Royal Navy, commanded by the superb Captain Philip Broke, which had already captured USS *Nautilus*.

Hull got *Constitution* turned around in good order, just in time for the wind to die, with them barely outside cannon range of His Majesty's ships. Hull took advantage of the fact that they were in fairly shallow water to try a trick that by all rights shouldn't have worked, but did; 'kedging' *Constitution* out of danger. This involved sending a boat over the side, lowering an anchor into it, then having it rowed out as far as the chain would go and dropped. *Constitution*'s crew would then pull their ship up to the anchor using the ship's windlasses. Far superior, and more importantly, far faster, than using the ships' boats to pull her out of danger, Hull opened the range between his ship and the British, to the point that when the wind finally did freshen, *Constitution* easily showed Broke her heels.

Skirting the coast all the way up to Boston, Hull replenished and then headed north for Nova Scotia to prey on British merchantmen in the Gulf of St. Laurence. It didn't take long for word of Hull's depredations to get back to Broke, who spun his command around to corner Hull against the coast of the Maritimes.

Broke wasn't quite fast enough. By the time he got there, Hull had sailed around him for Bermuda, the better to disrupt His Majesty's commerce. He probably would have made it, too, if it hadn't been for an unexpected development. Broke had detached the frigate HMS *Guerriere* to head back to Halifax. In that huge ocean the two ships managed to cross paths on August 19th, 1812.

Before the war, many of the frigate captains of both nations knew each other quite well. Hull and Captain James Dacres of *Guerriere* were no exceptions. In fact, by all accounts they were quite good friends and entertained one another as honored guests aboard each other's vessels whenever they met. At one of those meetings, the two men quite good-naturedly wagered each other's hats on the outcome of any combat between their ships. It is not hard to believe that when the two captains realized whom they were up against, the thought of that warm comfortable dinner echoed through their minds. But that

To Barbary's Far Shore

was in the past, a few years and a lifetime before. There was a job to do now and they would each do it to the best of their abilities; one for a flag and a belief that All Men Were Created Equal, the other for King and Crown.

For all the history that has since come down about this fight, it was actually much less even than it would seem. *Constitution* had a heavier broadside by about twenty percent and was more maneuverable than just about any other frigate on the planet. Dacres' only real hope was that his superbly trained gunners, the spiritual descendants of the men who had turned back the Armada, could fire with more speed than Hull's less experienced gun crews. It wasn't an entirely forlorn hope, but it didn't work. Speed the British gunners certainly had; *Guerriere* fired so rapidly that it looked like she was afire. But in moving as fast as they had ever moved in their lives ,they mistimed their shots, a fatal error in the days when a broadside had to be exquisitely timed with the movement of both the shooter and the target.

Guerriere's fire sprayed wildly across the waves. Hull roared out his commands and his crew responded a with timing and precision that would have been deemed impossible in peacetime practice. Hull danced *Constitution* nimbly through the waterspouts, except for one salvo that bored straight and true for her sides. Whether it was a matter of timing, poor powder or the angles at which the *Guerriere* fired and *Constitution* took the hit, no one would ever know. In a few seconds, it would never matter. The rounds thudded into her black oaken sides. Her crew flinched in expectation of what would come next, the bone-chilling CRACK of splintering timbers, the howls of injured and dying men, the bellows of infuriated commands.

They never came. Instead, the wicked black cannonballs literally bounced off *Constitution's* hull with hollow thumps, spinning like demented toys back into the water. A broadside that should, would, have crippled any frigate the Royal Navy had ever faced had done no more than scratch the American's paint. Both sides were stunned into momentary silence. Someone on *Constitution* spoke first, and when they did, they gave the frigate a name that has remained as deathless as her record.

"Hurrah!" called out one sailor, "Her sides are made of iron!"

From that moment on, *Constitution* would always and eternally be known as Old Ironsides. However, the fight wasn't over yet. After fifteen minutes of trading broadsides, *Constitution* had only very minor damage to some of her upper rigging. *Guerriere* was somewhat the worse for wear, with a disturbing number of her crew down with wounds from flying splinters of oak. Dacres wasn't ready to give in just yet, however, and tried to get his ship maneuvered out of the way. Hull wasn't having any of it though. His decades of time at sea honed his skills to an edge the Royal Navy hadn't believed possible and he saw his opening. A perfectly timed broadside brought down *Guerriere's*

Epilogue

mizzenmast, slowing her down. *Constitution* shot past her. Hull literally turned her on a dime, directly across *Guerriere*'s bows.

This was the moment every commander lived for, and Hull made the most of it. His gunners, now operating with a precision and speed that would do modern warships proud, got off two full broadsides, raking *Guerriere* from her bowsprit to her sternpost, the worst possible way for a sailing ship to be hit. The heavy oaken sides were intended to take heavy, ship killing blows, not the comparatively lightly built ends of the ship. Few ships could survive one blow like that, much less two. But Dacres was one of His Majesty's best, and he still had one trick up his sleeve. With his bow rigging now tangled in *Constitution's* mizzen rigging, Dacres saw that the gunners manning a bow chaser had somehow survived the hurricane of iron and fire that had cut down so many of their mates. Realizing it was their last chance, Dacres bellowed for them to fire.

That they were able to calmly and efficiently do so, almost within arms' reach of the marine sharpshooters in *Constitution's* fighting tops, is an everlasting testament to their training and courage. They got off one salvo from the bow chasers, which were too close to do anything but punch through even *Constitution*'s tough hide. As fate would have it, they landed squarely in Isaac Hull's cabin, the only serious internal damage the ship would ever take. But there were only two of them. Before the bow chaser crews could try again, Hull's gunners let go one more broadside that finally silenced them. Hull snarled one more order, and the helmsman put *Constitution's* wheel hard over. With an awful groan of fracturing wood and the crack of snapping rigging, *Constitution* tore free.

Hull and his crew were still very nearly at maximum combat efficiency, and *Constitution* was essentially unscathed. The same, however, could not be said for *Guerriere*. Her crew had literally been swept off the main deck, her rigging was mostly gone and what was left was hopelessly snarled. Because of that there was no possible way to bring her main battery to bear. James Dacres had done everything his King could have asked, and more beyond that. There was no possible dishonor in what he had to do next.

To the resounding cheers of *Constitution*'s crew, Dacres ordered the Red Ensign struck and surrendered.

A few minutes later, Isaac Hull stood on *Guerriere's* deck for what would be the last time, to accept his old friend's surrender. In keeping with tradition, Dacres came to attention on the quarterdeck with his surviving staff, bowed slightly to Hull and offered him his sword.

It would have been wonderful to see Isaac Hull place his hand on his old friend's shoulder in comfort and speak with dignity and respect. "Keep it, James. Anyone who fought as well as you did deserves to." What would have

To Barbary's Far Shore

been even more enjoyable to see would have been the grin on Hull's face, and Dacres', when Hull spoke again.

"But, about that hat . . . "

His Majesty's Frigate *Guerriere* was wounded beyond hope of recovery. Hull got her crew off, the wounded succor and the dead a decent burial at sea, where crews of two nations stood as one to hear Hull and Dacres read the immortal words of the Apostle Paul, reminding the grieving that their friends would one day be resurrected. Hull showed yet one more moment of the decency that marked his career; he ended his cruise and put about for Boston to land his prisoners. Isaac Hull was greeted as a conquering hero, a man for the ages. Even his enemies had to admit that his tactics and ship handling had been superior.

He would never have another fighting command in the United States Navy.

Hull was a brave, courageous warrior, and not a man who tolerated fools or stupidity. He spoke his mind plainly and honestly and had no use for anyone who did not. As other men of that breed have since discovered, skill and victory in combat do not always make up for a refusal to tolerate political games, and that is almost certainly what put Hull on the sidelines for the rest of his career. As the British blockade grew tighter and losses mounted, there were few commands in any event. Hull's undeniable skills at organization and training were invaluable in holding the battered remnants of the Navy together through the rest of the war. The fact that the USN was able to head back to the Mediterranean so quickly afterwards is in no small way due to his efforts.

Hull would live another three decades, serving for quite some time as a semi-retired Naval Commissioner with his friends Will Bainbridge and Stephen Decatur, and commanding the Pacific Squadron. He rests in Philadelphia, not far from Bainbridge. His last words: "I strike my flag."

Four destroyers have honored him, all decommissioned now. One of the new *Burke* class missile destroyers should rightfully bear his name but, until that time, his best memorial is a massive red brick wall at the ancient Washington Navy Yard on the corner of M Street SE across from the Metro station. Somewhat gothic in design and as solid as anything ever built in the better parts of the District, it seems heavy enough to hold down all of Anacostia, the blighted neighborhood that surrounds the old Yard. In black iron letters, impervious to time and vandals alike, it says:

ISAAC HULL GATE

As solid, forthright, and unbending as the man himself. Far worse men are commemorated by far grander monuments, but I, for one, believe Isaac Hull would have been satisfied with that simple, unadorned red brick wall.

Epilogue

Samuel Barron

Samuel Barron was, unfortunately, not one of the early US Navy's more inspired choices for command. In fairness, a great deal of his dithering and lack of initiative could be laid to his illness. After all, men in the best of health could find shipboard duty of that time difficult. Laid up with a debilitating disease and trying to run a war on the cheap four thousand miles from home could not have made things any better. Be that as it may, it seems that someone somewhere should have had the courage to stand up on this particular subject. It is unlikely that the Navy Department didn't know that there was a problem, and that Barron should have been relieved. It would not have been a black mark on his record, and it might have prolonged his life. In any event, Barron returned home to Norfolk in late 1805 and went to work alongside, of all people, Edward Preble on the gunboat project. By all accounts they got along correctly, if not warmly. However, Barron's health continued to steadily deteriorate until he died at the age of 45 in 1810.

William Eaton

Consul William Eaton returned to the United States aboard USS *Constellation* in late 1805. Like O'Bannon, he was regarded as a hero, but Eaton was still shocked and angry about how their mission had been undercut. Congress was certainly willing to reward him with land and Massachusetts certainly did, 10,000 acres worth. Eaton made no friends when he and O'Bannon started insisting very loudly and publicly that if they had been given the original contingent of marines Eaton had requested, they could have marched straight through to Tripoli to end matters once and for all.

Eaton himself, in public, actually accused Tobias Lear of treason, a much stronger word in a day when honor was the greatest component of a man's reputation. However, in an era where starlets have manned communist antiaircraft guns, ordained preachers visit genocidal dictators to 'negotiate,' California teenagers join in jihad against Western civilization and presidential candidates are found to have met with communist governments, 'treason' has lost its punch. In any event, Lear, however duplicitously, was only following his orders from Higher Authority, so nothing ever came of Eaton's charges. O'Bannon always supported Eaton's views, but he was in enough trouble with his own comments. Angry and frustrated, Eaton resigned from the diplomatic service, then crawled into a bottle and pretty much stayed there until his death on June 1, 1811. He had two sons, Nathaniel and Amos, both of whom graduated from West Point and served their country well. Nathaniel became a respected surveyor and professor, while Amos rose to the rank of Brevet Brigadier General by the end of the Civil War.

To Barbary's Far Shore

Eaton's final moment in the spotlight was in 1807, when Aaron Burr attempted to enlist Eaton in his infamous conspiracy. There does not seem to be much indication of why Burr wanted Eaton to join him, but it seems reasonable that Burr felt that Eaton had been wronged, as he himself had been, and would welcome the chance to revenge himself. William Eaton, proving, even at the end, that he was every bit as honest and loyal as his friends knew him to be not only turned down the tempting offer to help rule a new Empire in the North American wilderness, but then became a vitally important witness at Burr's trial. Whatever Will Eaton thought of what his country had done to and with him, he would be damned if he would betray her.

We really don't know what to make of William Eaton. Perhaps we should remember him differently. After all, he was one of the first of many Americans who believed that no problem was unsolvable and that, through sheer force of will alone, he could bend the rest of the world to his efforts. He was far from the last to discover that success did not necessarily mean victory or that what he had won in the field could be lost by men in quiet well-lit rooms far away from danger; men only interested in keeping things quiet. Had Eaton been less eager to press his point, he might have been able to parlay his newfound fame into political gain. After all, men have ascended to the Presidency since with far thinner military records.

It is interesting to speculate what a Secretary of War Eaton, or a Secretary of State Eaton, or, dare we say it, President Eaton would have done a few years later in the crises that led up to the War of 1812. It might not have made things any better, and, given Eaton's mercurial style, might have made them infinitely worse. But there is no question he impressed O'Bannon with his courage and Presley O'Bannon was not an easy man to impress. Let us say then that William Eaton saw a threat to his country and went after it with all the strength and bravery God gives a man, if not necessarily all the wisdom. In that, he still stands in honored company.

Tobias Lear

Tobias Lear remained in North Africa until 1812. He apparently channeled his predecessor and annoyed Yusef Karamanli to the point where the Dey made it exceptionally clear that he was no longer welcome in Tripoli. Given the frequency with which US envoys were PNG'd, this was not ordinarily a problem. In this case, it posed a serious risk for Consul Lear, as war with England had broken out. Lear and his wife made a dangerous trip home past the Royal Navy and reached port at Norfolk. The Lears had to take a tortuous route back to their home in New Hampshire to avoid the British, and got there just in time for President James Madison to call Lear back to Washington to serve as an assistant secretary in the War Department. Lear made the trip back to DC, where

Epilogue

he worked from an office near the President's House until the city was burned on August 24th, 1814.

Say what one might about Lear, it appears that he did his job quietly and well in the midst of the worst defeat that had yet been inflicted upon the young republic, and he should be remembered for that. Unfortunately, Lear's life began to spiral out of control not long afterwards. He was known to have suffered from what were almost certainly severe migraines and the horrifying depression that can accompany them. He was still being occasionally raked over the coals by the press regarding his dealings with George Washington's estate. On October 11th, 1816, Lear put a pistol to his head and pulled the trigger. The man known as a consummate administrator and record keeper left neither suicide note nor testament.

Eugene Leitensdorfer

Eugene Leitensdorfer, or whatever his name may have actually been, was paid off at Syracuse. At first, he decided to try and get to his home in Bavaria, but took an odd route home that somehow laid him across the path of Turkish pirates. Captured and sold into slavery, he helped nurse some injured and ill Turkish sailors back to health. It must have been a most impressive display, for his captors freed him on its account. It took him a while, four more years, to be precise. Years which included yet another marriage and desertion; one would truly like to know just how many Frau Leitensdorfers he ultimately left behind. But he finally landed upon the shores of the United States, where he promptly tracked down William Eaton and asked for a job.

By this point, Eaton was in no condition to offer much assistance, but he did write a few letters of introduction. Leitensdorfer put them to good use. Between the letters and Leitensdorfer's own gift of gab, he landed a genuine plum of a job. Benjamin Latrobe, Thomas Jefferson's chief architect, hired him as a night watchman in the growing Governmental district of Washington, D.C. This brought with it a fairly decent salary for the time, along with a place to live. Leitensdorfer, never one to miss the main chance, also branched out into sales of just about anything he could think of.

Finally, using contacts he'd made through William Eaton and on his own, he actually convinced Congress to pass legislation that not only got him a substantial land grant in the Missouri Territory, but also back pay for the Derna expedition at the grade of full colonel.

They threw in a mileage and travel allowance as well.

Leitensdorfer died in St. Louis, MO, in 1845, after at least one more marriage and four more children. He appears to have enjoyed every moment of his long, well deserved retirement.

To Barbary's Far Shore

Stephen Decatur

Stephen Decatur was something of a hotspur both before and after his raid on Tripoli. It would bring him both fame and tragedy. After all, his reputation as a midshipman was nothing to sneeze at; he was known to dive from jib booms and, at the age of fourteen, he defended his mother against a man who has come down through history to us only as 'a drunken ruffian,' He was appointed a midshipman at the age of twenty and a year later, a remarkably short period for the time, was commissioned a lieutenant. Five years later, he was leading *Intrepid* into Tripoli Harbor. From there on, he built a legend of daring, boldness, and aggressive leadership. He was, perhaps, a touch too aggressive at times. He rarely returned without a long casualty list and damage to his ship, but he always got results, and has gone down as one of the great figures of the Barbary Wars. From 1805 until the War of 1812, he had a fairly routine career, but he was then assigned to command the famed Old Waggon, *United States*. Decatur took the Waggon out on October 8, 1812 as if God had personally appointed him the scourge of Britannia, in company with *Chesapeake* and his old friend *Argus*.

Decatur made good time, arriving five hundred miles southwest of the Azores on October 25th, but they hadn't caught sight of anything worth hunting until that morning. A lookout called a sail off the starboard bow, just before nine o'clock. We can only imagine the thrill that went through Decatur's heart when he realized that it was His Majesty's Ship *Macedonian*, under Captain John Carden. Like so many of the frigate skippers of that time, Decatur and Carden knew one another well from before the war. Decatur certainly knew that *Macedonian* was only two years old and fresh from a refit, and had five guns on him. Carden too recognized his foe, and the hunt was on. After a few short minutes of maneuvering under the brisk wind, the fight started. Decatur shot first, but missed. Carden returned the compliment, bringing down part of the *United States'* masts. Decatur must have been apoplectic; this writer would give anything to have heard Decatur's bellowing at his gun captains.

They seem to have most assuredly heard every word. *United States*' second broadside looked to go high, but it was actually perfectly aimed at *Macedonian's* rigging. Tearing away a mizzen top, the plummeting wreckage took more of the intricate spider's web of masts and rigging with it. The result was that Carden's control over his ship was suddenly halved. Although the *United States* was slightly less maneuverable than her sisters, she could still dance pirouettes around *Macedonian*. Add to that sailors like Stephen Decatur, and his superbly trained crew, and there was only one way things could end. By eleven o'clock, *Macedonian* had been reduced to an uncontrollable, wallowing wreck. Carden knew it was over. He surrendered, having taken more than one hundred casualties. *United States* had just twelve, most of those injured. It had been a magnificent performance by Decatur, his ship, and his crew.

Epilogue

Six weeks later, Decatur hauled *Macedonian* into port, where she was repaired and purchased into the US Navy. Decatur, no fool, ordered *Macedonian's* colors transported to Washington by Lieutenant Hamilton, son of the Secretary of the Navy. Hamilton arrived in grand fashion, interrupting a formal ball to make his announcement and then presenting the bunting to Dolley Madison.

Transferring to *President*, Decatur's luck didn't quite hold. On the night of January 14, 1815, Decatur was attempting to lead his squadron out from New York City past the British blockade, when *President* grounded, badly. It took two nightmarish hours to get her clear; all the while, the frigate's copper-sheathed bottom crashed into a sand bar. When she finally did get clear, it was apparent from the way she was handling that *President's* cruise was over. Decatur reluctantly gave the order to put about and head for New York. When the sun came up though, even Decatur's heart must have skipped a beat in dread. A British squadron had spotted *President* in extremis on the sandbar, and quietly maneuvered itself into position between her and safety.

Decatur headed for the open ocean and made a run for it. It took eleven hours for the British to finally run Decatur down. He made them pay for their victory by savaging the frigate HMS *Endymion* to the point where she had to be abandoned. But just before midnight, Decatur was surrounded and almost out of ammunition. He may have been aggressive, but he was far from stupid, and he did care for his men. When he struck his colors, he did so in the knowledge that he had given far better than he'd gotten, and the rest of the squadron had made it to sea.

Returning to the US after the war, he was made a Commodore, and led the squadron that put paid once and for all to the Barbary pirates. Married and semi-retired after his return, and quite well off due to wise investments of prize money, Decatur built a home which still stands in Washington DC and helped lay the groundwork for the growing United States Navy. But in 1820, he managed to get himself into a fight ashore that would be his last. Years before, he had sat on a board of inquiry that had found Captain James Barron, Sam's brother, guilty of allowing the hard-luck *Chesapeake* to be boarded and members of her crew impressed by a British ship. Although there was plenty of blame to go around, Barron had been the captain. He had sailed *Chesapeake* in a condition that could only be described as scandalous, then made almost no effort to defend her. Barron was suspended from the USN for five years, and blamed Decatur for his troubles.

There was a foundation to Barron's grudge, Decatur had testified at the 1808 board of inquiry, and had not spoken well of Barron's abilities or judgment. In addition, Decatur had been the yard commander when *Chesapeake* sailed from Norfolk. If he was truly unaware of her condition, then Decatur

To Barbary's Far Shore

hadn't been doing his job either and bore at least some responsibility for what happened. On the other hand, Barron and his staff had made a sufficient mess of the whole thing on their own that Decatur's comments were probably only icing on the cake. It should be noted that the only officer on Barron's staff who wasn't court-martialed was, surprise, Will Allen. This feeling wasn't at all soothed when Decatur strongly opposed Barron's reinstatement at a point in time when Barron might very well have had an outside chance for a flag officer's billet upon his return. Neither man was likely to budge over matters like this, and it finally arrived at the point where Barron demanded satisfaction.

Early on the morning of March 22, 1820, Decatur and Barron met on the field of honor at Bladensburg, Maryland. Barron, much older than Decatur, had been ill for some time, and his eyesight was, to put it gently, poor. Decatur therefore decided to allow a distance of only eight paces instead of the traditional ten. He had previously told his seconds that he would not shoot to kill. Indeed, Decatur, no mean shot, put one neatly into Barron's thigh. Unfortunately, Barron either didn't feel the same way or, just as likely, was aiming to miss as well and inadvertently hit Decatur. The commodore died a short while later. Most of Washington turned out for his funeral. Barron's career, shaky to begin with, ended there and then.

Decatur is probably best known today not for his naval skills, though the USN still holds him up as an exemplar. Rather, it is for a declaration he made during the last Barbary expedition that he is most familiar. Standing to make a toast following the successful negotiation of a treaty with the Deys, Decatur raised his glass and intoned, "Our country! In her intercourse with foreign nations may she always be in the right; but our country right or wrong." This author has the feeling that were Decatur here to see the war we face now, and his namesake ship (DDG-73) head into combat, his feelings would be the same.

John Rodgers

John Rodgers remained in the Mediterranean until the summer of 1806, when he returned home to take command of the New York flotilla, as well as finding himself involved in President Jefferson's gunboat project. One can only imagine Rodgers' relief when he was assigned to actual patrol duty along the Atlantic coast in 1807. In early 1808, he was ordered to Norfolk to preside over the board of inquiry on James Barron's disastrous performance aboard *Chesapeake*. By all accounts, he was in remarkable form throughout, having worked himself into a truly monumental rage at that unlucky ship's officers. Barron and his staff, all, save one, either disciplined, suspended, or outright discharged from service, probably considered themselves fortunate indeed that Commodore Rodgers didn't decide to save the Navy the expense of the board and simply shoot them himself.

Epilogue

In 1811, Rodgers, in command of *President*, headed into the waters off New York, where the British frigate *Guerriere* had been busily impressing seamen from American merchant ships. Just after noon May 16th, a lookout reported sails on the southeast horizon. Rodgers, of course, gave chase believing it to be *Guerriere*. What happened next has been the subject of animated discussion for almost two centuries. The ship *President*'s lookout spotted wasn't *Guerriere*, but rather the very much smaller sloop HMS *Lille Belt*, Lieutenant Arthur Bingham commanding. Bingham may have commanded one of His Majesty's smaller warships, but say this for him, he knew his duty and intended to do it as smartly and as professionally as if he were Horatio Nelson himself aboard HMS *Victory*. Altering course to intercept *President*, Bingham made proper, businesslike signals asking her to identify herself. She didn't and Rodgers was always just a little evasive as to whether or not he ever actually saw the signals. In the days before radio and radar, when one relied on men with spyglasses in the foretops to read signals across open ocean, it was always very possible to miss something. Rodgers was probably relying on that fact.

In any event, Bingham closed far enough to see Rodgers' commodore's pennant at *President*'s masthead. He quite reasonably concluded that a) this was an American frigate, b) he was considerably outgunned, and c) discretion dictated that he resume his original course. However, Rodgers stated later that he believed at this point that the ship he was chasing was indeed HMS *Guerriere*. Being John Rodgers, he acted accordingly. Whether or not Rodgers actually believed that is, to put it gently, questionable. *Lille Belt* was far smaller than any frigate, and she was now easily within spyglass range. It seems unlikely that Rodgers, with his years of experience and knowledge, could have mistaken her for the *Guerriere* for more than a very brief period of time. By this time, Rodgers had put on every inch of sail *President* had. He was now knifing through the waves directly towards *Lille Belt*. His crew manned their battle stations with the cool speed and precision that Rodgers demanded.

At this point, poor Lieutenant Bingham was realizing he had a problem on his hands. *President* could easily outrun him and he seemed to have an accurate grasp of what kind of opponent he was up against, even if Rodgers didn't. Prudence dictated that even though there were no formal hostilities between the US and Great Britain, Bingham needed to be ready. He acted accordingly, sending his crew to action stations, double-shotting his guns, and running up the Red Ensign.

At 8:30 that night, Rodgers finally maneuvered *President* into firing position. There is no question whatsoever that by this point Rodgers knew he was up against a much smaller vessel; as he stated later, he could see the other ship's flag but could not make it out. In a moment that would have been funny if it hadn't been so potentially deadly, both captains called out simultaneously.

To Barbary's Far Shore

"What ship is that?"

Both captains claimed that they received no answer.

Again, both asked, "what ship is that?"

Both captains stated again that they received no answer.

Had it been anyone else but John Rodgers on *President*'s quarterdeck, this might have just ended in a tense face-off and both ships turning back to their original courses. But Rodgers had seen close up what had happened to the hard-luck *Chesapeake* and had no intention of letting it happen to him. British ships were stopping American shipping up and down the coast, and Rodgers had seen every report of their impressments of US citizens, something that could not have failed to enrage him. But, perhaps most importantly, he had sat in Norfolk and found James Barron guilty of negligence and dereliction of duty, with outright cowardice implied in every word of the charge. No one, no one, would ever say that about John Rodgers, whether he was up against a rowboat or a ship of the line.

According to Rodgers' report of the fight, before there could be a third hail the unknown vessel fired at *President*. Rodgers therefore returned the favor. For his part, Bingham always maintained that he had not fired, but rather Rodgers had fired first. Looked at logically, it would have required an utter lack of sanity and prudence on the part of Arthur Bingham to open fire on an American frigate, and so far Bingham had shown calm, professional judgment every step of the way. Be that as it may, the shooting had started and wasn't going to stop until one of them was unable to continue the fight.

Bingham handled *Lille Belt* quite skillfully, holding out against the larger ship for nearly an hour, but he never had much of a chance. Rodgers had every possible advantage, and he was using them with a vengeance. Finally, at around 10:00 PM, Rodgers ordered cease-fire. He asked again "what ship is that?"

This time Bingham answered. *Lille Belt* had been badly handled by *President*, with her sails and rigging almost gone and thirty-two casualties. At daylight, Rodgers sent one of his lieutenants over to offer assistance. Bingham's manners and tolerance must have been severely strained at that moment, but he remained polite and declined the offer. Within a few hours, Bingham had gotten a jury rig up and was headed back to Halifax.

Given the public uproar over the attack on *Chesapeake*, no one was terribly inclined to look closely at Rodgers' account of things. His Majesty's minister to the United States made a formal protest, was politely heard and then just as politely ignored. There was, of course, a formal inquiry at Norfolk, which found Rodgers blameless and Lieutenant Bingham entirely at fault. Within a few months, Rodgers' actions that night off New York would go a long way towards

Epilogue

helping the US and Great Britain into a state of war, though they were far from the only cause.

Rodgers took a full squadron out on the outbreak of war, with less than outstanding results. He would take *President* out on two more occasions, before handing her over to Stephen Decatur and being assigned to Baltimore. Most of the USN had been either sunk, captured, or blockaded by this point. For a man with John Rodgers' instincts combat was where you find it, and he found it in spades as the Royal Navy came up to attack Fort McHenry on September 13th, 1814. Without ships to command, Rodgers threw himself into helping organize the defense of Baltimore, and distinguished himself in keeping the embattled defenders at their posts and firing back. Of course, it is just as likely that the Americans were far more afraid of John Rodgers than they were of the British, but never mind.

Like many of the other legendary names of the early US Navy, Rodgers never had another fighting command after the War of 1812. He did serve for some years as a naval commissioner, returned to the Mediterranean briefly in the 1820s, and died in Philadelphia on August 1st, 1838. Presumably, Rodgers was, to the last, cursing Death to his face. Rodgers' son would become an honored flag officer during the Civil War, and his Great grandson, Admiral John Rodgers, would be an early pioneer in Naval Aviation with exploits that included one of the first flights from the continental United States to Hawaii before his tragic death in a plane crash in 1929. Six Navy ships have honored Rodgers and his descendants.

John Rodgers may well have been one of the most thoroughly dislikable men who ever commanded a ship of the United States Navy. On the other hand, more recent USN leaders have sometimes been less than pleasant men. William Halsey and Richmond Kelly Turner could be, and often were, hard drinking, belligerent, and temperamental, though Halsey's reputation is heavily cushioned by the adulation he received from the media. Fleet Admiral Ernest King, whose temper was short, vicious, and volcanic, is said to have grinned wickedly when notified that he would take over as Chief of Naval Operations after Pearl Harbor and said "when things get rough, they always call for the sons-of-bitches." Heaven knows that Ernie King would have known one when he saw one.

But John Rodgers was first and foremost a commander who absolutely refused to give up, whether afloat or ashore, was highly capable and skilled and refused to give an enemy any quarter. Those qualities will always make up for a multitude of sins, and they will always stand in his favor.

Edward Preble

Edward Preble returned home and was set to work on one of Thomas Jefferson's pet projects, a fleet of gunboats that he expected to take the place of

To Barbary's Far Shore

the big frigates and ships of the line in USN service. It was a good idea, build lots of smaller ships instead of the half dozen big frigates that cost so much to man and operate. Unfortunately, the ships were too small, poorly armed and not at all fit to sail much out of sight of shore. On top of that, this was one of the biggest defense projects in the short history of the Republic, which meant that Congress wanted to spread the wealth around as much as possible. As the majority of the US population was still concentrated within a hundred miles or so of the Atlantic Ocean, this meant that everyone could get a piece of the pie. In 1804, 278 gunboats, the number Congress authorized, was a very large and profitable pork pie. By way of comparison, the modern US Navy has roughly three hundred combat vessels.

Preble did his best, but there were problems from the start. Two different prototypes were ordered. One turned out to be nearly unseaworthy and many of those that were completed were never armed or even fitted out. Instead they were towed away to be placed in mothballs and never entered into service. Tired and ill, Preble passed away on 25 August 1807. He was still a young man – he had just turned 46 – but he had tuberculosis, and two centuries ago that was a death sentence.

Preble was of a type we would see much later in the USN's history. Not only a highly skilled warrior, but exceptionally capable as an administrator and a firm believer in training, training, and training. Uniquely among the early leaders of the USN, he also brought promising younger officers up through the ranks as well. They did not fail us either. Stephen Decatur, Charles Stewart, David Porter, Charles MacDonough, Isaac Hull and others eventually came to be known as 'Preble's Boys,' a title they bore proudly to the end of their days. The modern USN has not forgotten him. Two previous destroyers have borne his name, and the latest (DDG-88) is manned by the latest generation of Preble's Boys, and in recognition of modern sensibilities, some of Preble's Ladies as well. I'm not entirely sure what the Commodore would think of that, but, in the end, I believe he would be fine with it as long as they could fight.

Richard Somers

Richard Somers would come back to our attention once more, and sadly in a way neither fitting nor proper when compared to the man and his courage.

In the late 1830s, the US Navy decided to honor Lieutenant Somers, and accordingly named a spanking new brigantine being built at the New York Navy Yard for him. Commanded by one Alexander Slidell McKenzie, *Somers* headed out for Africa on a training cruise in September of 1843. While returning from their one and only port call at Monrovia, Liberia, something went dreadfully, terribly and horribly wrong. Although the particulars are still the subject of learned discourse, the end result is a matter of naval and legal history. Captain

Epilogue

McKenzie became convinced through a series of disturbing events that as many as twenty of his crew of 192 were planning a mutiny, led by one Midshipman Philip Spencer, who just happened to be the son of the Secretary of War. The final number actually charged with mutiny was eventually brought down to three, Midshipman Spencer and two others.

Spencer, it seems, was something of a handful to begin with. He'd faced disciplinary action for alcohol related incidents three times, on two different ships, before being sent over to *Somers*; all of this by the tender age of 19. Spencer also had what can be charitably described as an attitude problem, being less than respectful towards Captain McKenzie and flaunting the little luxuries his father made sure to send along without regard for the rest of the crew's welfare. Bad enough, certainly annoying, and definitely setting the touchy McKenzie's nerves on edge but nothing too outrageous, until the captain found out about midnight meetings, secret notes written in Greek, and conversations about taking the *Somers* a-pirating.

That did it for McKenzie; he went to his staff, along with three other midshipmen, and laid out the evidence. They in turn unanimously decided that Spencer and his friends, Seaman Elisha Small and Boatswain's Mate Samuel Cromwell, had indeed intended to seize the little ship and live the buccaneer's life. Spencer frantically, and without effect, claimed that the whole thing was an overblown fantasy. He, Small, and Cromwell were found guilty and hung from the yardarm on December 1st, 1843, and buried at sea that day. Spencer, in a final statement, retracted his claims of innocence and acknowledged his guilt.

McKenzie's actions have been debated ever since. When *Somers* reached New York two weeks later, McKenzie was immediately set before an inquiry, which completely exonerated him, as did a full-dress court martial a few months later. Unfortunately, both endorsements were, shall we say, less than ringing. The nation was deeply divided over the incident, and there was a fair amount of feeling inside and out of the Navy that McKenzie had badly overstepped his authority. Most who felt that way, however, had never heard of the old saying "ashore there is God and the Captain, but at sea there is only the Captain."

One saving grace came from the whole awful mess. The Navy realized that the sea-going midshipman system of educating future officers was quickly becoming hopelessly obsolete, so the United States Naval Academy was founded to take over their training. *Somers*, sadly, never saw the benefits the Academy would bring; she was run down by a squall in the Gulf of Mexico on December 8, 1846, taking more than thirty men with her.

Now, when Richard and his men died that awful night in Tripoli harbor, Yusef Bey did one of his few decent things in his long, indecent life; he ordered that Somers and his men be given a 'Christian' burial. The order was obeyed

To Barbary's Far Shore

with alacrity. For two hundred and eight years the last captain and crew of *Intrepid* have rested in Tripoli town.

There was some concern after the 1986 raids that Gaddafi, insanely desperate to get some kind of revenge on the Americans, had exhumed and desecrated the bodies. There was horrifying precedent. Gaddafi had a . . . 'thing' . . . about the remains of his enemies. At least one who was dead for more than a decade was found in a refrigerated morgue drawer in Gaddafi's home town of Sirte, and the remains of the two USAF aircrew killed over Tripoli were the focus of cruel ransom demands by Gaddafi's berserk sons. One airman, after a horrifying case of mistaken identity, was returned to the US; the other remains missing in action. There were reports that he had even ordered a search in the Libyan Desert for the missing remains of the crew of the *Lady Be Good*, a B-24 bomber that crashed there in 1943 and the source of one of the most tragic, and haunting, stories in the history of the United States Air Force.

However, after years of uncertainty, US authorities, including Secretary of Defense Leon Panetta, visited the graves of Somers and his crew in 2012. Two points should be noted: first, we have only the word of the current Libyan government that the remains . . . well, remain, but, for now, we shall take this on faith.

The second thing is that the graves lie in two different places. One overlooks the harbor and *Intrepid's* last resting place. The other, where Somers and seven others rest, are under the cool shade of an olive tree in Tripoli's Green Square, where Gaddafi and his loyal minions would gather on a regular basis to shout "Death To America!"

It must have driven Colonel al-Gaddafi just a bit more insane every time.

The Midshipmen

Unfortunately, the midshipmen for the most part disappear into history. This author was unable to find an indication of what happened to Midshipmen Danielson and Peck, though this is not necessarily a bad sign. Surprisingly few midshipmen went on to long naval careers, given the risks and dangers inherent in it at that time in history. In most cases, they were from fairly well to do families, so ending their careers and opting for something a bit safer was probably not a difficult choice.

We do know that Midshipman George Washington Mann left the Navy not long after the Derna raid due to an eye injury, which he may very well have received on the raid. However, in 1807 he volunteered to return to duty to fight French pirates who were then slipping into Chesapeake Bay to raise hell with the coastal shipping there. Finally receiving his commission as a lieutenant in 1809, he stayed on active duty for only two more years, resigning once and for

Epilogue

all in 1811. His family, which still lives near Annapolis, MD, has preserved, and more importantly, cherished his heritage.

Their professional and spiritual descendants today march forth every summer from Reserve Officer Training Corps programs at colleges all over the nation, and from the very heart of the USN's officer corps, the US Naval Academy at Annapolis, MD. They are probably the most rigorously selected and highly trained, skilled, and educated midshipmen in the world, and capable of doing things as a matter of daily routine that the midshipmen aboard *Argus* could have barely conceived; manning nuclear reactors, flying Mach 2 fighters, or diving to the bottom of the seas. It is a vastly more exciting, adventurous, and dangerous, world the new midshipmen sail into ever year, but they are up to the task.

The Marines

The men of the marine detachment are to an extent unknown to us. As mentioned in the Forward, not even the formidable resources of the USMC Historical Center were able to determine which of the men aboard USS *Argus* with Presley O'Bannon went ashore with him. We only know two beyond any question. John Whitten and Edward Stewart. Other authors may have nailed down other names, but it is not sure.

It is sad, I think, that we do not know with certainty who went ashore with O'Bannon that shimmering day in Egypt two hundred years ago. They were at the beginning of a tradition of going ashore, sometimes outnumbered, frequently outgunned, and not at all sure what lay over that next rise, but prepared to do their duty in any event, no matter the cost. They pressed on; not even sure if their orders were still in effect, but determined to follow them until told otherwise. And, like future marines in other places, they pulled it off magnificently.

We know Privates Stewart and Whitten still lie somewhere near Derna. British Commandos landed there during the North African campaigns to try and capture or kill Erwin Rommel, the dashing, dangerous Desert Fox. They failed, and their leader lies there still. Almost certainly, U.S. aircraft and missiles soared overhead in 1986 during Operation Eldorado Canyon, when the heir to the Deys of Tripoli was reminded that in the end America would only be pushed so far. Someday it would be fitting if Whitten and Stewart are brought home, perhaps to Arlington. They earned it.

As for the rest . . .

I hope that Edmund Bancroft truly did have a Rose to go home to. Bancroft deserved a retirement of peace and prosperity, but that would have been interrupted by the War of 1812 and his home would have stood very nearly at

To Barbary's Far Shore

what later generations would call Ground Zero. Knowing Bancroft, he would have stood at the door of his tavern that long summer night nine years later as the Royal Navy stood offshore to batter Fort McHenry into rubble. He would have seen the smoke and mist clear to reveal a ragged, battered banner still waving defiantly into the dawn. Eight generations of men like Bancroft have since marched into history with the Corps, and when most people think of marines, they see the sergeants; eternal, devoted to their nation, their Corps, and their marines. The sergeants have stood, Cerberus-like before those who would have harmed us and shepherded other O'Bannons through the flames. We should say a grateful prayer that they have done so without any other desire than to get their marines home.

And I do hope that Pritchard, Curtz, Duffy, and O'Brien went on to long, happy lives that ended quietly someday in the arms of their loving families. Those who go abroad in the service of their nation, in the end, ask for no more than that. They deserved no less, and ultimately far more. The heirs of those five marines report daily to a hot, sandy little island in South Carolina, where Sergeant Bancroft still stands in spirit, in flawlessly starched khakis and a campaign hat that has since become synonymous with the drill instructor's trade. He roars and thunders in incomprehensible fury at the God and the recruiters who sent him the worst excuses for marines he has ever seen, demanding in thundering frustration to know just what he is supposed to do with them. Eventually, eleven weeks later, he will change his opinion, but only because he has helped them find the spirit of those five and helped them find the strength and skill to put it to use. May they continue to do so forever.

USS *Argus*

Brig *Argus* continued on duty in the Mediterranean until 1806, when she was recalled to the United States. She was laid up 'in ordinary' for a year, then was refitted for patrol off the coast of America until the war of 1812. From October 8, 1812 to January 3, 1813 *Argus* was a literal terror afloat under the command of Lieutenant William Allen, and Her Majesty's Admiralty has the records to prove it. *Argus* captured six vessels and, on one memorable occasion, evaded an entire Royal Navy squadron during a three-day chase. As if that wasn't enough, one of *Argus'* prizes was taken while the Royal Navy watched in bitter frustration, unable to reach it in time. Returning home *Argus* was detailed with a vital diplomatic mission, getting the new US minister to France safely to his post through waters so thick with British ships that you could almost walk from Plymouth to the Continent without getting your feet wet.

Allen showed his usual skill by managing to neatly avoid the RN and deliver the ambassador to L'Orient on July 11th, 1813. From there, Allen was given an assignment he must have relished, a raiding cruise into the Royal Navy's home waters in the Irish Sea. In just thirty-one days on station, *Argus*

Epilogue

was able to rack up nineteen prizes, a record unmatched by any other American vessel during the war; one even more impressive when one considers that almost all of them were taken literally in sight of British territory.

Had Lieutenant Allen decided that discretion was the better part of valor and then headed home, no one could have faulted him. But devotion to duty, and perhaps just a little hubris, brought Nemesis down upon him in the form of HMS *Pelican* at just after six o'clock in the morning on August 14th, 1813. Allen had gone after his twentieth prize, a British merchantman, and was in the process of burning her off Saint David's Head when *Pelican* spotted him. *Argus'* crew fought with all the courage and skill that they had but the British had the wind, the size, and the armament.

Allen must have been a sight for the ages up there on that quarterdeck, for he was not a man to ever give in. Everything he had learned, everything his crew had trained for, would have been put to use that beautiful August morning. It wouldn't be enough. One of *Pelican's* rounds came sweeping across the deck and it struck Allen directly in his right thigh. The massive and comparatively low-powered projectiles of that day did horrifying damage to ships and worse to mere humans, and it was no different in this case.

When everyone looked up, Allen was there on the deck, his right leg gone. But true to the courage, bravery and skill he had shown that day, Allen stayed at his post. Two of his officers put a tourniquet on him and he kept giving orders. He didn't last long but he tried, God rest him. In a few minutes, he had fainted from shock and loss of blood, and was taken belowdecks. In that brief space of time, *Pelican* put two broadsides through *Argus'* rigging, and she was dead in the water. *Pelican* raked her at will, causing horrifying damage, then moved in for the kill.

"Away the boarding parties!"

It was all over in just under an hour, but the Fates gave one last kindness to Lieutenant Will Allen. He died belowdecks, aboard his own ship, and was spared having to surrender her. In a world where honor, bravery, and courage were watchwords for a life well lived, the British paid Lieutenant William Allen, USN, the ultimate compliment they could give to a sworn enemy, burial with full military honors at Plymouth, England. During William Allen's long rest there, future generations of American sailors have stood beside the Royal Navy and give their lives in England's defense. Overlords, adversaries, friends, and finally allies; the circle closes.

A new *Argus* had been laid down at the Washington Navy Yard, but she was still unfinished on the ways when the British Army came calling. She was burned where she sat almost a year to the day after her namesake's loss. There would not be another *Argus* until 1940, when a former German yacht was sold to the USN and given that name. She had a respectable and solid career, serving from San Francisco as a patrol ship until her decommissioning in 1946. She still survives today, restored to her peacetime yacht appearance.

As *Haida G*, she elegantly cruises the waters her old namesake patrolled with the Mediterranean Squadron. It might be fitting, perhaps, that one of America's next warships, be it an amphibious warfare vessel, or one of the new DDGs, or even a nuclear carrier, take the name of a vessel that always fought far above her weight, and left a lasting record to be eternally proud of.

USS *Constellation*

Frigate *Constellation* went on to a solid, if unspectacular career after the Barbary Wars. During the War of 1812, the Yankee Racehorse ended up on a series of fruitless patrols. These became less and less relevant as the British blockade of the Atlantic coast began to take effect. Eventually, outgunned and outnumbered at sea, *Constellation* headed home. Home, as it turned out, was where she was needed most. British coastal and riverine forces had pretty much stopped traffic on Chesapeake Bay, then a major commercial artery. By June of 1813, a handful of rickety gunboats with less than reliable crews were all that stood between America's most strategic seaports and the Royal Army, when British regulars from Bermuda arrived off Norfolk with the intent of taking the vital port city. That they did not can be ascribed to a single vessel, *Constellation*, ordered south to try and hold things together.

She did so admirably. Slipping past the British blockaders, *Constellation* anchored off Craney Island, VA, site of today's Norfolk Naval Base, and turned her guns on the invasion force. The gunboats and their reluctant crews, heartened by her arrival, turned to their task with renewed hope. Acting in concert with the big frigate and a scratch force of Virginia militia, they drove the invaders back. Norfolk and Hampton Roads were now safe from any further attempts at British occupation.

The British then turned their attention to the less easily defended northern Bay with a series of amphibious raids that culminated in the burning of Washington. *Constellation* and her crew had to watch in horrified frustration from afar as the new capital of their young nation was burned. For the rest of the war, one of the finest sail warships ever built served as a floating battery. The fighting edge of her crew slipped away as surely as that of the finest cutlass left in the weather to corrode and dull.

When the war ended on February 16th, 1815, *Constellation* really wasn't in fighting trim, but she was deployed anyway for one last trip to visit her old friends in North Africa. With the US Navy busy or trapped, the Dey of Algiers had decided it was quite safe to increase his harassment of US shipping. But President James Madison wasn't Tom Jefferson, always willing to give an old

Epilogue

reprobate one more chance. Madison may not have been able to take on the Royal Navy, but by God he could teach the Algerians a lesson they would never forget. Commodores Stephen Decatur and William Bainbridge took two powerful squadrons, including *Constellation*, back across to the Mediterranean.

On June 17th 1815, *Constellation* spotted a large ship in the distance, the Algerine frigate *Mashuda*. *Constellation*, along with two other ships, turned to cut her off. That is precisely what they did. Within minutes, *Mashuda's* captain was sliced in half by a forty-two pound shot from the brig *Epervier*. The terrified crew hauled down their colors, bleating for mercy. It was the only significant battle of the short, glorious little war. *Constellation* anchored in Algiers a few days later and the Dey was a most charming and gracious host, especially as most of the United States Navy was now in long cannon range of his castle. Less than two weeks later, on June 30th, Decatur signed the first of a series of treaties that effectively ended the Barbary wars, and *Constellation* returned home.

For the next thirty-five years, *Constellation* served admirably, showing the flag in a hundred different corners of the globe. However, by the 1850s she had deteriorated badly and it is at this point that her history takes an unusual turn. Of the 'Original Six' frigates, the USN's first major warships, *Chesapeake* and *President* had been captured by the British while *Congress* had been surveyed as nearly derelict a few years earlier and a new *Congress* commissioned. *Constitution* had barely escaped scrapping in the 1840s and was now on restricted training and patrol duty. She would soon be seen for the national symbol she is and would be preserved to become, as of this writing, the oldest commissioned warship afloat in the world. *United States* was a receiving hulk at Norfolk, having been tied up 'in ordinary' for decades with a skeleton crew and serving as a glorified dormitory for new sailors reporting to Norfolk. She would be survive to be captured by the Confederacy in 1861 and briefly rechristened CSS *Confederate States* before being scuttled by retreating CSA forces. She would be raised and towed dockside in the Elizabeth River to finally be scrapped in the mid 1860s. But, before her death, she would see another ship just a few yards away, the burned-out steam frigate *Merrimack*, metamorphose into the CSS *Virginia*, the first steam-powered, ironclad warship ever built in America, and head into an epochal gunfight with *Monitor* that changed naval combat forever.

In any event, *Constellation* put into Norfolk for inspection in late 1852 and was found to be unsafe to put to sea again. In particular, she was 'hogging,' a situation where the bow and stern actually sag down to a point below the keel amidships. In his work Fouled Anchors: The *Constellation* Question Answered, Dana Wegener states that *Constellation* was so far gone that, even with her original plans, yard workers had a hard time getting proper measurements of her distorted hull. Combined with extensive internal rotting, there was no choice.

To Barbary's Far Shore

Constellation was hauled ashore at Norfolk on February 22, 1853, and dismantled over the next seven months. By September 12th, she was gone. Within sight of *Constellation's* dismemberment, the USN used existing stockpiles of wood, including some intended for those 74-gun ships-of-the-line, to build a new vessel, also named *Constellation*. It should be pointed out here that it appears that no components or materials from the 1797 ship made it into the new vessel.

This *Constellation* was the last large, all-sail vessel built for the United States Navy. She was a sleek, beautiful ship, the ultimate expression of military sail technology in the USN. She got into active service just in time for the Civil War, and she served on anti-slavery patrol as well as hunting Confederate blockade-runners. She served in a variety of flagship and training duties through the end of the First World War, surviving into an age of the Dreadnoughts and naval aviation along with her distant cousin *Constitution* and David Farragut's old flagship *Hartford*. In the early 1920s Assistant Secretary of the Navy Franklin D. Roosevelt ordered her 'restored to her 1814 appearance,' under the impression that she was the 1797 frigate. Although, contrary to a well-known misconception, the USN never tried to hide the fact that she was not the 1797 vessel, no one ever seems to have enlightened Assistant Secretary Roosevelt as to the vessel's true status. *Constellation* was semi-properly altered to resemble the 1797 frigate. In the Second World War, *Constellation* served her country one last time as a headquarters flagship.

When the USN turned *Constellation* over to the city of Baltimore for preservation as a museum in the late 50s, less than diligent work by independent researchers led the city in all innocence to represent the ship as the 1797 frigate. This was questioned from the beginning, exacerbated by the USN's unwillingness to admit that it couldn't locate the records confirming the disposal of one of its most famous ships. When maritime historian Howard Chapelle published a paper in the early 70s proving otherwise, a controversy was started that still has its partisans. The argument grew so bitter that it started to resemble a Hollywood thriller, with conspiracies and forged, planted documents. Over time, however, the fight has quietly wound down and there is now no question of *Constellation's* provenance. She was built in Norfolk in the 1850s, within sight of her namesake.

Constellation was the last all-sail warship built in the US and the sole surviving USN combatant of the Civil War. As such is just as deserving of preservation. Baltimore has embraced her, and today, after a thorough restoration, she is lovingly maintained by a group of devoted professionals and enthusiastic volunteers, and she is expected to see her two hundredth birthday in 2056. Her name was carried on in USN service by the aircraft carrier USS *Constellation* (CV-64). The third ship to bear the name, she served off Vietnam. The first US ace of the Vietnam War, Randy Cunningham, flew from her deck.

Epilogue

She was on station when America struck back against Al-Queda. She was retired at the end of 2003, but she made the name *Constellation* once more the terror of those who would strike this nation with fire and fear.

USS *Enterprise*

Schooner *Enterprise* continued to serve in the Mediterranean Squadron until sent to Venice in 1804, where she was almost completely rebuilt. Remaining there until 1807, she fought several noteworthy battles, including more against the Barbary pirates, and one against a group of Spanish gunboats off Gibraltar. Returning the to US, she repaired and refitted until 1811, when she began operating from Savannah, Georgia and Charleston, South Carolina. At sea when the War of 1812 broke out, she engaged HMS *Boxer* in a legendary fight on September 5th, 1813. Although *Enterprise* was victorious, the cost was high; the commanders of both ships were killed and were buried alongside each other in Portland, Maine.

She returned to the Mediterranean in 1815 for one more tour, which included a spectacular running fight between *Enterprise* and six Spanish gunboats, which ended in the gunboats being run off. She then came back home to stay. In 1817, she was assigned to seaboard patrol, stopping smugglers, pirates, and other malefactors - and in the process, capturing thirteen vessels. On July 9th, 1823, *Enterprise* ran aground on Little Curacao Island in the West Indies. Although the crew escaped without injury, the ship was a total loss.

Her name has since been given to another sailing ship (1831-1844), the most decorated US ship of the Second World War (CV-6, 1938 - 1958), and the first nuclear-powered aircraft carrier (CVN-65, 1961 - 2012). During Operation Enduring Freedom in 2001, *Enterprise* led the US Navy against a new generation of pirates, ones far more vicious and cruel than those the first *Enterprise* fought. She had been on her way home after a long deployment when the first reports came in of the attacks on New York and Washington, and without orders, *Enterprise* reversed course and went to battle stations. And once again, the Big E lived magnificently up to a reputation that goes back nearly two hundred years against pirates, slavers, fascist and communist dictators alike, and stands for a true, genuine legend.

On December 1st, 2012, *Enterprise* was deactivated after a final combat deployment. For the first time in fifty-two years, the United States Navy was without the name that has come to symbolize courage, commitment, and dedication. Fortunately, despite a recent list of naming choices that can, at best, be considered ... unwise, Secretary of the Navy Ray Mabus announced that the third *Gerald R. Ford* class nuclear carrier would be named *Enterprise*, to join the fleet in 2025, assuming that, once again, the politicians do not decide that the world has become too peaceful or expensive for warships.

To Barbary's Far Shore

USS *Hornet*

Sloop *Hornet* continued to serve with the Mediterranean Squadron until June of 1806, when she was caught in a severe storm that took her topmast. The Navy decided to send her home, where she was surveyed and sold that September. There is some question as to exactly when she was decommissioned, for the authoriative Dictionary Of American Naval Fighting Ships has her decommissioned at Philadelphia on August 9th, 1806. Her replacement, a larger brig bearing the same honored name, is shown by DANFS as commissioning at Baltimore on October 18th, 1805, ten months earlier. As the US Navy has rarely, if ever, had two vessels in commission with the same name, this poses a minor quandary for the historian. One possible explanation is that the new brig ran into some kind of delay that kept her from deploying until the following March, when DANFS says she sailed for the Med. On the other hand, this has an interesting echo of the controversy over the frigate *Constellation*, which for many years was believed to have been built with funds and materials earmarked for repair and rebuilding of the original 1797 frigate. That is someone got a new ship named *Hornet* built by saying they were going to repair the original.

In the long run, it matters not; the new *Hornet* served well and honorably through the War of 1812, and then against pirates closer to home in the Caribbean. Sadly, she was lost during a ferocious storm there in the fall of 1829. The name has lived on in a schooner (1813-20), a captured Confederate blockade runner (1864-69), and two famed aircraft carriers, the *Yorktown* class CV-8 (1941-42), and the *Essex* class CV-12 (1943-70), now a museum in honored retirement in California.

USS *Intrepid*

There have been three *Intrepids* since 1804. The first, a remarkable 'torpedo ram,' was built at Boston Navy Yard in 1874. She was something of a technological dud, and was quickly put into a retirement that lasted 18 years before she was sold for scrap. The second *Intrepid* was built as a dedicated training ship and served from 1907 to 1921, with thousands of future sailors and officers crossing her brow. She was a lovely and graceful vessel, and didn't deserve the fate she ultimately met, sold for conversion to a civilian barge just after her decommissioning. The Navy got her back in 1942 and turned her into *YSR-42,* a sludge removal barge. Let it be said though that even in this humble duty, the ex-*Intrepid* provided great service. Sent to Pearl Harbor, her services were invaluable in pumping out and refloating the sunken ships there so they could return to service and extract a nation's vengeance against the Japanese Empire. There is, however, a picture of *YSR-42* moored alongside the shattered *Oklahoma* (BB-37), looking truly sad and forlorn as she helped bring the sunken battleship back from her shallow grave. After World War Two she returned to civilian ownership and was later lost off the Columbia River.

Epilogue

And of course, there is probably the most well known of her line, the *Essex* class carrier CV-11 (1942 – 74). She is now the magnificent centerpiece of the Intrepid Sea-Air-Space Museum in New York City. *Intrepid* likely took more repeated punishment than any other carrier in the Pacific, and her crew got quite skilled at extricating her from tight corners. In one memorable instance, with her steering gone from a Japanese torpedo, the crew rigged a canvas hatch cover as a makeshift sail and helped steer her to safety. Later she served as an attack carrier in Vietnam, an anti-submarine carrier and the prime recovery vessel for the Mercury *Aurora 7* and Gemini 3 missions before her decommissioning in 1974, followed by her rebirth as an immensely popular museum.

There is, though, a picture of that awful wall of smoke and dust moving out from what had been the World Trade Center towards *Intrepid* standing firm at her moorings, the retired destroyer USS *Edson* beside her, offering shelter and protection as strongly and securely as her namesakes had. When a ship is named *Intrepid*, she can do no less.

USS *Nautilus*

Schooner *Nautilus* went home with the rest of the Mediterranean Squadron in early 1812 and immediately commenced patrols off the northeast coast under Lieutenant William Crane, looking for the enemy wherever he might be found. On July 17th Crane found them off the coast of New Jersey, and in spades: HMS *Shannon*, thirty-eight guns, HMS *Guerriere*, thirty-eight guns, HMS *Africa*, sixty-four guns and HMS *Aeolus*, thirty-two guns. Outnumbered four to one and outgunned by twice that, Crane knew he didn't have a chance of taking on the British squadron, so he put on full sail and tried to get home. It was close, but, after a seven-hour chase, *Shannon* came up close aboard, easily in a position to reduce *Nautilus* to flinders. William Crane was brave enough, but he was no fool and his crew's life meant far more to him than a glorious end in combat. Striking her colors and surrendering, *Nautilus* became the first warship lost by either side during the war. Her name has since come to be associated with submarines - first, one of the largest conventional submarines built by the USN, SS-168 (1927-45), and the first nuclear submarine SSN-571 (1952-70), now a museum in Groton, Connecticut.

USS *Constitution*

USS *Constitution* has earned the title of 'legend' several times over, but her survival was a very close-run thing. *Constitution* got out of Boston one last time, repaired and refitted but barely supplied, during the War of 1812. Under the command of Charles Stewart, she escaped through the British blockade early on December 18th, 1814. Stewart proceeded to take Old Ironsides on a cruise that would have brought a grin to a buccaneer's face. Stewart first captured the merchant *Lord Nelson* on Christmas Eve, and using the merchantman's stores

To Barbary's Far Shore

and cargo to provide a proper Christmas dinner for the Americans. On February 8, 1815, Stewart was off the coast of Spain when he was told of the Treaty of Ghent, ending the War of 1812. Stewart, however, was not going to let a little thing like a peace treaty stand in the way of glory. Declaring that since the Treaty had not yet been ratified, the treaty was signed the day Stewart took the *Lord Nelson*, and could admittedly not yet have been presented to and debated by Congress, as far as he was concerned, the war was still on.

On the 16th, *Constitution* captured another merchantman, and then on the 20th, her lookouts spotted not one, but two masts on the horizon. They were His Majesty's Ships *Cyane* and *Levant*, headed back to Great Britain. Showing his usual zeal for combat, Stewart ordered the helm to head directly for them, and then proceeded to put on a bravura performance of what Americans could do to the Royal Navy.

Outnumbered, and outgunned, Stewart concentrated his fire on *Levant*, forcing her to move off for repairs. He then focused all his wrath on the *Cyane*. The *Cyane* took only a few shots before her skipper realized just what he was up against and struck her colors. *Levant* then came back into the fight. *Levant's* captain, seeing the *Cyane* now boarded by marines and firmly under American control, decided he'd had enough of the fight and tried to make a run for it. Charles Stewart would have none of that. He brought *Constitution* about and went after the *Levant*, his superior speed and maneuverability making the difference. A few broadsides were all it took to convince the *Levant* that discretion was the better part of valor, and she too struck her colors. While Stewart was supervising repairs to all three ships, *Constitution* had taken no serious damage and her fighting efficiency was unimpaired, he discovered twelve cannonballs embedded in her timbers. Old Ironsides had lived up to her nickname a second time.

Stewart and his prizes then moved out for the Cape Verde Islands, followed by a large and extremely angry British squadron. Unable to handle all three vessels in a fight, Stewart ordered them to scatter. *Cyane* made it back to the states in April and was purchased into the Navy, but *Levant* was recaptured. *Constitution* pressed on, this time looking for a Royal Navy frigate that rumor said was carrying gold. Stewart never did find her, instead putting in to Brazil to resupply. There he got word that the Treaty of Ghent had been ratified. It wasn't absolute confirmation, but there was so much information in the report that even Charles Stewart had to admit that this time, it looked like peace, damn the politicians. A quick stop in Puerto Rico got final, undeniable proof that the war was over, and the *Constitution* headed for home. Her final combat deployment was over on May 15th, when she pulled into New York for a hero's welcome.

Constitution was in rough shape, having only gotten enough work over the last few years to keep her in fighting trim. Accordingly, she was placed in

Epilogue

ordinary in 1816, missing out on one last visit to her old friend the Dey of Tripoli. In 1820, Isaac Hull supervised a thorough overhaul and refit which brought her up to speed with the latest in naval design. She also tried out a bizarre experiment where two hand-cranked paddlewheels were briefly fitted to see if they could assist in getting away from enemies when becalmed. Hull and her new skipper, Captain Jacob Jones, are reported to have been doubtful about their use, and the paddles disappear into history.

Jones then took her out in May of 1821, back to her old haunts in the Mediterranean. The cruise was quiet in terms of combat, but not in terms of her crew's behavior. Apparently the men of Old Ironsides not only exercised a sailor's right to occasionally debauch himself on liberty, they abused the hell out of it. The Navy finally had enough and ordered Jones home just a bit early. Tom MacDonough, supervised by a thoroughly unamused John Rodgers, took command and, not incidentally, got the crew's attention on matters of deportment and discipline. MacDonough took her back to the Med, with Rodgers close behind in the ship-of-the-line *North Carolina*, and had a blessedly quiet deployment until February of 1826. Unfortunately, she was starting to show her age, and returning to Boston in 1828, she was once again placed into ordinary.

Constitution was now thirty-one years old, by that era's reckoning more than twice the age most warships were designed to simply survive, much less serve. It was going to cost a great deal of money to get her back into shape, along with quite a few other vessels in the same situation. So accordingly, the Secretary of the Navy directed that she be surveyed, in the sense of the word meaning, 'inspection,' to see exactly what it would take to get her shipshape again. However, a reporter in Boston got hold of the order and misinterpreted it to understand 'survey' in its naval sense; that is, mark a ship for disposal.

Never has a mistake by the media been so helpful to history. The citizens of Boston had a fierce affection for Old Ironsides, and their attitude spread, inspiring a young doctor named Oliver Wendell Holmes to write a poem defending the old girl. Most people remember the poem itself, but little of it save the last stanza:

> "... *Oh, better that her shattered hulk*
> *Should sink beneath the wave;*
> *Her thunders shook the mighty deep,*
> *And there should be her grave;*
> *Nail to the mast her holy flag,*
> *Set every threadbare sail,*

To Barbary's Far Shore

And give her to the god of storms,
The lightning and the gale!"

That, most mercifully and wonderfully, did it. She was rebuilt and refitted, the first of several rebuilds so far, and has served well and honorably since, in ways far better documented in fascinating detail by others. Suffice it to say here that United States Ship *Constitution* is the oldest commissioned warship afloat in the world; the Royal Navy's noble HMS *Victory* is older, but confined to a concrete drydock. She is manned to this day by a highly trained, professional Navy crew, which now includes some women, surely to the occasional wonder of those holystoned decks. She is considered a prize assignment. I shall only add that I hope my great grandchildren will see her on her yearly trips out into Boston Harbor and marvel that such ships once ruled the seas . . . and in some ways, still do.

Hamet Bey

Hamet Bey, of course, did not rule long or well over the little bit of Tripoli O'Bannon and Eaton won for him. It was not sure whether or not his family would ever actually rejoin him. Yusef was required to release them by the treaty he signed with Tobias Lear but, as we have since discovered to our dismay, absolute rulers in that part of the world do not always trouble themselves with such things as commitments and treaties. However, Yusef was a practical man when it suited him to be. He had no desire to be embarrassed again by the Americans quite so soon, so there was at least some hope that Hamet did see his family again, which he finally did in 1807.

Once in Sicily, no one was quite sure what to do with him. As far as President Jefferson, and therefore, Congress, was concerned, the matter had been closed some time before when Lear paid off the Dey once again. The men who tried to restore him to his throne, however, felt a distinct responsibility to him, including a Dr. George Davis, who was at one point Consul to the Dey of Tunisia and later succeeded Lear at Tripoli. He took up Hamet's cause as his own and eventually convinced, or perhaps more appropriately, shamed, Congress into providing some money for Hamet and his family, along with a move to Morocco.

But in 1808, Hamet received a message from Yusef; "Come home, all is forgiven," Dr Davis had convinced Yusef that Hamet was just the man to take charge at Derne, then in another one of its periodic uprisings. The idea probably made sense. The citizens liked Hamet, and with a little careful planning he could keep the locals under control and a close eye on his brother. That lasted two short years before Yusef grew suspicious; with or without reason, we do not know. Hamet, this time with his family firmly in tow, ran for safety in Egypt, where he died shortly thereafter. His memory has faded into the mists of history,

Epilogue

a cautionary but ultimately forgotten lesson in the wisdom of trusting the United States to guard one's throne.

Yusef Bey

Brother Yusef Bey, on the other hand, seems to have enjoyed life for almost three more decades. He did, pretty much, keep to his word on not molesting US shipping; though, with Yusef, one could never be quite sure. There were some fairly substantial challenges to his rule; most notably by two plucky relatives named Mehmed and Mehmed ibn' Ali, who between them took a crack at Yusef an astounding five times. On the other hand, Yusef doesn't seem to have been too hard on them. It was not the usual way one dealt with recalcitrant family members in that part of the world, but Yusef seems to have taken it in stride.

By 1819 though, the corsairs were almost all gone and this would have put something of a crimp in Yusef's ability to maintain the lifestyle to which he had become accustomed. Searching for new sources of income, Yusef hit upon one that, sadly, may still be found in some parts of the world, the slave trade; in particular, the sub-Saharan variant of that vile custom. Yusef was no fool. Although the United Kingdom and the United States had banned the slave trade in 1807 and 1808 respectively, in Africa the horrors continued well into the twentieth century. In some North African countries, the practice is officially outlawed but continues in 'secret,' Yusef, however, had no real problems profiting from the misery of others. It was, as the saying goes, nothing personal.

Business was good for about ten more years, and then things started to finally and truly go wrong for the Dey of Tripoli. First, the British, who were asserting themselves in Africa, declared the slave trade piracy in 1827, punishable, by the way, by death, and in 1833 banned slavery outright. The Royal Navy grimly hunted down slave traders foolish enough to take their cargo to sea, and even Yusef knew better than to challenge the might of the most powerful fleet on the planet. Slaves were still bought and sold to Tripoli's south, but less and less money could be gained from that particular market. The game was not quite up for Yusef, but he could see it coming, assuming he wanted to, which apparently he did not.

The silks, the jewels, and the palaces stayed, but the glory of Tripoli fled. The Tripolitian economy began to nosedive, aggravated by Yusef's unwillingness to acknowledge the obvious. By 1830, Yusef was now in his mid-60s, no longer the terror he had been thirty years before. He refused to admit the world had changed. Rulers in that position tend to discover that there are wolves circling, and Yusef was no different. And in the best traditions of that sort of thing, the wolves were already in his fold.

Yusef had three sons, who were probably every bit as cruel, despotic and hedonistic as their father. And each of them was quite sure that he deserved to

To Barbary's Far Shore

sit on the throne of Tripoli, spending a great deal of time and theoretically secret effort on making sure they were the ones who prevailed. As with most despotic courts, each one had a small group of followers who eventually coalesced into three factions that were gunning to make their man Dey. The intrigues and plots would have been fascinating to observe, were it not for the fact that the losers would have at best found themselves in exile or at worst found themselves kneeling before a chopping block in Tripoli town while a herald reminded the onlookers of the fate of traitors and a hooded executioner stood behind a scimitar.

By 1832, Yusef seems to have finally realized that the old days were gone for good. If he allowed his sons to get into an uncontrolled fight for what was left, he might very well be the first casualty. So, at the age of 66, Yusef pulled off what he hoped would be his greatest coup. He abdicated and appointed his son Ali II as Dey of Tripoli, Ali apparently having impressed his father with his skill, charisma, and cunning, not to mention the almost certain promise that Yusef would be allowed to live out his life in unmolested peace.

All in all, not a bad plan on Yusef's part. Had things been slightly different in that part of the world, he might have pulled it off. Unfortunately, Ali's siblings did not take the news of their brother's rise to power with the equanimity and decency that Yusef had hoped. They rose up against Ali and a full-blown civil war broke out, aided and abetted by the two Mehmeds who had so bedeviled Yusef in the past. What happened next is slightly unclear, but troops from the Ottoman Empire, then undergoing one of its periodic moments of lucidity during its long decline, came in 1835 to 'keep the peace,' There is the possibility that Ali got just a little too clever for his own good and invited the Ottomans in, thinking that they would be kind enough to conquer his country back for him. If so, it was a bad idea.

The Ottomans promptly deposed Ali and set up their own governor, writing finis to the saga of the Karamanli dynasty. Yusef himself survived for three more years, finally dying quite peacefully and in bed in 1838. One may safely assume that there were a great many sailors and marines who were not at all unhappy to see him off to commune with the Prophet.

Tripoli is today officially known as Libya, but for forty-some years before that it was known as The Great Socialist People's Libyan Arab Jamahiriya, a grandiose title that meant little to anyone except the homicidal, Napoleonic madman who created it, but we shall return to that momentarily. Little, however, has ever truly changed on that beknighted square of sand and scrub on the southern shore of the Mediterranean Sea.

When the Ottomans returned in the 1830s, the disintegrating rule of Yusef Bey and his highly competitive brood was replaced by a combination of corrupt, numbing boredom and low-level civil war that would occasionally flare up into

Epilogue

the 1850s. By that time, the Ottomans had actually gotten a fair handle on things and life settled down to that of a slowly decaying backwater province. Not much would happen there for nearly seventy years until the great European powers started dismantling the dying Ottoman Empire for their own benefit.

At this point, the Italians actually invaded across the Mediterranean and took possession of Tripoli, turning it into three separate colonies. In 1934, the Italians combined them into a single one known as Libya, from the ancient Greek name for North Africa, and declared that it was now and forever part of Benito Mussolini's glorious new Imperium. Sadly the newly christened Libyans had something to say about that and started a determined resistance against the Italians. It would be a few years before the Italians would formally ally themselves with Nazi Germany, but they learned some lessons quickly and early from Hitler's growing nightmare. The Italians started their own private and little-known holocaust against the resistance that may have killed off as much as fifty percent of the Bedouin tribes that were its backbone with concentration camps and disease.

Libya became a battlefield again as Montgomery and Rommel sparred across its empty dunes during the great desert campaigns of the Second World War, and accordingly fell under British rule during the long retreat of die Deutsches Afrika Korps. As was their wont, the British administered Libya well and fairly until 1951, when they granted Libya its independence with UN backing. Interestingly, as kingdoms and monarchies were retreating all over the world, the Libyans chose a monarchy for their government. Their first king was Idris I, AKA Sayyid Muhammad Idris bin Muhammad al-Mahdi as-Senussi. His Majesty was a member of the Senussi, a Muslim sect that sought to cleanse Islam through banning fanaticism and taking vows of poverty. Given those who have claimed to want to purify Islam since, the goal seemed reasonable enough and Idris seems to have been sincere at first. The Libyan Constitution implemented under his rule was a solid document that would have warmed the heart of the men who wrote our own, and though Libya was poor it was reasonably happy and an American ally to boot, with Wheelus AFB in the northwest corner of the country.

All good things however must come to an end. They did in Libya with the discovery in 1959 of another good thing, vast quantities of oil; deposits still estimated to be the ninth largest in the world. As is sadly the case when these sort of things happen, most of the money that came pouring in ended up concentrated in the hands of a small elite and King Idris, whose vows of poverty seem to have taken a back seat to more realistic concerns. Pretty much on schedule, some people started to realize that others were getting rich while they weren't. When combined with the tide of Arab nationalism that was then sweeping the Middle East, things started to go south for good King Idris. They got worse when a young Libyan Army officer named Muammar al-Gaddafi

To Barbary's Far Shore

became infatuated with the pan-Arab screeds of Gamal Nasser of Egypt. He decided that he would be the one to lead the Arab people to greatness after Nasser's humiliation at the hands of the Israelis in 1967.

Gaddafi led a coup overthrowing King Idris in 1967, and Libya has never been the same since. Nor has that part of the world. Gaddafi swung wildly across the world stage. He started several wars, and lost each and every one of them. He bounced between Islam and sectarianism. He was soundly beaten on at least two occasions by the United States and then got his revenge by blowing Pan Am Flight 103 out of the sky along with 259 innocent human beings. Finally, he made a mockery of himself and his nation by prancing about with his Amazonian virgin guards while wearing uniforms that frankly would have made Michael Jackson or Elton John blush. He banned dissent, the teaching of foreign languages, and is known to have broadcast executions live on television. He ordered the execution of Libyan dissidents overseas and actually killed at least twenty-five of them, according to Amnesty International. He started his own nuclear weapons program, which was rather far along when the United States overthrew Saddam Hussein. This lead him to what was probably the only second thought he ever had and promptly turned everything over to the Americans.

According to Freedom House, a respected international organization that monitors freedom of the press around the world, Libya was the most censored nation in the Middle East and North Africa; considering the governments of that part of the world, that is a grim honor indeed. In February 2011, Libyans, inspired by the 'Arab Spring' that had ejected other tyrants, began demonstrations in Benghazi. That quickly escalated into a full-dress civil war, with the rebels aided and abetted by the United Nations,for all practical purposes, the US and the United Kingdom, and damned little of that.

The fight see-sawed back and forth for months until the limited air support for the rebels began to tell. Gaddafi's forces, bolstered by vicious mercenaries and goaded on by his demented sons, who behaved and spoke in ways that would have made Yusef Bey's lads step back in horror, began to disintegrate. Tripoli fell in August and Gaddafi vanished underground into his hometown of Sirte, leading a flagging resistance from there. When the end came, it was ugly - Gaddafi and a small band of fanatic loyalists held out until October 20th when the dictator was making a break for it and was caught by a NATO airstrike. Abandoning the convoy, he was captured, appropriately hiding in a sewer pipe. He was brought out, assisted by someone with a bayonet who kept stabbing him in the buttocks. Gaddafi lasted but a few minutes before someone in the crowd put him out of their misery. His final words were "do you not know right from wrong?"

Clearly, someone did.

Epilogue

When this book was written, the new ruling National Transitional Council was trying to get things up and running again, but they were having some problems. The old tribal ways came back with a vengeance once Gaddafi's heavy hand was gone. The psychotic minions of Al-Queda are now circling the crippled nation, a situation not helped by the apparent willingness of the NTC to embrace Sharia and some of the other more unpleasant aspects of Islamic society. The UN has pretty much left Libya to its own devices and, as long as the oil flows, it's likely to stay that way. The United States, always willing to believe that things have changed in that part of the world, was so willing to believe the best that they allowed the Ambassador, Chris Stevens, to be guarded by people who had known ties to Islamic extremist groups . . . an action that resulted in an attack on the Ambassador and the Benghazi consulate on September 11th, 2012. The result was the death of the Ambassador and two civilian security specialists, probably at the hands of al-Queda or a closely linked group.

The US response would probably have stunned the men who faced the Dey two hundred and eight years before; utter confusion, bizarre claims that the attack, with all its symbolism, was actually a response to an amateurish video insulting Islam made by a con man in California, absolute disbelief that a terror attack could occur on September 11th, denials that any of this had to do with the fact that very dangerous men hate the United States and everything that it stands for, and finally an admission that the murder of our envoys and his men simply weren't worth the effort to avenge them. The innocent were punished, the guilty allowed to walk away. Yusef Bey is probably smiling. In a timeless land, his people have not changed, nor have ours.

The Mameluke Sword

According to tradition, Hamet Bey presented O'Bannon with a Mameluke sword before the Lieutenant and his men boarded *Argus* to leave Derna. The sword became the pattern for the US Marine Corps' officers' sword to this day.

There is, however mystery as to the fate of the actual sword itself. O'Bannon may have given a sword to George Mann, a sword which now rests in the US Naval Academy. If he did, according to Mann family tradition, it was not Hamet's sword, but rather the one surrendered by the Governor of Derna. However, diligent research by Brigadier General Edwin Simmons, USMC (Ret) indicates that this may not be the case. In addition, O'Bannon is known to have received a blade from the Viceroy of Egypt before the Derna expedition. A rare surviving letter from O'Bannon to Isaac Hull confirms this, plus the fact that all who were at that dinner with Kourshek Pasha received one. That, of course, includes George Mann, which could also explain the sword at the Naval Academy. O'Bannon also received one more from his home state of Virginia (with his name misspelled) when he returned to the United States.

To Barbary's Far Shore

General Simmons honestly points out that, despite almost two centuries of tradition, there is simply no mention in any of the surviving USMC records of Hamet Bey actually giving Presley O'Bannon a blade that long-ago day in Tripoli. There is tradition, however, of O'Bannon turning a sword over to another Marine legend, Archibald Henderson. Henderson was a brand new second lieutenant when O'Bannon arrived home in the summer of 1805. Henderson, in turn, used that as the pattern for the new officer's sword. Unfortunately, General Simmons' research says this cannot be true, as the sword O'Bannon gave to Henderson was a standard US pattern blade. According to General Simmons, Henderson directed that the Marine sword have a 'Mameluke hilt,' but there is no sure way to know where this decision came from. It is unlikely at best that it was O'Bannon; the two men do not seem to have known one another well and only served together for nine months.

In any event, far be it from this writer to question tradition, especially Marine tradition. But even if, at the end of the day, there was no sword, what of it? It is not so much a matter of a length of steel and ivory and silken tassel, but rather what the Sword stands for. For two hundred years, the Mameluke Sword and its bearers have stood for bravery against great odds, leadership in the grimmest circumstances and resolute courage when all seems lost. One need only look at the long line of names, Tripoli, Chapultepec Castle, Belleau Wood, Tarawa, Iwo Jima, Chosin, Khe Sanh, Kuwait, and now Baghdad and Fallujah, to know that the Sword has done its job though eight generations of Marines. Eight generations, hence the Sword will remain a symbol of defense to those without, of victory to those alone, outnumbered and outgunned, and righteous justice to those resisting tyranny. In short, a symbol of everything the Marine Corps has stood and fought for.

The Mamelukes

The Mamelukes are still synonymous with courage, bravery, skill, and ferocity in battle, but they met a grim end worthy of Greek tragedy. After the Ottomans shakily took power again, the surviving Mameluke chieftains tried to take their old positions back as well, but they no longer had the military strength needed to keep power. An Ottoman pasha named Mehemet Ali rose to power in Egypt, playing the Mameluke leaders off one another as he consolidated his own position. It took him more than a decade, but he finally got the Mamelukes to accept and support him as Viceroy of Egypt. In the best traditions of that part of the world, Mehemet made plans to properly reward the Mamelukes for their support.

In 1811, Mehemet declared a feast and celebration in honor of his Mameluke friends, who gathered in the Bab-al-Azab neighborhood of Cairo, near Saladin Square. When the festivities were over, the Mamelukes saddled up and prepared to go home only to find the square blocked. Thousands of

Epilogue

Mehemet's troops suddenly rose from their hiding places overlooking the narrow streets. Between bullets, arrows, and even rocks, the Mamelukes were literally cut down to the last man. According to tradition, only one survived, a clan chief called Amim Bey. Legend says that Amim Bey managed to get his horse out of the killing grounds and onto the ramparts of Cairo's citadel. There he jumped the horse over Mehemet's men and out to safety. It was a thirty-foot drop that killed the noble steed, but Amim was able to make a run for it and survived to escape into Asia. For six more days, the massacre continued against men, women, and children alike. When it was over, the Mamelukes were no more. In reality, there were almost certainly a few pitiful survivors who faded into the mist. Even in the well-planned and efficient genocides of recent memory, the killing has been far from complete. But however many were left, it was nowhere sufficient for the Mamelukes to ever reestablish themselves as a distinct minority. They have since been subsumed into the larger Egyptian population, gone except for the legends of their bravery.

Derne

Derne, Libya, would be fought over a few more times, most notably during the Second World War. Australian forces would take it in January of 1941, and the Germans would take it back in April and then give it up once and for all in November of 1942. Before and after that, Derne had a reputation, one to be proud of or not, depending on your particular outlook, as the most devout city in Libya. Now, religious fervor in that part of the world tends to be a harbinger of worse things, especially when one is a dictator of the stripe of Muammar al-Gaddafi, who knew perfectly well what would happen if said fervor was turned against him.So he struck upon the brilliant idea of convincing these zealots to fight the only people they disliked more than him, the Americans. After the US invasion of Iraq in 2003, jihadi recruiters routinely signed up dozens of Derne residents. According to one captured roster, more than half of the Libyans on it came from Derne. Needless to say, this kind of social and religious outlook worked against Gaddafi when the final challenge to his rule began in 2011. Derne was the first big city to join the Libyan rebels, and it was never in danger of falling to loyalist forces. It today remains a major trading city and source of Libyan religious thinking.

The United States Marine Corps

The Corps has been, is, and shall remain, Always Faithful. And really, nothing more needs be added to that proud motto.

M.J.K

March, 2013

REFERENCES AND BIBLOGRAPHY

Allen, Gardner W. Our Navy and the Barbary Corsairs. Houghton, Mifflin, and Co., New York, 1905

Denyer, Simon. 'Remains Of First 'Navy SEALS' Lie In Tripoli', *The Washington Post*, May 29th, 2011

Fowler, William M. Jack Tars and Commodores: The American Navy, 1783-1815. Houghton, Mifflin, and Co., New York, 1984.

Oakes, Lorna, and Gahlin, Lucia. Ancient Egypt: An Illustrated Reference To The Myths, Religions, Pyramids And Temples Of The Land Of The Pharaohs. Anness Publishing, London, UK, 2003.

Simmons, Edwin H., Brigadier General, USMC (Ret). "O'Bannon's Sword?" From *Fortitudine, The Newsletter Of The Marine Corps Historical Program*, Vol. XIV, Summer 1984, Volume 1.

Towers, John (printer). DOCUMENTS, OFFICIAL AND UNOFFICIAL, RELATING TO THE CASE OF THE CAPTURE AND DESTRUCTION OF THE FRIGATE PHILADELPHIA AT TRIPOLI ON THE 16TH FEBRUARY 1804. Washington, DC, 1850.

Whipple, A.B.C. To The Shores Of Tripoli. Naval Institute Press, Annapolis, MD, 1991.

Wegner, Dana. Fouled Anchors: The *Constellation* Question Answered. David Taylor Research Center, Bethesda, MD, 1991.

Zacks, Richard The Pirate Coast: Thomas Jefferson, the first Marines, and the Secret Mission of 1805. Hyperion, New York, 2005.

ONLINE SOURCES

http://africanhistory.about.com/od/militaryhistory/ss/Yusuf-Karamanli-1.htm for the life and ultimate fate of Yusef Bey Karamanli.

References & Bibliography

http://www.navsource.org/archives/09/86/86idx.htm Brief but highly informative pages, with photographs, of many of the 'Old Navy' ships and their descendants that served during the Barbary Wars.

http://www.history.navy.mil/danfs/ The Dictionary Of American Naval Fighting Ships - the official USN histories of almost every warship that has ever served in the fleet. Constantly being revised and updated by the Naval History and Heritage Center.

To Barbary's Far Shore

APPENDIX 1

Treaty of Peace and Amity, signed at Tripoli June 4, 1805

Treaty of Peace and Amity, signed at Tripoli June 4, 1805 (6 Rabia I, A. H. 1220). Original in English and Arabic. Submitted to the Senate December 11, 1805. Resolution of advice and consent April 12, 1806. Ratified by the United States April 17, 1806. As to the ratification generally, see the notes. Proclaimed April 22, 1806.

The English tent of the copy of the treaty, signed by Tobias Lear, follows; to it is appended the receipt for the $60,000 ransom paid on June 19, 1805 (21 Rabia I, A. H. 1220), as written in the same document; then is reproduced the Arabic text of that paper, in the same order as the English. Following those texts is a comment, written in 1930, on the Arabic tent.

Treaty Of Peace and Amity between the United States of America and the Bashaw, Bey and Subjects of Tripoli in Barbary.

ARTICLE 1st

There shall be, from the conclusion of this Treaty, a firm, inviolable and universal peace, and a sincere friendship between the President and Citizens of the United States of America, on the one part, and the Bashaw, Bey and Subjects of the Regency of Tripoli in Barbary on the other, made by the free consent of both Parties, and on the terms of the most favoured Nation. And if either party shall hereafter grant to any other Nation, any particular favour or priviledge in Navigation or Commerce, it shall immediately become common to the other party, freely, where it is freely granted, to such other Nation, but where the grant is conditional it shall be at the option of the contracting parties to accept, alter or reject, such conditions in such manner, as shall be most conducive to their respective Interests.

ARTICLE 2d

The Bashaw of Tripoli shall deliver up to the American Squadron now off Tripoli, all the Americans in his possession; and all the Subjects of the Bashaw of Tripoli now in the power of the United States of America shall be delivered up to him; and as the number of Americans in possession of the Bashaw of Tripoli amounts to Three Hundred Persons, more or less; and the number of Tripolino Subjects in the power of the Amelicans to about, One Hundred more or less; The Bashaw of Tripoli shall receive from the United States of America,

Appendix I

the sum of Sixty Thousand Dollars, as a payment for the difference between the Prisoners herein mentioned.

ARTICLE 3rd

All the forces of the United States which have been, or may be in hostility against the Bashaw of Tripoli, in the Province of Derne, or elsewhere within the Dominions of the said Bashaw shall be withdrawn therefrom, and no supplies shall be given by or in behalf of the said United States, during the continuance of this peace, to any of the Subjects of the said Bashaw, who may be in hostility against him in any part of his Dominions; And the Americans will use all means in their power to persuade the Brother of the said Bashaw, who has co-operated with them at Derne &c, to withdraw from the Territory of the said Bashaw of Tripoli; but they will not use any force or improper means to effect that object; and in case he should withdraw himself as aforesaid, the Bashaw engages to deliver up to him, his Wife and Children now in his powers

ARTICLE 4th

If any goods belonging to any Nation with which either of the parties are at war, should be loaded on board Vessels belonging to the other party they shall pass free and unmolested, and no attempt shall be made to take or detain them.

ARTICLE 5th

If any Citizens, or Subjects with or their effects belonging to either party shall be found on board a Prize Vessel taken from an Enemy by the other party, such Citizens or Subjects shall be liberated immediately and their effects so captured shall be restored to their lawful owners or their Agents.

ARTICLE 6th

Proper passports shall immediately be given to the vessels of both the contracting parties, on condition that the Vessels of War belonging to the Regency of Tripoli on meeting with merchant Vessels belonging to (citizens of the United States of America, shall not be permitted to visit them with more than two persons besides the rowers, these two only shall be permitted to go on board said Vessel, without first obtaining leave from the Commander of said Vessel, who shall compare the passport, and immediately permit said Vessel proceed on her voyage; and should any of the said Subjects of Tripoli insult or molest the Commander or any other person on board Vessel so visited; or plunder any of the property contained in the full complaint being made by the Consul of the United States America resident at Tripoli and on his producing sufficient proof substantiate the fact, The Commander or Rais of said Tripoline Sh or Vessel of War, as well as the Offenders shall be punished in the most exemplary manner.

All Vessels of War belonging to the United States of America meeting with a Cruizer belonging to the Regency of Tripoli, and having seen her passport and

To Barbary's Far Shore

Certificate from the Consul of t] United States of America residing in the Regency, shall permit her to proceed on her Cruize unmolested, and without detention. No pas port shall be granted by either party to any Vessels, but such as are absolutely the property of Citizens or Subjects of said contracting parties, on any presence whatever.

ARTICLE 7th

A Citizen or Subject of either of the contracting parties having bought a Prize Vessel condemned by the other party, or by any other Nation, the Certificate of condemnation and Bill of Sale she be a sufficient passport for such Vessel for two years, which, considering the distance between the two Countries, is no more than a reason able time for her to procure proper passports.

ARTICLE 8th

Vessels of either party, putting into the ports of the other, and having need of provisions or other supplies, they shall be furnish at the Market price, and if any such Vessel should so put in from disaster at Sea, and have occasion to repair; she shall be at liberty to land and reimbark her Cargo, without paying any duties; but in no case shall she be compelled to land her Cargo.

ARTICLE 9th

Should a Vessel of either party be cast on the shore of the other all proper assistance shall be given to her and her Crew. No pillar shall be allowed, the property shall remain at the disposition of its owners, and the Crew protected and succoured till they can be sent to their Country.

ARTICLE 10th

If a Vessel of either party, shall be attacked by an Enemy within Gun shot of the Forts of the other, she shall be defended as much as possible; If she be in port, she shall not be seized or attacked when it is in the power of the other party to protect her; and when she proceeds to Sea, no Enemy shall be allowed to pursue her from the same port, within twenty four hours after her departure.

ARTICLE 11th

The Commerce between the United States of America and the Regency of Tripoli; The Protections to be given to Merchants, Masters of Vessels and Seamen; The reciprocal right of establishing Consuls in each Country; and the priviledges, immunities and jurisdictions to be enjoyed by such Consuls, are declared to be on the same footing, with those of the most favoured Nations respectively.

ARTICLE 12th

Appendix I

The Consul of the United States of America shall not be answerable for debts contracted by Citizens of his own Nation, unless, he previously gives a written obligation so to do.

ARTICLE 13th

On a Vessel of War, belonging to the United States of America, anchoring before the City of Tripoli, the Consul is to inform the Bashaw of her arrival, and she shall be saluted with twenty one Guns, which she is to return in the same quantity or number.

ARTICLE 14th

As the Government of the United States of America, has in itself no character of enmity against the Laws, Religion or Tranquility of Musselmen, and as the said States never have entered into any voluntary war or act of hostility against any Mahometan Nation, except in the defence of their just rights to freely navigate the High Seas: It is declared by the contracting parties that no pretext arising from Religious Opinions, shall ever produce an interruption of the Harmony existing between the two Nations; And the Consuls and Agents of both Nations respectively, shall have liberty to exercise his Religion in his own house; all slaves of the same Religion shall not be Impeded in going to said Consuls house at hours of Prayer. The Consuls shall have liberty and personal security given them to travel within the Territories of each other, both by land and sea, and shall not be prevented from going on board any Vessel that they may think proper to visit; they shall have likewise the liberty to appoint their own Drogoman and Brokers.

ARTICLE 15th

In case of any dispute arising from the violation of any of the articles of this Treaty, no appeal shall be made to Arms, nor shall War be declared on any pretext whatever; but if the Consul residing at the place, where the dispute shall happen, shall not be able to settle the same; The Government of that Country shall state their grievances in writing, and transmit it to the Government of the other, and the period of twelve callendar months shall be allowed for answers to be returned; during which time no act of hostility shall be permitted by either party, and in case the grievances are not redressed, and War should be the event, the Consuls and Citizens or Subjects of both parties reciprocally shall be permitted to embark with their effects unmolested, on board of what vessel or 1Vessels they shall think proper.

ARTICLE 16th

If in the fluctuation of Human Events, a War should break out between the two Nations; The Prisoners captured by either party shall not be made Slaves; but shall be exchanged Rank for Rank; and if there should be a deficiency on either side, it shall be made up by the payment of Five Hundred Spanish Dollars for each Captain, Three Hundred Dollars for each Mate and Supercargo and One hundred Spanish Dollars for each Seaman so wanting. And it is agreed that Prisoners shall be exchanged in twelve months from the time of their capture, and that this Exchange may be effected by any private Individual legally authorized by either of the parties.

ARTICLE 17th

If any of the Barbary States, or other powers at War with the United States of America, shall capture any American Vessel, and send her into any of the ports of the Regency of Tripoli, they shall not be permitted to sell her, but shall be obliged to depart the Port on procuring the requisite supplies of Provisions; and no duties shall be exacted on the sale of Prizes captured by Vessels sailing under the Flag of the United States of America when brought into any Port in the Regency of Tripoli.

ARTICLE 18th

If any of the Citizens of the United States, or any persons under their protection, shall have any dispute with each other, the Consul shall decide between the parties; and whenever the Consul shall require any aid or assistance from the Government of Tripoli, to enforce his decisions, it shall immediately be granted to him. And if any dispute shall arise between any Citizen of the United States and the Citizens or Subjects of any other Nation, having a Consul or Agent in Tripoli, such dispute shall be settled by the Consuls or Agents of the respective Nations.

ARTICLE 19th

If a Citizen of the United States should kill or wound a Tripoline, or, on the contrary, if a Tripoline shall kill or wound a Citizen of the United States, the law of the Country shall take place, and equal justice shall be rendered, the Consul assisting at the trial; and if any delinquent shall make his escape, the Consul shall not be answerable for him in any manner whatever.

ARTICLE 20th

Should any Citizen of the United States of America die within the limits of the Regency of Tripoli, the Bashaw and his Subjects shall not interfere with the property of the deceased; but it shall be under the immediate direction of the Consul, unless otherwise disposed of by will. Should there be no Consul, the effects shall be deposited in the hands of some person worthy of trust, until the party shall appear who has a right to demand them, when they shall render an account of the property. Neither shall the Bashaw or his Subjects give hindrance in the execution of any will that may appear.

Appendix I

Whereas, the undersigned, Tobias Lear, Consul General of the United States of America for the Regency of Algiers, being duly appointed Commissioner, by letters patent under the signature of the President, and Seal of the United States of America, bearing date at the City of Washington, the 18" day of November 1803 for negotiating and concluding a Treaty of Peace, between the United States of America, and the Bashaw, Bey and Subjects of the Regency of Tripoli in Barbary-

Now Know Ye, That I, Tobias Lear, Commissioner as aforesaid, do conclude the foregoing Treaty, and every article and clause therein contained; reserving the same nevertheless for the final ratification of the President of the United States of America, by and with the advice and consent of the Senate of the said United States.

Done at Tripoli in Barbary, the fourth day of June, in the year One thousand, eight hundred and five; corresponding with the sixth day of the first month of Rabbia 1220.

[Seal] TOBIAS LEAR.

Having appeared in our presence, Colonel Tobias Lear, Consul General of the United States of America, in the Regency of Algiers, and Commissioner for negotiating and concluding a Treaty of Peace and Friendship between Us and the United States of America, bringing with him the present Treaty of Peace with the within Articles, they were by us minutely examined, and we do hereby accept, confirm and ratify them, Ordering all our Subjects to fulfill entirely their contents, without any violation and under no pretext.

In Witness whereof We, with the heads of our Regency, Subscribe it.

Given at Tripoli in Barbary the sixth day of the first month of Rabbia 1220, corresponding with the 4th day of June 1805.

(L. S.) JUSUF CARAMANLY Bashaw

(L. S.) MOHAMET CARAMANLY Bey

(L. S.) MOHAMET Kahia

(L. S.) HAMET Rais de Marino

(L. S.) MOHAMET DGHIES First Alinister

(L. S.) SARAH Aga of Divan

(L. S.) SEEIM Hasnadar

(L. S.) MURAT Dqblartile

To Barbary's Far Shore

(L. S.) MURAT RAIS Admiral

(L. S.) SOEIMAN Kehia

(L. S.) ABDAEEA Basa Aga

(L. S.) MAHOMET Scheig al Belad

(L. S.) ALEI BEN DIAB First Secretary

[Receipt]

We hereby acknowlidge to have received from the hands of Colonel Tobias Lear the full sum of sixty thousand dollars, mentioned as Ransum for two hundred Americans, in the Treaty of Peace concluded between Us and the United States of America on the Sixth day of the first Month of Rabbia 1220-and of all demands against the said United States.

Done this twenty first day of the first month of Rabbia 1220.

(L. S.) Signd (JOSEPH CARMANALY) Bashaw

LION BY LION
PUBLISHING

Growing Up In A Pennsylvania Steel Town During The Great Depression by Edward Nebinger. This personal memoir looks back at the years before World War II. They represented a time when the U.S. was struggling through the Depression, but people never gave up and instead made the best of what they had. Above all, the Bethlehem Steel Company was the second largest steel producer in the world, and the U.S. was the leading industrial power on earth. Now that great industrial base is largely gone, having moved to other Continents, with Bethlehem Steel among the first of the industrial dominos to fall - an event of tragic proportions. 238 pp, paperback $22.95, E-book, $9.95

The Big One by Stuart Slade. Europe is being torn apart by a war which nobody can win. Nazi Germany occupies Europe from the Pyrenees to the Volga. In the East, Russian and American troops fight to stop the German Army from breaking through. In the West, American carriers prowl the Atlantic, hurling their midnight-blue fighter-bombers against any target they find. Nothing can stop the madness. America has one last hand left, a plan to bring the war to an end in a single terrible blow. 213 pp, paperback $19.95, E-book, $9.95

Winter Warriors by Stuart Slade. One great German offensive has broken through the Russian defenses leaving an Allied army trapped in the frozen waste land of the Kola Peninsula. While the armies try to survive the bitter cold opposing ski-troops fight a vicious private war to dominate the ground between by their armies. Desperate to break the deadlock, the German Navy sets sail in an effort to destroy the convoys that keep the allied troops on Kola alive. And so, an epic naval battle brews in the icy waters of the North Atlantic. In the midst of the fighting, a U.S. Navy railway gun crew, Russian railway engineers and Siberian ski-troops come together in a desperate battle to save the great guns from the advancing German troops. Behind the scenes, in a war-weary America, another political battle is being fought, one in which a supposed friend can be as deadly an enemy as any found on the Kola Peninsula.

High Frontier by Stuart Slade. Living and working in space is man's greatest challenge. The conquest of space will guarantee man's survival. Yet, there may not be enough time left. On Earth, one empire is collapsing under its own weight and there are those in its government who would prefer to suffer utter destruction rather than defeat. Another is trying to repair the damage from previous blunders and rebuild its relations with the rest of the world. Will humanity have time to scale The High Frontier? 236pp, paperback $22.95, E-book, $9.99

A Mighty Endeavor by Stuart Slade. The unthinkable has happened and a theoretical possibility has become an ugly reality. Britain is out of the war. The Commonwealth is on its own. How can it survive when its military, economic and political center has been stripped away? In a world that is suddenly filled with unexpected enemies and unlikely friends, the Commonwealth has a desperate struggle on its hands. Just to survive will be hard enough. To survive and win is truly A Mighty Endeavor. 420 pp, paperback $24.95, E-book $9.95

Conrad's Eye by Stuart Slade. Conrad Lorenz, Inquisitor - A soul eternally damned, doomed to wander the Earth until he had saved enough of those wrongly accused to redeem his soul from the guilt of the innocents he had condemned. For those wrongly accused and in desperate need, there is one last hope for justice. That Conrad will cast his eye upon their case. 464 pp, paperback $29.95, E-book, $9.95

All these books are available from our website

www.lionpubs.com

www.ingramcontent.com/pod-product-compliance
Lightning Source LLC
Chambersburg PA
CBHW020940230426
43666CB00005B/104